Insecticide Action

From Molecule
to Organism

Insecticide Action

From Molecule to Organism

Edited by

Toshio Narahashi

Northwestern University Medical School
Chicago, Illinois

and

Janice E. Chambers

Mississippi State University
Mississippi State, Mississippi

PLENUM PRESS • NEW YORK AND LONDON

Library of Congress Cataloging-in-Publication Data

Insecticide action : from molecule to organism / edited by Toshio Narahashi and Janice E. Chambers.
p. cm.
"Proceedings of the symposium on insecticide action: from molecule to organism, held as part of the 196th National Meeting of the Agrochemicals Division of the American Chemical Society, September 25-30, 1988, in Los Angeles, California"--P.
Includes bibliographical references.
ISBN-13: 978-1-4684-1326-7 e-ISBN-13: 978-1-4684-1324-3
DOI: 10.1007/978-1-4684-1324-3
1. Insecticides--Toxicology--Congresses. 2. Insecticides--Physiological effect--Congresses. I. Narahashi, Toshio. II. Chambers, Janice E. III. American Chemical Society. Division of Agrochemicals. IV. American Chemical Society. Meeting (196th : 1988 : Los Angeles, Calif.)
RA1270.I5I57 1990
668'.651--dc20 89-26595
CIP

Proceedings of the symposium on Insecticide Action: From Molecule to Organism, held as part of the national meeting of the Agrochemicals Division of the American Chemical Society, September 25–30, 1988, in Los Angeles, California

Softcover reprint of the hardcover 1st edition 1989
A Division of Plenum Publishing Corporation
233 Spring Street, New York, N.Y. 10013

PREFACE

Intoxication of humans and animals has become increasingly important in recent years as has contamination of the environment by a variety of chemicals. In order to develop effective means by which such intoxication and contamination can be properly handled, it is imperative to know how these environmental agents act in humans and animals. Despite studies conducted by various investigators, the mechanisms of action of these environmental agents have not been fully elucidated. Insecticides are by no means an exception in terms of the seriousness of the problem and of the urgency of the need for such information.

In order to complete a picture of the effects of any particular insecticide, it is of utmost importance that its actions at various levels ranging from those of molecules to whole animals be analyzed and synthesized. To understand the toxicological action on animals or humans, it is not sufficient to know the action at each level only. The actions at various levels must be integrated to construct a picture of the toxic effect on the intact organism. However, in spite of the large body of information that has been accumulated during the past few decades, little or no attempt has been made to integrate experimental data obtained at the molecular, cellular, organ, and animal levels together in order to define the whole picture of insecticidal action.

This book is such an attempt, with contributions by researchers actively involved in insecticide toxicology at the molecular, cellular and whole organism level. For example, the sodium channel of nerve membrane has been identified as the major target site of pyrethroids and DDT through advanced electrophysiological experiments and measurements of sodium flux. This modification of the sodium channel subsequently causes an increase in transmitter release from nerve terminals, which can then have large effects on overall physiology or behavior. However, some pyrethroids appear to inhibit the GABA receptor-channel complex, and this alternative action, if significant, could modify the organism's overall response. Similarly, the organophosphorus insecticides, by inhibiting acetylcholinesterase, cause an accumulation of acetylcholine in synapse, and subsequently hyperactivity of cholinergic pathways, thus impacting physiology and behavior. However, some organophosphates have been shown to affect glutamate- and acetylcholine-activated ion channels, and this alternative action could also modify the overall response. A variety of animal defense mechanisms have proven to be critically important in detoxifying insecticides and contributing to resistance to insecticides. For example, microsomal enzyme systems play an important role in the detoxification of various insecticides, and certain physiological processes and reactions are effective means by which animals are protected against the toxic action of insecticides. These defenses, then, have the opportunity to attenuate the organism's response to the toxic effects.

The information presented in this book has allowed a clearer picture to be developed regarding the overall impact of insecticides on organisms. This information was presented in a symposium entitled Insecticide Action: From Molecule to Organism in the program of the Agrochemicals Division of the American Chemical Society meeting in Los Angeles, California, in September, 1988.

The Editors express appreciation to the contributors for their articles and their participation in the symposium. We are also grateful to the Agrochemicals Division of the American Chemical Society for its support of the symposium, and to the following companies for their financial support to the symposium: CIBA-GEIGY Corporation, Dow Chemical U.S.A., E.I. du Pont de Nemours and Company, Hoechst-Roussel Agri-Vet Company, and Lilly Research Laboratories. Thanks are due to Sherry L. Manick for editorial assistance and Vicky James-Houff for secretarial assistance.

Toshio Narahashi
Janice E. Chambers

CONTENTS

INSECTICIDE ACTIONS ON GABA RECEPTORS AND VOLTAGE-DEPENDENT CHLORIDE CHANNELS

Mohyee E. Eldefrawi and Amira T. Eldefrawi

Department of Pharmacology and Experimental Therapeutics
University of Maryland School of Medicine
Baltimore, MD 21201

ABSTRACT

γ-Aminobutyric acid (GABA) is the major inhibitory neurotransmitter. The $GABA_A$ receptor is a primary target for cyclodiene insecticides and a secondary target for several other insecticides. The $GABA_A$ receptor has a high affinity (K_d in nM) for cyclodienes and binds them at the site that binds the convulsant t-butylbicyclophosphorothionate (TBPS). There is an excellent correlation between the toxicities of cyclodienes and their potencies in inhibiting receptor binding and function in mammalian brain and insect neurons. Of four hexachlorocyclohexane (BHC) isomers, only the insecticide γ-isomer inhibits the $GABA_A$ receptor. Other insecticides that inhibit this receptor are the pyrethroids, with type II more potent than type I, and a few organophosphate anticholinesterases (e.g. leptophos and EPN) with IC_{50} values above 1 μM. Voltage-dependent chloride channels are also targets for insecticides, possibly primary targets for avermectins. Also γ-BHC, is a potent inhibitor of a voltage-dependent chloride channel binding, even more so than of the $GABA_A$ receptor. However, the $GABA_A$ receptor binding site is much more stereospecific than that of the chloride channel.

INTRODUCTION

The chloride (Cl^-) ion is essential for normal function of excitable tissues in both vertebrates and invertebrates. Cell membranes are fairly permeable to Cl^-. This negatively charged ion is transported across cell membranes via Cl^- selective transport proteins of which there are two major classes: The chemically-gated and the voltage-gated Cl^- channels (Hille, 1984). The first class is transmitter operated channels, which are receptors that are activated by their respective transmitters to open their Cl^- channels. These include the two best known inhibitory neurotransmitter receptors for GABA and glycine. The second class is the voltage-gated Cl^- channels which play important roles in maintaining the resting membrane potential of muscle and nerve. These channels are operated by a change in membrane potential. Thus far, more attention has been given to chloride channels of muscles rather than those of nervous tissue.

There are two classes of GABA receptors. $GABA_A$ is the major inhibitory receptor of the brain, and is associated with a Cl^- channel which is operated by binding of GABA, while the $GABA_B$ receptor is believed to couple to Ca^{2+} and K^+ channels via GTP-binding proteins (Enna and Carbon, 1987; Bormann, 1988). The $GABA_A$ receptor is now recognized as the molecular target for many prescribed drugs (e.g. sedative barbiturates and the anxiety-relieving benzodiazepines) as well as drugs of abuse (e.g. alcohol) (Enna, 1983; Suzdak and Paul, 1987). Recently, it has been shown that $GABA_A$ receptors are also targets for different insecticides and other toxicants (Eldefrawi and Eldefrawi, 1987). This presentation will focus on evidence from our laboratory that $GABA_A$ receptors are involved in the toxic reaction to several kinds of insecticides. In addition, preliminary data on the possible involvement of voltage-dependent (VD) Cl^- channels in insecticide toxicity will be discussed.

MOLECULAR AND PHARMACOLOGICAL PROPERTIES OF $GABA_A$ RECEPTORS AND VD Cl^- CHANNELS

The $GABA_A$ receptor of mammalian brain has been purified and its gene cloned (Sigel and Barnard, 1984; Schofield et al., 1987). It is a protein composed of two distinct subunits: α and β (53 and 57 kilodaltons, respectively). The monomer is believed to have an $\alpha_2\beta_2$-subunit structure. The binding site of GABA is located on the β-subunit and is recognized by structural analogs of GABA (e.g. muscimol and isoguvacine), but not by baclofen, which is a specific activator of the $GABA_B$ receptor. Bicuculline is the best known competitive antagonist of the $GABA_A$ receptor, and binds to the GABA recognition site on the β subunit. Many different classes of drugs are known to modulate $GABA_A$ receptor responses. Classical anxiolytic drugs (e.g. diazepam) bind to a specific site known as the benzodiazepine receptor, which is located on the α subunit of the GABA receptor. These drugs potentiate GABA responses by increasing the frequency of opening of the GABA-gated Cl^- channels. The sedative hypnotic barbiturates (e.g. pentobarbital) also potentiate GABA responses by increasing the open time of Cl^- channels. They bind to a site associated with the Cl^- channel which is made of all 4 subunits. Convulsant drugs like picrotoxinin and TBPS are believed to block the Cl^- channel. The binding sites for agonists, benzodiazepines, barbiturates and convulsants are allosterically linked so that binding of ligands to one site modulates the affinities of the other sites (Olsen, 1982).

By contrast, we know very little about the molecular properties of VD Cl^- channels. There is no *a priori* reason why drugs and toxicants that bind to the $GABA_A$ receptor would also bind to the VD Cl^- channel, except for possible molecular similarities in their ionic channels. $[^{35}S]$TBPS was also found to bind to the VD Cl^- channel, albeit with a 27-fold lower affinity than to the $GABA_A$ receptor. However, the high concentration of VD Cl^- channels in the non-innervated membrane of the electric organ of the electric ray, *Torpedo* sp., has made it possible to detect $[^{35}S]$TBPS binding to these channels (Abalis et al., 1985a; Matsumoto et al., 1988). The VD Cl^- channels are detected best by monitoring electrophysiologically the Cl^- currents in excitable tissues such as squid giant axon (Inou, 1986), cultured rat muscle (Blatz and Magleby, 1986) and *Torpedo* electric organs (Tank et al., 1982). The lack of knowledge of their pharmacology, shortage of high affinity radioactive probes and unavailability of functional biochemical assays have made it difficult to study these VD Cl^- channels.

CYCLODIENE INSECTICIDES

Cyclodiene insecticides were recognized as neurotoxicants that cause hyperexcitability and convulsions in vertebrates and invertebrates (Metcalf, 1955), but only in the last few years has the $GABA_A$ receptor been identified

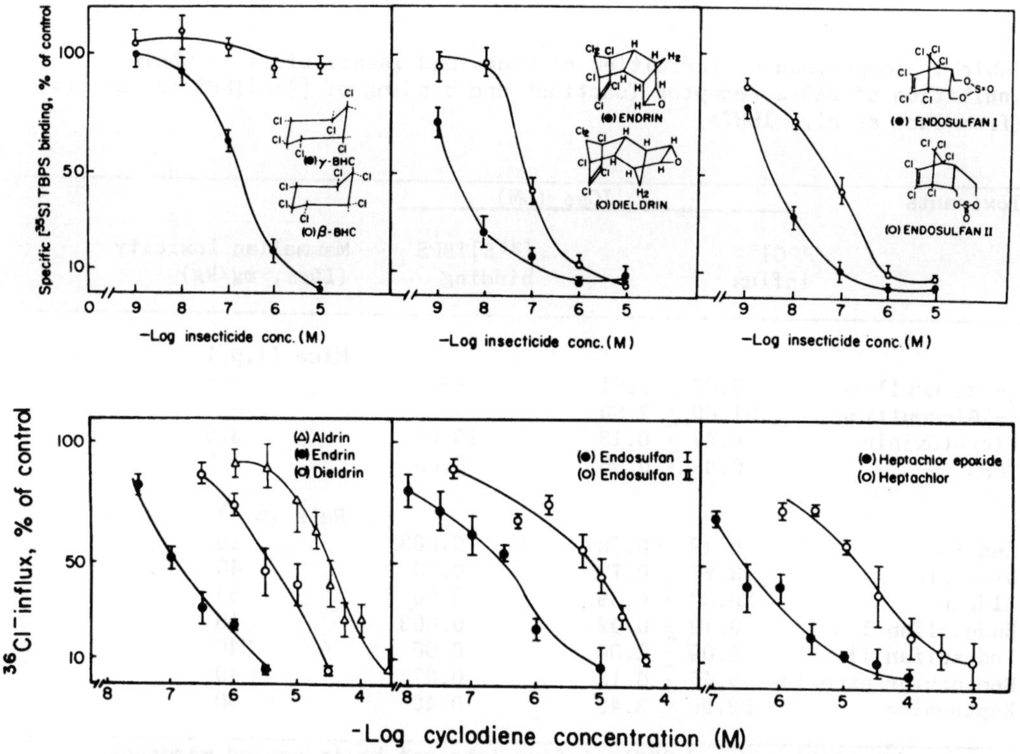

Fig. 1. The effects of isomers of BHC and cyclodiene insecticides on rat brain $GABA_A$ receptor. <u>Top</u> Inhibition of the specific binding of [^{35}S]TBPS (2 nM) to $GABA_A$ receptors. <u>Bottom</u> Inhibition of $^{36}Cl^-$ influx that is induced by 100 μM GABA. This assay of $^{36}Cl^-$ influx into resealed membranes is a biochemical correlate of the physiological function of the $GABA_A$ receptor (from Abalis et al., 1985b and Gant et al., 1987a).

as their molecular target (Lawrence and Casida, 1984; Abalis et al., 1985b; Bloomquist and Soderlund, 1985; Gant et al., 1987a). Cyclodienes bind with high affinity and stereospecificity to $GABA_A$ receptors of rat brain (Fig. 1). They displace binding of the convulsant drug [^{35}S]TBPS competitively Fig. 2), suggesting that cyclodienes have a picrotoxin-like action on $GABA_A$ receptors. They also inhibit $GABA_A$ receptor function as measured by GABA-induced $^{36}Cl^-$ influx into rat brain microsacs. The more toxic epoxides are more potent in inhibiting $GABA_A$ receptor binding and function than their parent compounds (Fig. 1; Table 1).

Cyclodienes have similar action on the GABA receptor of insects as they do on the vertebrate brain receptor, even though the insect GABA receptor is not inhibited by bicuculline. Endrin inhibits GABA responses of the D_f motoneuron of cockroach ganglia noncompetitively (Fig. 3). Thus, the convulsions produced by these insecticides in insects and mammals can be explained by their picrotoxin-like action on $GABA_A$ receptors.

Table 1. Comparison of toxicities of drugs and insecticides to their inhibition of $GABA_A$ receptor function[a] and binding of [^{35}S]TBPS in rat brain (from Gant et al., 1987a)

Toxicants	IC_{50} (μM)		
	$^{36}Cl^-$ influx	[^{35}S]TBPS binding	Mammalian toxicity (LD_{50}, mg/kg)
			Mice (i.p.)
(+)Bicuculline	3.07 ± 1.21	---	---
(-)Bicuculline	31.60 ± 7.60	---	---
Picrotoxinin	0.41 ± 0.13	10.00	3.0
TBPS	0.43 ± 0.05	0.05	1.1
			Rats (p.o)
Endrin	0.19 ± 0.06	0.003	10
Dieldrin	3.27 ± 0.72	0.10	46
Aldrin	26.30 ± 0.39	0.50	55
Endosulfan I	0.19 ± 0.07	0.003	18
Endosulfan II	8.09 ± 2.00	0.06	240
Heptachlor expoxide	0.45 ± 0.13	0.07	40
Heptachlor	22.90 ± 3.40	0.40	90

[a]Measured as GABA-induced $^{36}Cl^-$ influx into rat brain sealed membranes (microsacs).

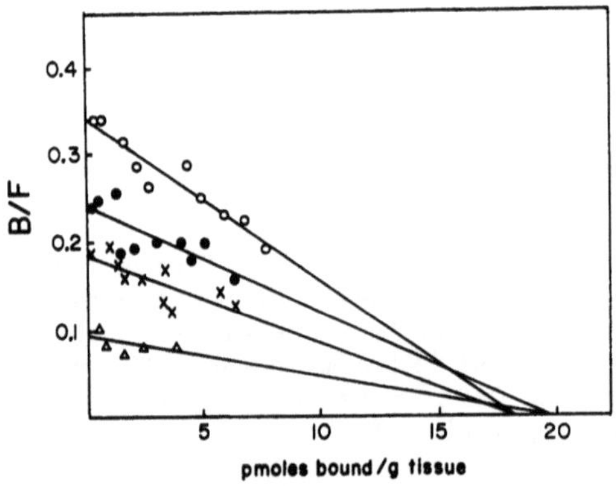

Fig. 2. Scatchard plot of the binding of [^{35}S]TBPS to the ionic channel site of the GABA receptor of rat brain in the absence (O) and presence of 1 nM (●), 3 nM (x), and 10 nM (Δ) endrin. B, amount bound in pmol/g tissue; F, free [^{35}S]TBPS concentration (from Abalis et al., 1985b).

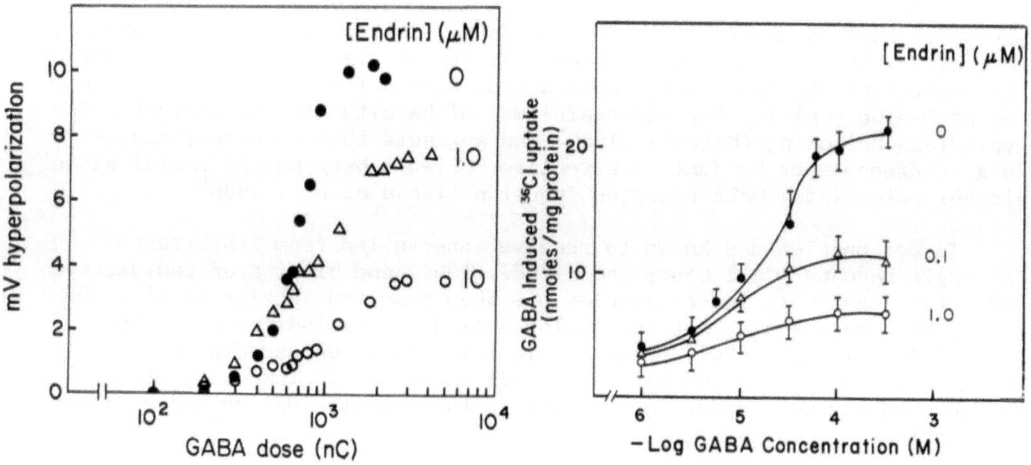

Fig. 3. Dose-dependent inhibition by endrin of: ***left***- GABA-induced hyperpolarizations of cockroach (*Periplaneta americana*) fast coxal motoneuron in ganglia; and ***right***- uptake of $^{36}Cl^-$ into rat brain microsacs induced by 100 μM GABA (from Wafford et al., 1989).

Cyclodienes also displace [^{35}S]TBPS binding to the "putative" VD Cl- channel in *Torpedo* sp. (Fig. 4). However, the affinity of cyclodienes for this Cl^- channel is 10-50 times lower than that for the $GABA_A$ receptor and the stereospecificity is also lower. Since inhibition of VD Cl^- channels could contribute to increased neural excitability and would potentiate the excitation which results from inhibition of GABA receptors, it is possible that cyclodiene toxicity may involve these VD Cl^- channels as well.

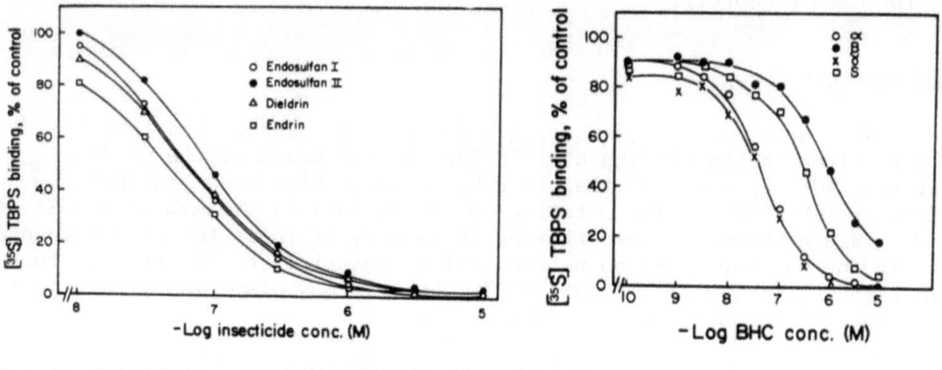

Fig. 4. Inhibition of [^{35}S]TBPS binding (2 nM) to putative Cl^- channel of *Torpedo* electric organ by: ***left***- cyclodiene insecticides; and ***right***- four BHC isomers (from Matsumoto et al., 1988).

By inhibiting $GABA_A$ receptor function, cyclodienes induce neural excitability as a result of silencing the inhibitory input of GABA at inhibitory synapses. Accordingly, activated excitatory neurons, which normally receive inhibitory GABAergic input as well, become more active in the presence of cyclodienes. This explains earlier experimental findings such as the increased transmitter release induced by cyclodienes, which results in elevated spontaneous activity in isolated cockroach nerve cord (Schroeder et al., 1977) and the marked increase in parasympathomimetic activity, that was most likely due to central stimulation of vagal centers (Metcalf, 1955). The increased vagal activity is likely to be the result of the removal of the inhibitory influence of GABA. The action on the GABA receptor also explains the successful use of barbiturates as antidotes for cyclodiene poisoning (Metcalf, 1955) and suggests that a benzodiazepine (e.g. diazepam) may be just as effective. Even a beer may be useful since alcohol potentiates GABA receptor function (Ticku et al., 1986).

Insect muscles are known to receive innervation from inhibitory GABAergic nerouns (Cull-Candy and Miledi, 1981) and binding of radioactive GABAergic ligands to insect muscles has been reported (Abalis et al., 1983). It has also been shown that heptachlor epoxide, a cyclodiene and γ-BHC inhibit GABA-induced $^{36}Cl^-$ uptake in insect muscle (Ghiasuddin and Matsumura, 1982). The inhibition of GABA receptors on insect muscle by cyclodienes may be a major contributor to the insecticidal action of these insecticides.

HEXACHLOROCYCLOHEXANE

Hexachlorocylohexane (BHC) isomers differ significantly in their toxicity to vertebrates and invertebrates. The γ isomer (lindane) is a convulsant; while the α isomer has no effect; the β isomer acts as an anticonvulsant delaying rates of kindling acquisition (Stark et al., 1986) and the δ isomer is comparatively inactive. It is possible that the convulsant action of lindane is due to its inhibition of the $GABA_A$ receptor. The difference in toxicity of the γ and β isomers is matched by their ability to displace [^{35}S]TBPS binding to rat brain (Fig. 1). Lindane also displaces competitively the [^{35}S]TBPS binding to the putative VD Cl^- channel of *Torpedo* (Fig. 5). Interestingly, the affinity of lindane for these putative VD Cl^- channels is higher ($K_d \simeq 30$ nM) than that for the $GABA_A$ receptor (K_d 150 nM). On the other hand, binding of the different BHC isomers to this putative VD Cl^- channel does not show the stereospecificity seen for GABA receptors.

MIREX AND CHLORDECONE

Neither mirex or chlordecone, two chlorinated hydrocarbon insecticides has any effect on the binding of [^{35}S]TBPS to rat brain or the GABA-regulated $^{36}Cl^-$ uptake. On the other hand, only chlordecone inhibits binding of [^{35}S]TBPS to the putative VD Cl^- channel of *Torpedo* with a Ki of 4.6 μM. Since these two insecticides share many of their toxicity symptoms with lindane and mirex may be metabolically converted to chlordecone, then it is tempting to speculate that VD Cl^- channels may also play a role in the toxic reaction to these insecticides.

PYRETHROIDS

The poisoning symptoms of pyrethroids in mammals are of two types (Gray, 1985). Type I pyrethroids produce a tremor syndrome and type II pyrethroids produce a syndrome characterized by profuse salivation followed by spontaneous sinuous writhing (choreoathetosis) and chronic/tonic convulsions. In insects, Type I pyrethroids generally produce restlessness, incoordination, prostration and paralysis (Gammon et al., 1981; Casida et al., 1983), while type II pyrethroids produce incoordination, convulsions and intense hyperactivity (Gammon et al., 1982). The primary molecular target for pyrethroids is the VD Na^+ channel. Pyrethroids interfere with the inactivation mechanism of the Na^+ channel, thereby keeping the activated channel open longer, and increasing the level of depolarization, which lead

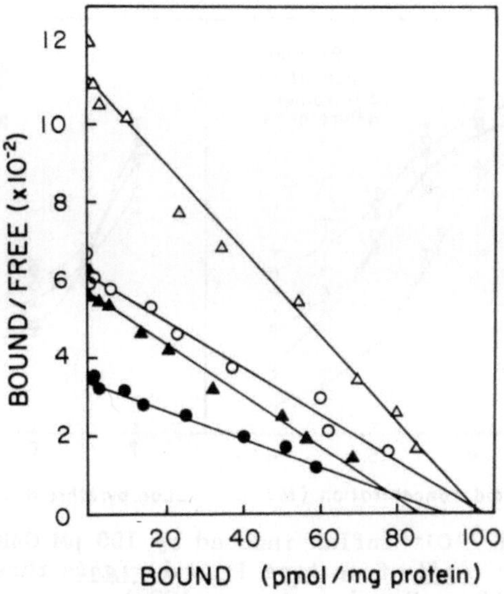

Fig. 5. Scatchard plot of specific [^{35}S]TBPS binding (at 2 nM) to the putative Cl^- channel in *Torpedo* membranes in absence (o, •) and presence (Δ, ▲) of 150 mM KCl. Closed symbols represent binding in presence of 30 μM γ-BHC (from Matsumoto et al., 1988).

to multiple discharges in response to a single stimulus. Type II pyrethroids are more potent than type I, though both types have the same mechanism of action on Na^+ channels and comparable to that of DDT (Narahashi, 1985). Pyrethroids also act on the $GABA_A$ receptor, inhibiting stereospecifically its [^{35}S]TBPS binding (Lawrence and Casida, 1983; Seifert and Casida, 1985) and the GABA-regulated $^{36}Cl^-$ influx (Fig. 6). Type II pyrethroids are much more potent. However, the higher the concentration required for inhibition of the $GABA_A$ receptor than the Na^+ channel and the poorer correlation with toxicity, suggest that the $GABA_A$ receptor is generally a secondary target for pyrethroid toxicity. The $GABA_A$ receptor is more important for the action of type II pyrethroids, which produce convulsions. The action on $GABA_A$ receptors could potentiate the effect of pyrethroids on Na^+ channels resulting in higher levels of excitation.

AVERMECTIN

The anthelmintic and insecticidal avermectin B_{1a} (AVM) also affects GABA receptor function (Egerton et al., 1979; Ostlind et al., 1979; Pong et al., 1982). It binds to the recognition site of $GABA_A$ receptor as evidenced by its ability to displace [^{3}H]muscimol binding competitively and to potentiate [^{3}H]flunitrazepam binding (Fig. 7) (Abalis et al., 1986). Like agonists, AVM also inhibits [^{35}S]TBPS binding to rat brain $GABA_A$ receptor. However, AVM inhibits the GABA induced $^{36}Cl^-$ uptake, suggesting that it binds to the $GABA_A$ receptor and possibly acts as a partial GABA agonist and also a blocker of its Cl^- channel. Since activation of the $GABA_A$ receptor leads to hyperpolarization, then AVM may be a hyperpolarizing blocker. A

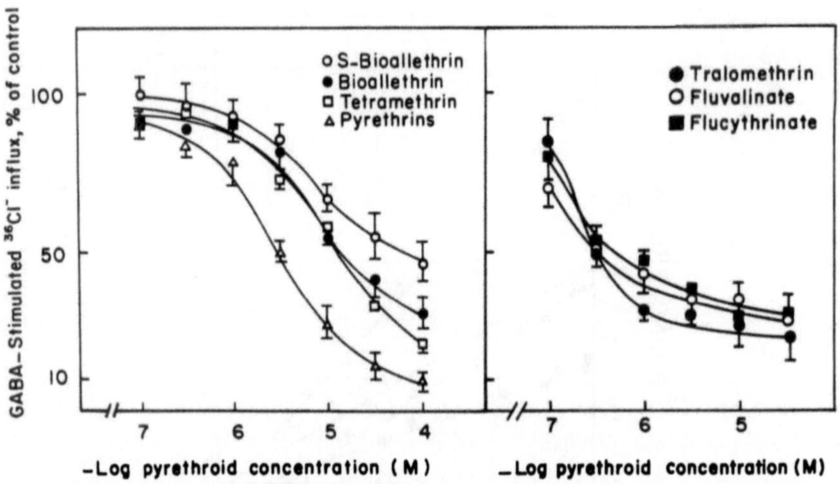

Fig. 6. Inhibition of $^{36}Cl^-$ influx induced by 100 μM GABA into rat brain microsacs by: ***left***- four type I; and ***right***- three type II pyrethroids (from Ramadan et al., 1988).

second possible target for AVM in the nervous system, may be non-GABAergic Cl^- channel. The findings of Duce and Scott (1985) that AVM irreversibly increases Cl^- conductance in insect muscles known to be void of GABA receptors, raises the possiblity that such action is a result of activation of VD Cl^- channels. AVM also increases $^{36}Cl^-$ efflux from rat brain synaptoneurosomes and this increase is insensitive to GABAergic drugs, suggesting that it may be the result of activation of VD Cl^-channels (Matsumoto and Eldefrawi, unpublished). However, this AVM-induced $^{36}Cl^-$ efflux is not blocked by Cl^- channel blockers. Also, AVM has no effect on [^{35}S]TBPS binding to the putative VD Cl^- channel of *Torpedo* electric organ (Matsumoto et al., 1988). An alternative explanation is that AVM by itself may form anion channels in the membrane allowing Cl^- flux. Such a possibility should be easy to test by studying the effect of AVM on $^{36}Cl^-$ efflux from $^{36}Cl^-$ loaded liposomes made from pure lipids.

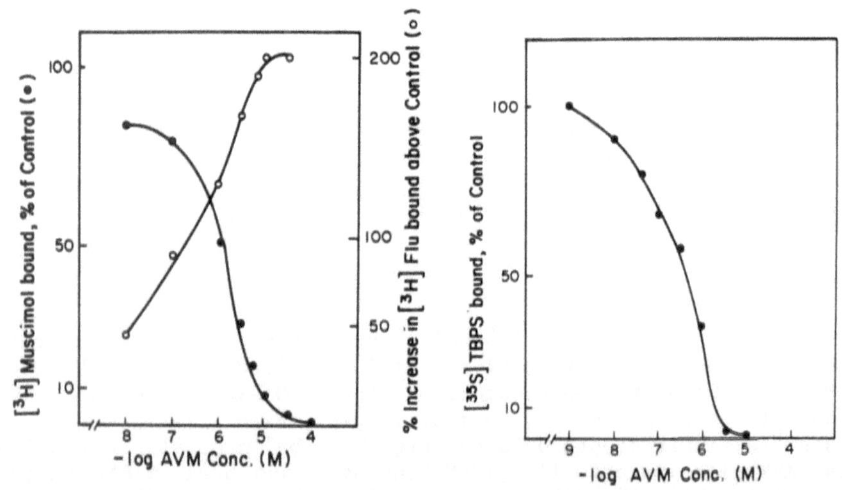

Fig. 7. The dose-dependent effect of AVM on binding to rat brain membranes of: ***left***- [^{3}H]muscimol (4 nM) (•); and [^{3}H]flunitrazepam (2 nM) (o) and ***right***- [^{35}S]TBPS (2 nM). Control binding, in the absence of AVM, is considered as 100% for [^{3}H]muscimol and [^{35}S]TBPS and 0% for [^{3}H]flunitrazepam (from Abalis et al., 1986).

ORGANOPHOSPHATES

The highly toxic bicycloorganophosphates (e.g. TBPS) are potent $GABA_A$ receptor blockers, but poor cholinesterase inhibitors. Organophosphate anticholinesterases are also highly toxic to mammals and induce seizures and convulsions. However, it is believed that their toxicity is due mainly to modulation of cholinergic transmission as a result of their potent and irreversible inhibition of acetylcholinesterase. A few such organophosphates were also found to inhibit $GABA_A$ receptor binding and function but only in relatively high concentrations (Table 2) (Gant et al., 1987b). Some compounds also react with the putative VD Cl^- channel of *Torpedo* electric organs with even higher affinity as evidenced by the lower IC_{50} values on the [^{35}S]TBPS binding (Table 2). There is lack of correlation between the affinities of these organophosphates and their potencies in inhibiting binding by the GABA receptor and VD Cl^- channel and their potencies in inhibiting acetylcholinesterase or in inducing delayed neurotoxicity. It suggests that neither of these two proteins is a primary target for the toxicity of organophosphate anticholinesterases although a few may act on Cl^- channels as a secondary target.

Table 2. Effects of organophosphates on GABA induced $^{36}Cl^-$ influx and [^{35}S]TBPS binding to *Torpedo* voltage-dependent chloride channels (from Gant et al., 1987b)

Organophosphate	IC_{50} (μM)[a] GABA-induced $^{36}Cl^-$ influx[b]	[^{35}S]TBPS binding[c] GABA receptor	[^{35}S]TBPS binding[c] Cl^- Channel
EPN	75.8 ± 15.4	52.2 ± 3.0	0.3 ± 0.1
Leptophos	253.0 ± 120.1	35.6 ± 1.5	8.7 ± 1.2
Leptophos oxon	89.6 ± 30.4	13.4 ± 0.1	2.4 ± 0.1
TOCP	14.2 ± 3.4	17.6 ± 0.1	1.5 ± 0.5
Triphenyl phosphate	18.2 ± 6.5	12.9 ± 0.6	10.3 ± 1.9

[a] Values are means ± standard errors of three separate experiments.
[b] GABA used at 100 μM to induce maximal $^{36}Cl^-$ influx.
[c] At 1 nM.

CONCLUDING REMARKS

Activation, inhibition, or modulation of the $GABA_A$ receptor can have a variety of pronounced effects on nervous function. These effects are represented by convulsion, depression, anxiolysis, hypnosis, and/or tremorgenesis. The $GABA_A$ receptor is the primary molecular target for many therapeutics as well as toxicants. Of all the insecticides, the $GABA_A$ receptor is suggested to be the primary target for the action of cyclodienes and possibly γ-BHC, and an important target for AVM and possibly type II pyrethroids. It may also contribute to the toxicity of certain organophosphate insecticides. VD Cl^- channels may prove to be primary molecular targets for insecticides such as AVM and γ-BHC and secondary target for certain organophosphates and chlordecone. It should be feasible to utilize the differences in the molecular structures of the GABA receptors of insects and vertebrates to develop more insect-selective toxicants.

ACKNOWLEDGEMENTS

This research was supported in part by NIH grant ES 02594. We thank Mrs. Sharon Mills for word processing.

REFERENCES

Abalis, I. M., Eldefrawi, M. E. and Eldefrawi, A. T., 1983, Bichemical identification of putative GABA/benzodiazepine receptors in housefly thorax muscles, Pestic. Biochem. Physiol., 20:39.

Abalis, I. M., Eldefrawi, M. E. and Eldefrawi, A. T., 1985a, Binding of GABA receptor channel drugs to a putative voltage-dependent chloride channel in *Torpedo* electric organ membranes, Biochem. Pharmacol., 34:2579.

Abalis, I. M., Eldefrawi, M. E. and Eldefrawi, A. T., 1985b, High affinity stereospecific binding of cyclodiene insecticides and γ-hexachlorocyclohexane to γ-aminobutyric acid receptors of rat brain, Pestic. Biochem. Physiol., 24:95.

Abalis, I. M., Eldefrawi, A. T. and Eldefrawi, M. E., 1986, Actions of avermectin B_{1a} on the γ-aminobutyric acid receptor and chloride channels in rat brain. J. Biochem. Toxicol., 1:69.

Blatz, A. L. and Magleby, K. L., 1986, Quantitative description of three modes of activity of fast chloride channels from rat skeletal muscle, J. Phýsiol. (London), 378:141.

Bloomquist, J. R. and Soderlund, D. M., 1985, Neurotoxic insecticides inhibit GABA-dependent chloride uptake by mouse brain vesicles, Biochem. Biophys. Res. Comm., 133:37.

Bormann, J., 1988, Electrophysiology of $GABA_A$ and $GABA_B$ receptor subtypes, Trends in Neuro., Sci. 11:112.

Bretag, A., 1987, Muscle chloride channels, Physiol. Rev., 67:618.

Casida, J. E., Gammon, D. W., Glickman, A. H. and Lawrence, L. J., 1983, Mechanisms of selective action of pyrethroid insecticides, Ann. Rev. Pharmacol. Toxicol., 23:413.

Cull-Candy, S. G. and Miledi, R., 1981, Junctional and extrajunctional membrane channels activated by GABA in insect locust muscle fibres, Proc. R. Soc. Lond., B211:527.

Duce, I. R. and Scott, R. H., 1985, Actions of avermectin B_{1a} on insect muscle, Br. J. Pharmac., 85:395.

Egerton, J. E., Ostlind, D. A., Blair, L. S., Eary, C. H., Suhayda, D., Cifelli, S., Piek, R. F. and Campbell, W. C., 1979, Autimicrab, Agents Chemother, 15, 327.

Eldefrawi, A. T. and Eldefrawi, M. E., 1987, Receptors for γ-aminobutyric acid and voltage-dependent chloride channels as targets for drugs and toxicants, FASEB J., 1:262.

Enna, S. J., 1983, GABA Receptors, In "The GABA Receptors" (Enna, S.J., ed), p.1, Humana, Clifton, N.J..

Enna, S. J. and Karbon, E. W., 1987, Receptor regulation: evidence for a relationship between phospholipid metabolism and neurotransmitter receptor-mediated cAMP formation in brain, Trends in Pharmacol. Sci., 8:21.

Gant, D. B., Eldefrawi, M. E. and Eldefrawi, A. T., 1987a, Cyclodiene insecticides inhibit $GABA_A$ receptor-regulated chloride transport, Toxicol. Appl. Pharmacol., 88:313.

Gant, D. B., Eldefrawi, M. E. and Eldefrawi, A. T., 1987b, Action of organophosphates on $GABA_A$ receptor and voltage-dependent chloride channels, Fund. Appl. Toxicol. 9:698.

Gammon, D. W., Brown, M. A. and Casida, J. E., 1981, Two classes of pyrethroid action in the cockroach, Pestic. Biochem. Physiol. 15:181.

Gammon, D. W., Lawrence, L. J. and Casida, J. E., 1982, Pyrethroid toxicologies: Protective effects of diazepam and phenobarbital in the mouse and the cockroach, Toxicol. Appl. Pharmacol., 66:290.

Ghiasuddin,S. M. and Matsumura,F., 1982, Inhibition of gamma-aminobutyric acid (GABA)-induced chloride uptake by gamma-BHC and heptachlor expoxide, Comp. Biochem. Physiol. 73C:141.
Gray,A. J., 1985, Pyrethroid structure-toxicity relationships in mammals, Neurotoxicol. 6(2):127.
Hille,B., 1984, "Ionic channels of excitable membranes", Sinauer Assoc. Inc., Suderland, Mass.
Inoue,I., 1986, Characterization of chloride channel of the squid axon membrane: its role in the resting potential generation, Biomed. Res. 7(Suppl):43.
Lawrence, L.J. and Casida, J.E. (1983) Stereospecific action of pyrethroid insecticides on the γ-aminobutyric acid receptor-ionophore complex. Science 221:1399.
Lawrence,L. J. and Casida,J. E., 1984, Interactions of lindane, toxaphene and cyclodienes with brain specific *t*-butylbicyclophosphorothionate receptor. Life Sci., 35:171.
Matsumoto,K.,Eldefrawi,M. E. and Eldefrawi,A. T., 1988, Action of polychlorocycloalkane insecticides on binding of [^{35}S] *t*-butylbicyclophosphorothionate to *Torpedo* electric organ membranes and stereospecificity of the binding site, Tox. Appl. Pharmac. 95:220 .
Metcalf,R. L., 1955, "Organic insecticides". pp. 392 Interscience Publishers Inc. New York.
Narahashi,T., 1985, Nerve membrane ionic channels as the primary target of pyrethroids, Neurotoxicol. 6(2):3.
Olsen,R. W., 1982, Drug interaction at the GABA receptor-ionophore complex. Ann. Rev. Pharmacol. Toxicol. 22:245.
Ostlind,D. A.,Cifelli,S. and Land, R., 1979 Insecticidal activity of the antiparasitic avermectins, Vet. Rec. 105, 168.
Pong,S.-S.,Dehavan,R. and Vange,C. C., 1982, A comparative study of avermectin B_{1a} and modulations of the γ-aminobutyric acid receptor chloride ion channel complex, J. Neurosci. 2:966.
Schofield,P.R.,Darlison,M. G.,Fujita,N.,Burt,D. R.,Stephenson,F. A., Rodriguez,H.,Rhee,L. M.,Ramachandran,J.,Reale,V.,Glencourse,T.,Seeburg, P. H. and Barnard,E. A., 1987, Sequence and functional expression of the $GABA_A$ receptor shows a ligand-gated receptor super-family, Nature 328:221.
Schroeder,M. E.,Shankland,D. L. and Hollingworth,R. M., 1977, The effects of dieldrin and isomeric aldrin dials and transmision in the American cockroach and their relevance to dieldrin poisoning, Pestic. Biochem. Physiol., 7:403.
Seifert,J. and Casida,J. E., 1985, Regulation of [^{35}S] *t*-butylbicyclophosphorothionate binding sites in rat brain by GABA, pyrethroid and barbiturate, Eur. J. Pharmacol., 115:191.
Sigel,E. and Barnard,E. A., 1984, A γ-aminobutyric acid/benzodiazepine receptor complex from bovine cerebral cortex. Improved purification with preservation of regulatory sites and their interactions, J. Biol. Chem. 259: 7219.
Stark,L. G.,Albertson,T. E. and Joy,R. M., 1986, Effects of hexachlorocyclohexane iosmers on the acquisition of kindled seizures, Neurobehav. Teratol., 8:487.
Suzdak,P. D. and Paul,S. M., 1987, Ethanol, membranes and neurotransmitters: Novel approaches to modifying behavioral actions of alcohol, Psychopharm. Bull., 23:445.
Tank,D. W.,Miller,C. and Webb,W. W., 1982, Isolated-patch recording from liposomes containing functionally reconstituted chloride channels from *Torpedo* electroplax, Proc. Natl. Acad. Sci. USA, 79:7749.
Ticku,M. K., Lowrimore,P. and Lehoullier,P., 1986, Ethanol enhances GABA induced ^{36}Cl-influx in primary spinal cord cultured neurons, Brain Res., 17:123.
Wafford,K. A.,Sattelle,D. B.,Gant,D. B.,Eldefrawi,A. T. and Eldefrawi,M. E. 1989, Noncompetitive inhibition of GABA receptors insect and vertebrate CNS by endrin and lindane, Pestic. Biochem. Physiol., 33:313.

ANTAGONISM OF INSECT MUSCLE GLUTAMATE RECEPTORS -

WITH PARTICULAR REFERENCE TO ARTHROPOD TOXINS

Peter N.R. Usherwood and Ian S. Blagbrough*

Departments of Zoology and Pharmaceutical Sciences*
The University of Nottingham, University Park
Nottingham NG7 2RD, United Kingdom

ABSTRACT

A model of the cation-selective channel gated by the quisqualate-sensitive glutamate receptor (GluR) of locust muscle is described. This model is based, in part, upon the vertebrate, nicotinic acetylcholine receptor channel with which the GluR channel has many properties in common. The GluR channel is c. 140Å in length and an integral part of the proposed tetrameric glutamate receptor protein. It has an outer vestibule, 30Å in diameter at its extracellular face, which narrows at the channel gate to a c. 45Å long ion-selectivity filter terminating on the cytoplasmic face of the GluR. The channel is lined by fixed -ve charges to which permeating cations bind and has a hydrophobic pocket just external to the selectivity filter. Chlorisondamine, an organic cation with a maximum dimension of 9.8Å, passes through the channel at high membrane potentials. This fixes the minimum dimension of the selectivity filter at c. 10Å. The argiotoxins and philanthotoxin should permeate the channel, but perhaps only at high membrane potentials, but the high affinity of these flexible, polycationic molecules for the fixed charges on the channel wall makes them potent open channel blockers. If these toxins compete with divalent cations for the fixed charges on membrane proteins, then competition between Joro spider toxin for Ca^{2+}-binding sites on the crustacean muscle GluR could cause closed channel block since Ca^{2+} is a requirement for channel gating in this system, but not for the locust GluR. Interactions with acidic groups of membrane phospholipids may lead to changes in membrane flexibility which could account for some of the effects of these toxins on GluR.

There have been many studies of antagonism of the quisqualate-sensitive receptor (GluR) of locust skeletal muscle, but, until recently, these have provided little insight into the molecular pharmacology of this receptor and its associated ionic channel. Antagonism by indolalkylamines of postjunctional receptors at excitatory synapses on locust leg muscle was first suggested in a study by Hill and Usherwood (1961) undertaken a few years before the discovery that glutamate was the transmitter at these sites (Usherwood and Machili, 1968; Usherwood, 1969). The indolalkylamines 5-hydroxytryptamine (see Fig. 1) and tryptamine were shown to block neuromuscular transmission at these synapses, a finding subsequently found to involve the postjunctional receptors (Usherwood and Machili, 1968). In a later, more extensive study of the neuromuscular blocking action of indolalkylamines in a locust, Clements and May (1974) concluded the N-phenoxycarbonyltryptamine was the most active compound tested. They suggested that

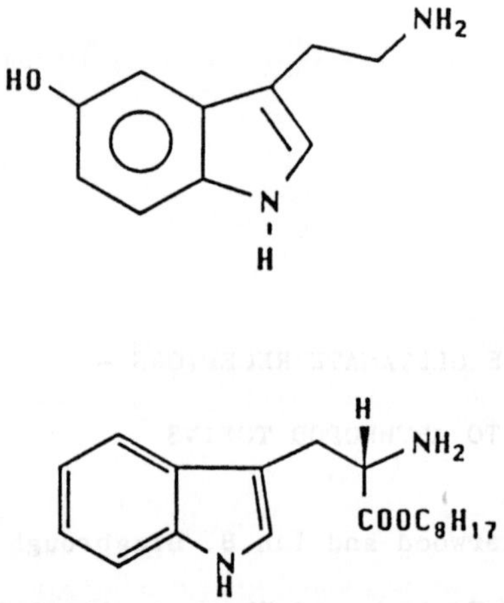

Fig. 1. 5-Hydroxytryptamine (5-HT) (top) is a possible competitive antagonist or weak agonist (high affinity; low efficacy) of locust muscle GluR, although its potential as a channel blocker has not been evaluated. L-Tryptophan octyl ester (bottom) is more potent than 5-HT in blocking glutamatergic nerve-muscle transmission in the locust.

the indolalkylamines, like some β-carbolines (e.g. harmine and harmol) also tested in their study may be partial agonists of GluR, but there remains the possibility that these compounds may also block the channel gated by GluR and perhaps even bind to other types of channel in locust muscle membrane. Unfortunately, quantitative investigations of these compounds using modern electrophysiological techniques have not yet been undertaken.

In principle, most organic compounds which are positively charged at neutral pH should be considered as possible open channel blockers of GluR. The surface membranes of excitable cells contain a majority of acidic groups attached to their extracellular faces and there is no reason to suppose that insect muscle is any different in this respect. It follows, therefore, that a negative surface potential must exist near the receptor channel (assuming of course, as we do in this chapter, that the channel of the GluR is an integral part of this membrane protein). In addition, it seems reasonable to assume that there are acidic groups within the channel which determine its selectivity for cations. Therefore, one might envisage all or part of the open channel as a chain of cation-binding sites extending across the membrane, although basic sites may also be present in part of the channel. Unfortunately, we do not have data on the anatomy and morphology of GluR, although biophysical studies undertaken by Gration et al. (1981a,b, 1982) and Kerry et al. (1987a, 1988a) suggest that it might be a tetrameric protein, perhaps with four identical subunits. Nevertheless, the channel of GluR shares many physiological properties in common with that of the vertebrate nicotinic acetylcholine receptor (AChR), for which there is good structural information together with some useful ideas on its quarternary morphology.

The vertebrate AChR is a pentameric protein which, in the electric organ of *Torpedo marmorata*, extends beyond the extracellular and cytoplasmic faces of the lipid bilayer in which it is suspended. The five protein sub-units surround a central water-filled area considered to be the receptor channel. There is an approximately 5-fold axis of symmetry along the channel axis over a large part of the molecule (Brisson and Unwin, 1985). This AChR is

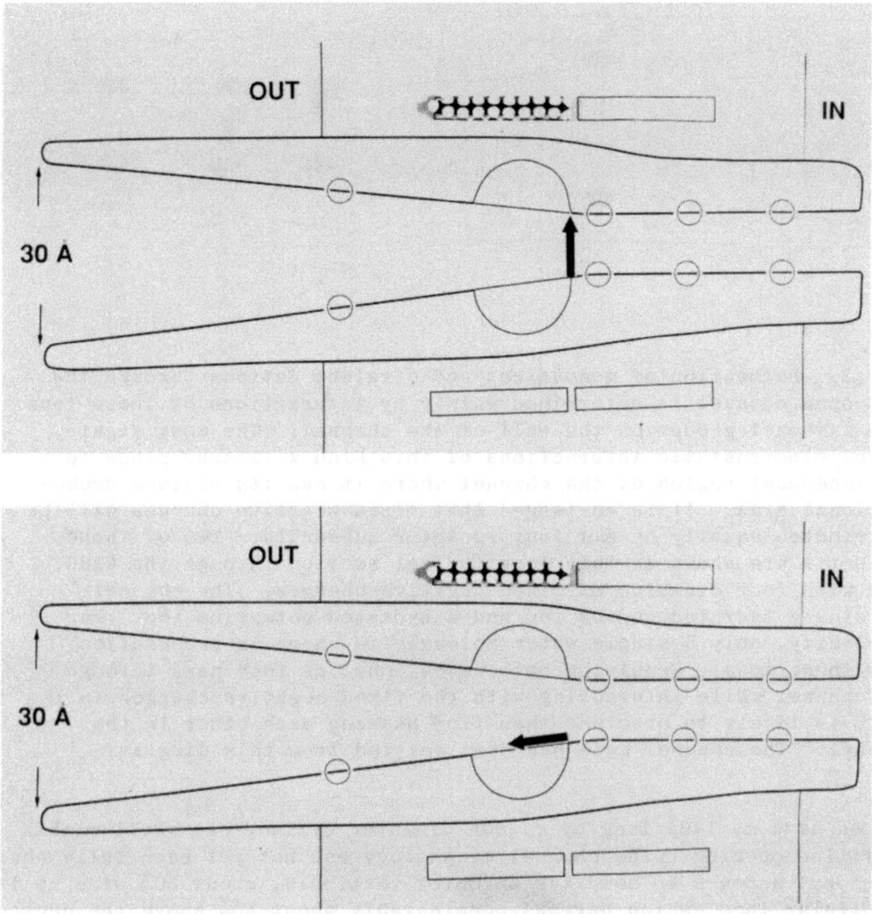

Fig. 2. Hypothetical longitudinal section of closed (top) and open (bottom) channel of GluR drawn to scale and based in part on the postulated structure of AChR (Brisson and Unwin, 1985). The lipid component of a membrane phospholipid molecule is shown [modelled using Corey-Pauling-Koltum (CPK) space filling molecular models]. The polar head groups of the phospholipids are not shown, although the approximate boundaries of the phospholipids are indicated. The GluR channel narrows to a constant cross-section at a point close to the mid-line of the membrane. It is at this point that the channel gate has been located (denoted by arrow). Carboxylic acid groups of aspartic acid or glutamic acid side chains, for example, present on the four receptor protein sub-units line the walls of the channel to provide an array of fixed negative charges (Barry and Gage, 1984). Just external to the channel gate the channel contains a hydrophobic pocket or pockets. The narrow region of the channel is the so-called selectivity filter. See text for further explanation.

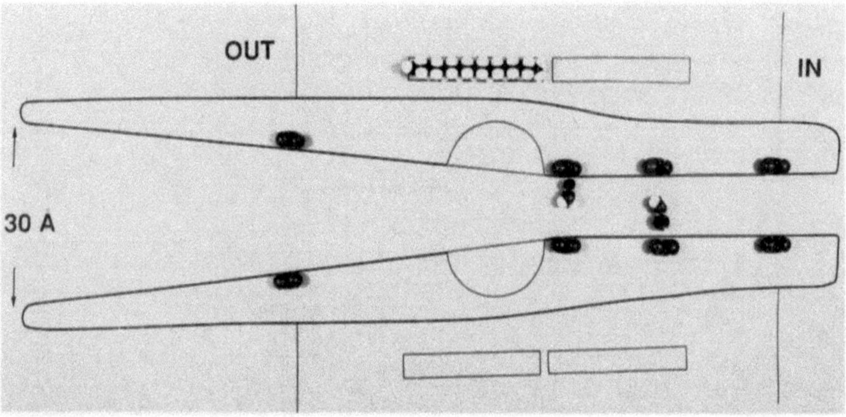

Fig. 3. Permeation of monovalent and divalent cations through the GluR open channel is determined mainly by interactions of these ions with carboxyl groups on the wall of the channel. The most significant electrostatic interactions of this kind will take place in the innermost region of the channel where it has its minimum cross-sectional area. It is envisaged that fixed negative charges will be contributed equally by all four receptor sub-units. Two of these sub-units are shown in this longitudinal section through the GluR, each with four examples of fixed negative charges. The channel contains a hydrated sodium ion and a hydrated potassium ion. For simplicity, only a single water molecule is shown in association with these ions. Repulsion between cations, as they pass through the channel while interacting with the fixed negative charges in its wall, is likely to preclude them from passing each other in the channel. The channel gate has been omitted from this diagram.

envisaged as a c. 140Å long by c. 80Å diameter cylindrical shell with a water-filled opening. The channel morphology has not yet been fully characterized, but appears to comprise an outer vestibule, about 30Å wide at its extracellular face, which narrows considerably about 15Å above the centre of the lipid bilayer. The postulated selectivity filter, which is thought mainly to determine which ion species cross the membrane through the AChR channel, is presumed to occur in this region, i.e. about two-thirds of the way into the receptor protein (Armstrong, 1971). The channel gate may also be present in this region.

THE GluR CHANNEL

Fig. 2 shows a hypothetical GluR based partly upon the AChR model. The outer part of the GluR channel narrows to a constant cross-section about mid-way across the membrane lipid. We have located the channel gate at this point, which is about two-thirds of the way in from the outside of the channel. Given the large volume occupied by the outer part of the GluR channel, the channel vestibule is unlikely to exert much influence over the movement of small inorganic and organic cations, although the presence of fixed negative charges in the channel wall in this region may concentrate cations in the vestibule. Large organic molecules could conceivably block the open channel by binding to acidic groups in the vestibule. On its cytoplasmic side the GluR channel has a relatively small, constant cross-sectional area and is lined by acidic groups. It is this part of the channel, the selectivity filter, which determines the permeability properties of the GluR, presumably on the basis of either unhydrated or hydrated ion size and ion charge. We shall assume that it is the unhydrated ion which permeates. From a purely physical perspective, electronegative sites are required within the channel to provide an energetically suitable environment for permeant cations and to account for the very low or zero permeability of anions (Anwyl and Usherwood, 1974; Anwyl, 1977). The binding of ions to fixed negative charges within the GluR channel slows permeation and those ions that dissociate slowly from these sites will block the channel. The reason for this is that fluxes of ions through membrane channels exhibit competition kinetics (Hille, 1984; Barry and Gage, 1984). This deviation from the independence principle and the known saturation of ion fluxes through membrane channels can be readily explained by assuming that ions compete for binding sites or fixed charges in the channel as part of the permeation process and that a site can only bind one ion at a time. A simple view then is that when the GluR channel opens sodium, potassium and calcium compete for sites in the channel beyond the channel gate on its cytoplasmic side, i.e. in the selectivity filter (Fig. 3). It follows, therefore, that organic cations which can enter the selectivity filter will compete with these inorganic ions. If the organic ions bind more tightly to the channel than their inorganic counterparts, channel block will ensue since permeation requires vacant electronegative sites within the channel. In principle, polycations with flexible structures should be potent open channel blockers of GluR because they may bind simultaneously to more than one fixed charged site in the channel.

Table 1 Maximum cross-sectional dimensions of some unhydrated inorganic [from Pauling (1960) radii] and organic ions (from Corey-Pauling-Koltum space-filling molecular models)

	M.Wt	Max. dimen. (Å)	Permeant
Na^+	23	1.9	Yes
K^+	39	2.66	Yes
Rb^+	85	2.96	Yes
Cs^+	133	3.38	Yes
Ca^{2+}	40	1.98	Yes*
Cl^-	35	3.62	No
NH_4^+	18	3.7	Yes
Trimetaphan	381	8.4	Yes*
Chlorisondamine	358	9.8	Yes*
Ketamine	238	7.5	? *
(+)-Tubocurarine	615	>10.0	No*
PhTx-433	453	7.5†	Yes*
ArgTx-636	636	6.5†	? *

*Also blocks the GluR channel
†Maximum cross-section, assuming linear conformation

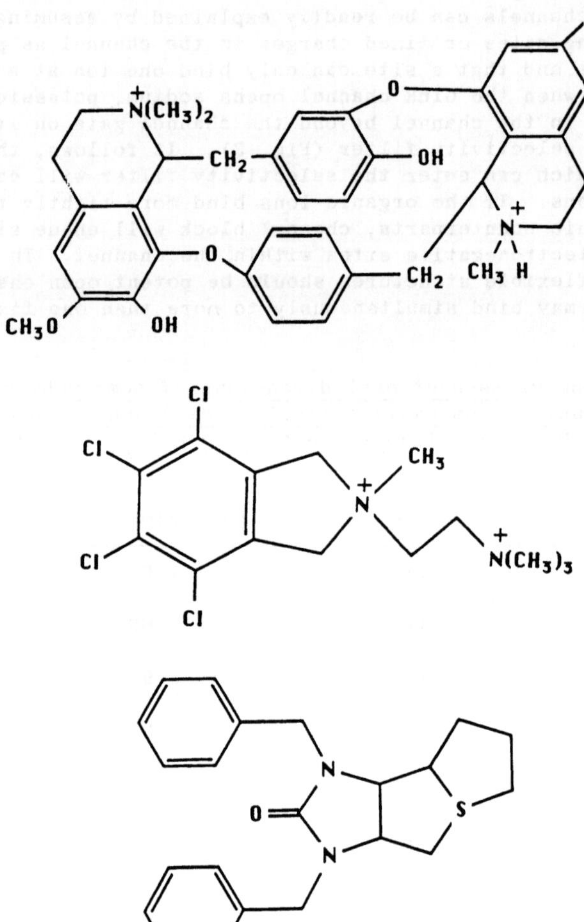

Fig. 4. Open channel blockers of locust muscle GluR. Top; (+)-tubocurarine; middle; chlorisondamine; bottom; trimetaphan.

The molecular basis for gating of the GluR channel, and for that matter other receptor channels, remains hypothetical. It probably involves tertiary or quartenary conformational changes of the channel protein either locally in the receptor protein, such as to produce a swinging door which opens and closes the channel, or more generally completely to occlude all or part of the channel. In our model of the GluR channel we have opted for the former scheme, but our proposals for channel permeation and block will accommodate either mechanism equally well.

Studies of non-competitive antagonism and permeation of AChR by a wide range of inorganic and organic cations has led to an estimate of the minimum dimensions of the channel gated by this receptor. Its permeability sequence agrees well with a frictional pore model for permeation through a circular cylinder with cross-sectional dimensions of 6.5 x 6.5Å with 0.5Å cut off at each corner (Dwyer et al., 1980). Like the AChR channel the GluR channel should be blocked by most organic ions bearing a net positive charge, although factors such as size and the presence of hydrophobic moieties in the blocking compounds may influence their potency. Although studies of permeation and block of GluR by organic cations has involved only a few compounds (e.g. Anwyl and Usherwood, 1974; Ashford et al., 1988) it is possible to gain some insight into the minimum dimensions of the open channel gated by this receptor.

Chlorisondamine (Fig. 4) is the largest ion known to permeate the GluR channel. Voltage clamp and patch clamp studies of the blocking action of this ion have shown that at high membrane potentials (i.e. > -120mV) open channel block is relieved, presumably because the ion is forced through the channel into the muscle fibre. Given that the maximum dimension of the unhydrated chlorisondamine molecule is c. 10Å (Table 1) it is reasonable to assume that the minimum dimension of the GluR channel is of the same order (Fig. 5). It is possible, of course, that the dimensions of the open channel are variable, but assuming that they are fixed and that the channel has a circular cross-section then it follows that the minimum open channel diameter is c. 10Å. This estimate has been incorporated in the channel model

Table 2 Hypothetical minimum cross sections of ionic channels

	Dimensions (Å)	Area (Å²)	Conductance (pS)	Å²/pS
K^+ (frog nerve)	3.3 x 3.3 (a)	8.6	4 (c)	2.15
Na^+ (frog nerve)	3.1 x 5.1 (a)	15.8	7 (d)	2.3
AChR (frog muscle end-plate)	6.5 x 6.5 (a)	40.3	27 (e)	1.5
GluR (locust muscle excitatory synapse)	9.8 diam. (b)	96	120 (f)	0.8

a, Dwyer et al. (1980); b, Ashford et al. (1988); c, Begenisch and Stevens (1975); d, Conti et al. (1976); e, Neher and Stevens (1977); f, Patlak et al. (1979).

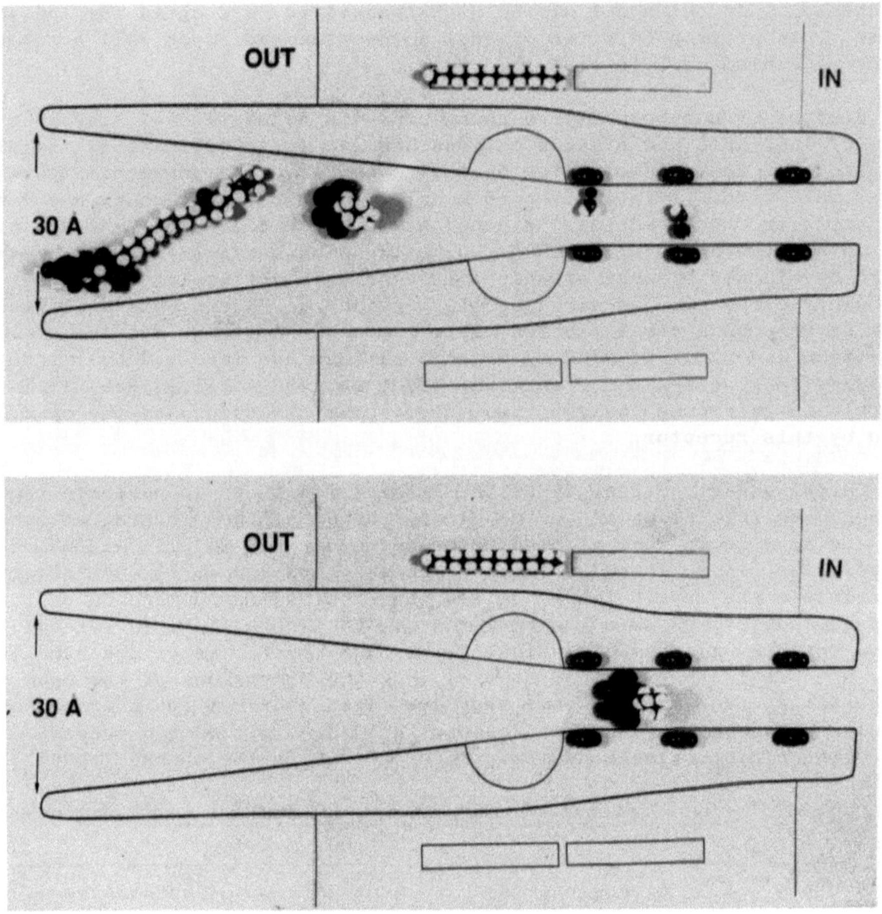

Fig. 5. Top; Chlorisondamine and ArgTx-636, as unhydrated molecules, located in the outer region of the GluR channel. Although fixed electronegative sites located on the wall of this part of the channel may favour accumulation of cations they are unlikely to have much direct influence on the permeation of small cations through the GluR channel. Large organic cations, particularly polycations, might co-operatively block the channel by binding to these sites and even single large, positively charged organic molecules may possibly block the lower reaches of the vestibule.

Bottom; An unhydrated chlorisondamine molecule fits tightly into the selectivity filter of the GluR channel. The maximum dimension of this molecule has provided an estimate of the minimum cross-sectional dimension of this region of the GluR channel.

illustrated in Fig. 2. The conductance of the open GluR channel is c. 120pS whereas that of AChR is only 25-40pS. One might anticipate from this difference that the GluR channel is much bigger than the AChR channel, but experience has shown that there is no simple relationship between open channel conductance and channel size [Hille (1984) and Table 2].

With our model of the GluR channel we are now in a position to predict which organic molecules will be permeant. This will include all positively charged and neutral compounds with maximum cross-sectional dimensions less than c. 10Å. Hence long molecules in linearly extended conformation will permeate if they are cationic (or neutral also perhaps) and their 'cylindrical' diameter does not exceed c. 10Å. We make the assumption that these ions can pass through the narrowest part of the GluR channel as unhydrated species. Channel blockers such as trimetaphan (Fig. 4) will be expected to permeate the channel, an expectation borne out by experiment when the locust muscle fibre is clamped at high membrane potentials (Ashford et al., 1987). (+)-Tubocurarine [(+)-TC] (Fig. 4) is another blocker of the open GluR channel (Kerry et al., 1987b). Like that of chlorisondamine and trimetaphan, (+)-TC block of GluR is voltage-dependent (Yamamoto and Washio, 1979; Cull-Candy and Miledi, 1983), but, unlike the block caused by chlorisondamine and trimetaphan, (+)-TC block is not relieved at high membrane potentials. To reach binding sites within the GluR channel all three compounds need to cross part of the membrane field, a factor which gives rise to their voltage-dependence, but the block by (+)-TC is dependent on voltage only at high membrane potentials. The weak voltage dependence of (+)-TC block and the absence of relief from block by this cation at high membrane potentials suggest that this compound binds to sites in the outer part of the GluR channel. The fact that it is too large to enter the selectivity filter supports this contention and leads one to conclude that it is an impermeant cation.

Other open channel blockers of locust muscle GluR which are currently receiving much attention are certain low molecular weight toxins isolated from venoms of spiders (Bateman et al., 1985; Magazanik et al., 1986; Adams et al., 1987) and of wasps (Clark et al., 1982; Eldefrawi et al., 1988), the argiotoxins (ArgTx-) and the philanthotoxins (PhTx-) respectively. The properties of these toxins have been recently reviewed by Jackson and Usherwood (1988). Both groups of compounds contain polyamines, which make the toxins polycationic (at physiological pH) and may contribute, thereby, in a major fashion to their action as open channel blockers of GluR. The rest of this Chapter will concentrate on these compounds, but to set the scene it may be useful to review briefly our knowledge of the natural polyamines.

NATURAL POLYAMINES

Polyamines with their point charges and very flexible polymethylene chains can adopt a wide variety of conformations in space and are well suited, thereby, to binding to rigid molecular frameworks of the type one might expect to find in the wall of the GluR channel. Tam and Williams (1984) have shown that the separation in space of charges on molecular frameworks is important in the association of organic ions, some cations having greater affinity for rigid anionic frameworks than could be expected from the charge on the rigid ion.

Polyamines are non-protein nitrogenous bases which are widely distributed in procaryotic and eucaryotic organisms. The high p*K*s of these compounds (Table 3) mean that at neutral pH they exist as polycations with each primary amino group, in the case of putrescine and cadaverine and, in addition, each secondary amino group in the case of spermidine, spermine

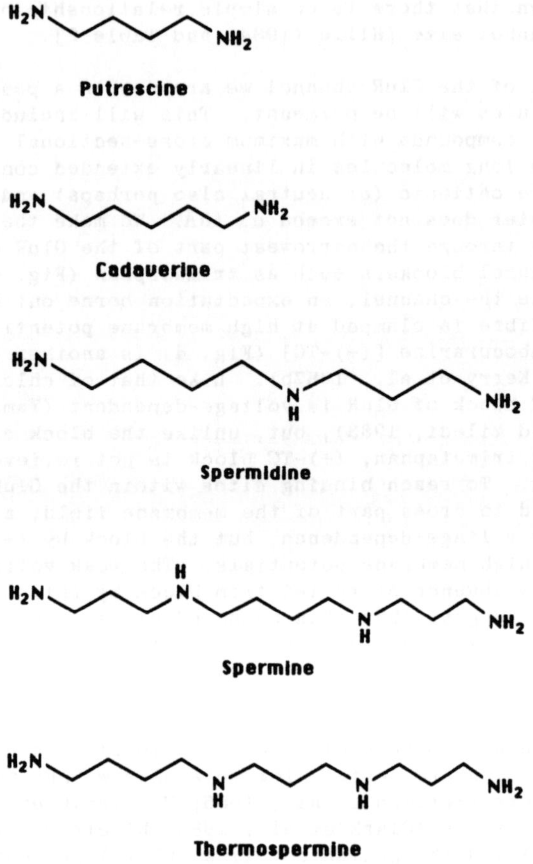

Fig. 6. Some natural polyamines. At neutral pH all primary and secondary amino groups will be protonated.

Table 3. pK values of natural polyamines in aqueous solution (c.1M). From Takeda et al. (1983) and Long et al. (1961).*

	pK_1	pK_2	pK_3	pK_4
Spermine	11.50	10.95	9.79	8.90
Spermidine	11.56	10.80	9.52	-
Thermospermine	11.62	10.57	9.35	8.31
Putrescine	10.50*	9.04*	-	-

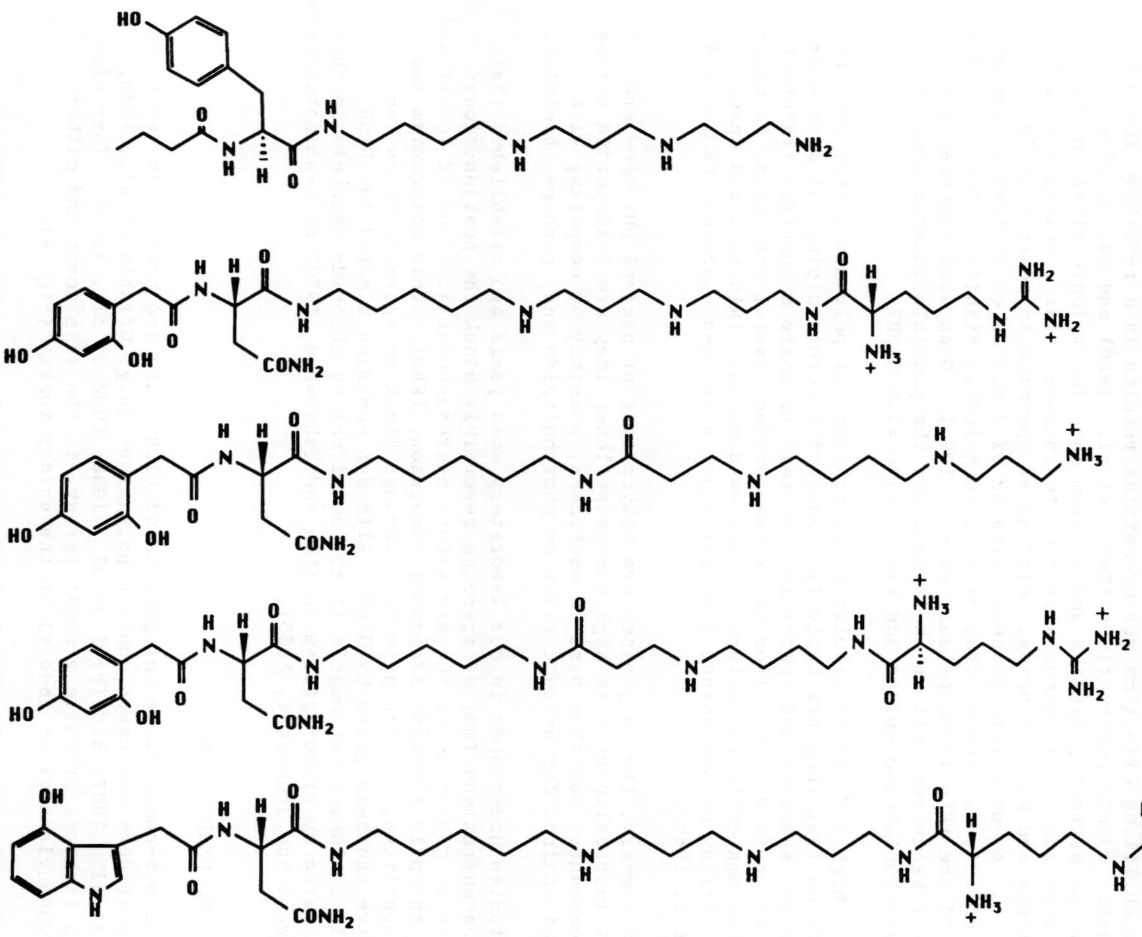

Fig. 7. Chemical structures of, top to bottom: wasp toxin (PhTx-433) and spider toxins (ArgTx-636; JSTx-3*; NSTx-3* and ArgTx-659). *From Hashimoto et al., 1987; Teshima et al., 1987. See text for further explanation.

and thermospermine (Fig. 6) protonated (or at least partially protonated). It follows, therefore, that through protonation, association and dissociation of polyamines with biological macromolecules may be controlled dynamically in correspondence with small changes in the local environment, such as ionic strength, pH, dielectric constant and temperature.

Polyamines bind to cell membranes through ionic interactions with component acidic phospholipids and proteins. Interactions with phospholipids are thought to result in membrane stabilization with concomitant reduction in the lateral mobility of membrane-bound proteins (Yung and Green, 1986). Their stabilizing effect on rat myometrium results in a reduction in spontaneous muscle contractions (Keiji et al., 1986) and may involve competition between polyamine and calcium ions for anionic sites on the muscle membrane. Polyamines also bind to intracellular constituents such as nucleotides and nucleic acids. Binding of spermine to yeast tRNA is a sequential, co-operative process; binding of 4 molecules or less of spermine has no effect, but the binding of a fifth molecule stabilises the anticodon region of the tRNA (Nothig-Laslo et al., 1981). Does such remarkable co-operative behaviour tell us anything about the possible interactions of polyamine-containing spider and wasp toxins with GluR?

The high (mM) intracellular concentrations of polyamines have lead to suggestions that they are actively transported across biological membranes. Putrescine, spermine and spermidine are said to share a specific transport system which is distinct from any known nutrient system and which, in mouse neuroblastoma cells in culture, is sodium-dependent (Rinehart and Chen, 1984). Efflux of polyamines from cells has also been observed (Nadler and Takahashi, 1985).

In summary, the polyamines are polycations at neutral pH; they are present intracellularly at high concentrations; they are transported across cell membranes; and they bind to membranes, possibly by competing with divalent cations for anionic sites on phospholipids and, perhaps, proteins.

Studies undertaken in our laboratory some years ago established that high concentrations (mM) of spermine reversibly block the postjunctional response to glutamate of the excitatory neuromuscular junction of locust and larval sheepfly (*Lucilia sericata*) (Robinson, 1980). This antagonism was non-competitive, arising primarily through block by spermine of cation-selective channels gated by GluR. Although spermine blocked the GluR channels of locust leg muscle it had no effect on chloride channels in this system gated by either glutamate (Lea and Usherwood, 1973) or γ-aminobutyrate (Usherwood and Grundfest, 1965).

SPIDER AND WASP TOXINS

The non-competitive antagonists of GluR which are present in venoms of certain spiders and wasps contain polyamines (e.g. Grishin et al., 1986; Adams et al., 1987; Eldefrawi et al., 1988; Budd et al., 1988). They also possess terminal aromatic groups; tyrosyl in the wasp toxin and either hydroxyphenylacetyl or indolyl in the spider toxins (Fig. 7).

PhTx-433 and ArgTx-636 block the open GluR channel presumably by competing with inorganic cations for binding sites in the channel. Given the relatively small size of these molecules it seems unlikely that singly they cause channel block by binding to anionic sites in the wall of the vestibule although co-operative binding of a significant number of toxin molecules in this region might be sufficient to 'occlude' the channel. Additionally these cationic molecules will be swept by the transmembrane potential into the

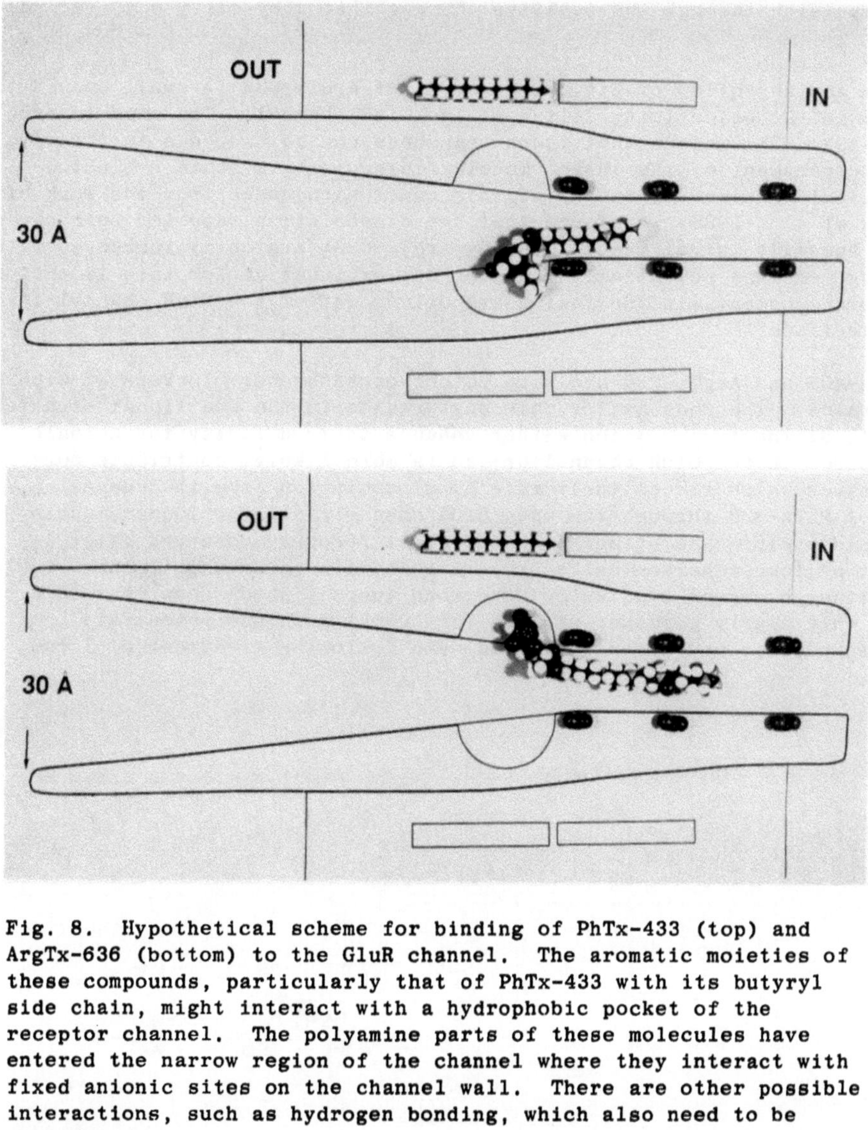

Fig. 8. Hypothetical scheme for binding of PhTx-433 (top) and ArgTx-636 (bottom) to the GluR channel. The aromatic moieties of these compounds, particularly that of PhTx-433 with its butyryl side chain, might interact with a hydrophobic pocket of the receptor channel. The polyamine parts of these molecules have entered the narrow region of the channel where they interact with fixed anionic sites on the channel wall. There are other possible interactions, such as hydrogen bonding, which also need to be considered.

selectivity filter of the open GluR channel where they will bind to sites on the channel wall (Fig. 8). The flexibility of the polyamine moieties in these toxins and their many protonated sites at physiological pH leads one to suppose that binding to the GluR channel might be quite tight, which could account for their high affinity (Kerry et al., 1988b). Given the dimensions of the fully extended toxin molecules (Table 1), it is conceivable that more than one molecule of toxin could exist side by side in the narrow part of the GluR channel so that binding in this region may also be co-operative, but mutual repulsion through the positive charges that they carry may preclude this.

From an inspection of Fig. 9, we see that ArgTx-636 is small enough to pass through the selectivity filter of the GluR channel. The same is true for PhTx-433. This means that these compounds can gain access to the cytoplasmic compartment of the insect muscle fibre via this route. Electrophysiological evidence in favour of this conclusion comes from the work of Magazanik et al. (1986) who found that the dissociation rate for open channel block of housefly larval muscle GluR by ArgTx-636 (argiopin) increases with increasing membrane potential. The simplest explanation for this is that at high membrane potentials the toxins are driven out of the GluR channel into the muscle fibre.

PhTx-433 and ArgTx-636 are more potent open channel blockers of GluR than spermine. The reasons for this must reside in the additional structural components of the toxins which either enhance their affinity for anionic binding sites in the GluR channel or, as is more likely, contribute more binding sites which reduce their rate of dissociation from the channel. Passage of PhTx-433 through the open GluR channel into the locust muscle fibre could explain the stimulation-dependent recovery observed after treatment of locust nerve-muscle preparations with this toxin (Clark et al., 1982) [although recent studies by Usherwood (unpublished) show that this toxin is only highly permeant at high (> -100mV) membrane potentials]. It could also explain why after prolonged wash following treatment of locust

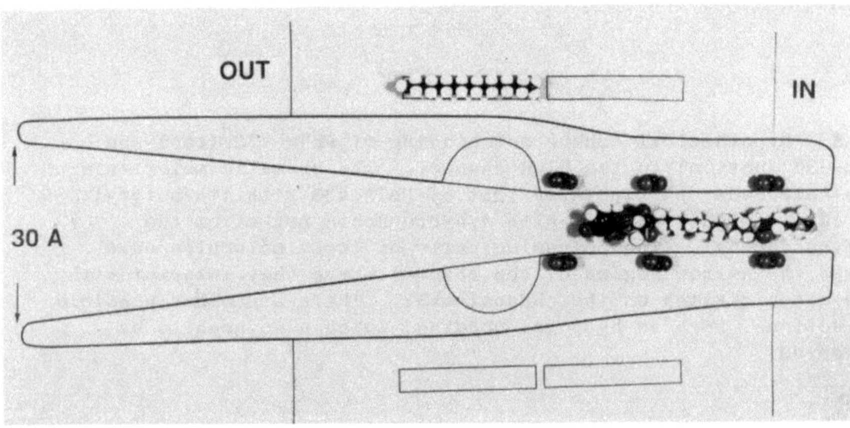

Fig. 9. ArgTx-636 is small enough to enter the narrow region of the GluR channel and to pass through to the intracellular compartment of the muscle fibre. This model assumes a linear conformation for the toxin.

nerve-muscle preparations with ArgTx-636 it is still possible to invoke use-dependent reduction in muscle twitch amplitude because of open channel block of GluR (Usherwood et al., 1984; Bateman et al., 1985; Usherwood and Duce, 1985). Of course, it is possible that these toxins gain entry to the muscle cytoplasm via a polyamine uptake system, but such a system has not been demonstrated for locust skeletal muscle and it is not known whether such a system (Fig. 10), if present, would transport these toxins.

We can now provide explanations for some, hitherto, puzzling observations on the effects of PhTx-433 and ArgTx-636 on GluR. It now seems clear that these toxins block the open GluR channel and that they can enter the muscle via the open channel route, particularly at high membrane potentials. It still remains to be established whether these hydrophilic toxins can cross the lipid bilayer by other routes and whether they can block the GluR channel by gaining access via the membrane lipid (Jackson and Usherwood, 1988) as has been proposed for channel block of AChR by some local anaesthetics (Hille, 1984). There is no evidence to suggest that they are competitive antagonists of GluR. Also, there is no evidence that they affect GluR channel gating by binding to membrane phospholipids, i.e. through membrane stabilization, although this deserves further investigation, particularly in terms of possible interactions of these toxins with the closed channel GluR.

This leaves one outstanding problem of significance to comparative toxicology. Toxins similar in structure to PhTx-433 and ArgTx-636 have been isolated from venoms of Japanese (Joro, JSTx-) and New Guinean (Nephila spp., NSTx-) spiders (Fig. 7). It has been argued that the toxin in the venom of the Japanese spider *Nephila clavata* is a competitive antagonist of

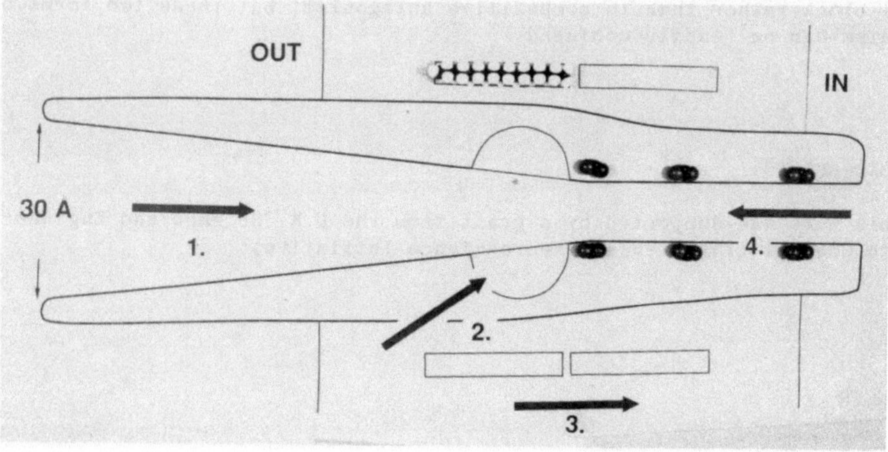

Fig. 10. Possible pathways for interaction of ArgTx-636 with locust muscle GluR.
1. via the extracellular environment of the muscle fibre to block the open channel;
2. via the membrane lipid non-competitively to block channel gating;
3. across the muscle membrane, e.g. via a polyamine transporter and then via the open GluR channel from the intracellular side of the channel;
4. from the muscle fibre interior to block the open channel.

postjunctional glutamate receptors at excitatory synapses on crayfish skeletal muscle (Abe et al., 1983). There are also arguments that the low-molecular weight toxins found in orb-web spider venoms are competitive antagonists of non-NMDA-type glutamate receptors in vertebrate central nervous systems (Jackson and Usherwood, 1988) although there is conflicting evidence on this score (Clark et al., 1982; Usherwood, 1987).

What could account for the apparent differences in mode of action between, for example, ArgTx-636 on insect muscle on the one hand and JSTx-3 on crayfish muscle on the other hand? Given the very close structural relationships demonstrated within this group of polyamine toxins, these differences are unlikely to be related to the chemical properties of the molecules, so one must look for a physiological explanation.

One major difference between insect and crustacean muscle GluR that gate cation channels is that in the insect [at least in the locust (Dudel et al., 1988)] channel gating will occur in the presence of glutamate, but in the absence of extracellular calcium, whereas in the crayfish calcium is essential either for binding of glutamate to the receptor or for coupling such binding to channel gating (Hatt et al., 1988). The dependences of channel activation in the crayfish on the extracellular Ca^{2+} concentration and glutamate concentration are identical (Hatt et al., 1988). This led Hatt et al. (1988) to conclude that in this system Ca^{2+} binds to a site on the crayfish postjunctional GluR which either makes the glutamate binding site available to exogenous glutamate or couples glutamate binding to channel gating, as mentioned previously. Other multivalent ligands e.g. Ni^{2+} can displace Ca^{2+} from its binding site and, thereby, inhibit receptor activation. It may be the case, therefore, that JSTx- and NSTx- displace Ca^{2+} from its binding site on the GluR and, thereby, inhibit either glutamate binding or channel gating. Such antagonism would, nevertheless, be akin to closed channel block rather than to competitive antagonism, but these two forms of antagonism can be readily confused.

ACKNOWLEDGEMENTS

This work was supported by a grant from the U.K. Science and Engineering Research Council (Invertebrate Neuroscience Initiative).

REFERENCES

Abe, T., Kawai, N. and Miwa, A., 1983, Effects of a spider toxin on the glutaminergic synapse of lobster muscle, J. Physiol., 339:243-252.

Adams, M.E., Carney, R.L., Fu, E.T., Jamera, M.A., Li, J.P., Schooley, D.A., Shapiro, M.J. and Venema, V.J., 1987, Structures and biological activities of three synaptic antagonists from orb weaver spider venom, Biochem. Biophys. Res. Commun., 148:678-683.

Anwyl, R., 1977, Permeability of the postsynaptic membrane of an excitatory glutamate synapse to sodium and potassium, J. Physiol., 273:367-388.

Anwyl, R. and Usherwood, P.N.R., 1974, Voltage clamp studies of glutamate synapse, Nature (Lond.), 252:591-593.

Armstrong, C.M., 1971, Interaction of tetraethylammonium ion derivatives with the potasisum channels of giant axons, J. Gen. Physiol., 58:413-437.

Ashford, M.L.J., Boden, P., Ramsey, R.L., Shinozaki, H. and Usherwood, P.N.R., 1987, Effects of trimetaphan on locust muscle glutamate receptor, J. exp. Biol., 130:405-424.

Ashford, M.L.J., Boden, P., Ramsey, R.L., Sansom, M.S.P., Shinozaki, H. and Usherwood, P.N.R., 1988, Voltage-dependent block of locust muscle glutamate channels by chlorisondamine, J. exp. Biol., 134:131-154.
Barry, P.H. and Gage, P.W., 1984, Ionic selectivity of channels at the end plate, in "Ion Channels: Molecular and Physiological Aspects", W.D. Stein, (Ed.), Academic Press, N.Y., pp. 1-51.
Bateman, A., Boden, P., Dell, A., Duce, I.R., Quicke, D.L.J. and Usherwood, P.N.R., 1985, Postsynaptic block of glutamatergic synapses by low molecular weight fractions of spider venom, Brain Res., 339:237-244.
Begenisch, T. and Stevens, C.F., 1975, How many conductance states do potassium channels have?, Biophys. J., 15:843-846.
Brisson, A., and Unwin, P.N.T., 1985, Quarternary structure of the acetylcholine receptor, Nature (Lond.), 315:474-477.
Budd, T., Clinton, P., Dell, A., Duce, I.R., Johnson, S.J., Quicke, D.L.J., Taylor, G.W., Usherwood, P.N.R. and Usoh, G., 1988, Isolation and characterisation of glutamate receptor antagonists from venoms of orb-web spiders, Brain Res., 448:30-39.
Clark, R.B., Donaldson, P.H., Gration, K.A.F, Lambert, J.J., Piek, T., Ramsey, R.L., Spanjer, W. and Usherwood, P.N.R., 1982, Block of locust muscle glutamate receptors by δ-philanthotoxin occurs after receptor activation, Brain Res., 241:105-114.
Clements, A.N. and May, T.E., 1974, Pharmacological studies on a locust neuromuscular preparation, J. exp. Biol., 61:421-442.
Conti, F., Hille, B., Neumcke, B., Nonner, W. and Stampfli, R., 1976, Conductance of the sodium channel in myelinated-nerve fibres with modified sodium inactivation, J. Physiol. (Lond.), 262:729-742.
Cull-Candy, S.G. and Miledi, R., 1983, Block of glutamate-activated synaptic channels by curare and gallamine, Proc. Roy. Soc. B., 218:111-118.
Dudel, J., Franke Chr., Hatt, H., Ramsey, R.L. and Usherwood, P.N.R. 1988, Rapid activation and desensitization by glutamate of excitatory, cation-selective channels in locust muscle, Neurosci. Lett., 88:33-38.
Dwyer, T., Adams, D.J. and Hille, B., 1980, The permeability of the end-plate channel to organic ions in frog muscle, J. Gen. Physiol., 75:469-492.
Eldefrawi, A.T., Eldefrawi, M.E., Konno, K., Mansour, N.A., Nakanishi, K., Oltz, E. and Usherwood, P.N.R., 1988, Identification and synthesis of a potent non-competitive glutamate receptor antagonist in wasp venom, Froc. Natl. Acad. Sci. USA, 85:4910-4914.
Gration, K.A.F., Lambert, J.J., Ramsey, R.L., Rand, R.P. and Usherwood, P.N.R., 1981a, Agonist potency determination by patch clamp analysis of single glutamate receptors, Brain Res., 230:400-405.
Gration, K.A.F., Lambert, J.J., Ramsey, R.L. and Usherwood, P.N.R., 1981b, Non-random opening and concentration-dependent life times of glutamate-gated channels in muscle membrane, Nature (Lond.), 291:423-425.
Gration, K.A.F., Lambert, J.J., Ramsey, R.L., Rand, R.P. and Usherwood, P.N.R., 1982, Closure of membrane channels gated by glutamate receptors may be a two-step process, Nature (Lond.), 295:599-601.
Grishin, E.V., Volkova, T.M., Arseniev, A.S., Reshetova, L.G., Onoprienko, V.V., Magazanik, L.G., Antonov, S.M. and Fedorova, I.M., 1986, Structure-functional characterization of argiopine an ion channel blocker from the venom of spider *Argiope lobata*, Bioorg. Khim., 12:1121-1124.
Hashimoto, Y., Endo, Y., Shodo, K., Armaki, Y., Kawai, N. and Nakajima, T., 1987, Synthesis of spider toxin (JSTx-3) and its analogs, Tetrahedron Letters, 28:3511-3514.
Hatt, H., Franke, Chr. and Dudel, J., 1988, Calcium-dependent gating of the glutamate activated excitatory synaptic channel in crayfish muscle, Pflugers Arch., 441:17-26.
Hill, R.B. and Usherwood, P.N.R., 1961, The action of 5-hydroxytryptamine and related compounds on neuromuscular transmission in the locust *Schistocerca gregaria*, J. Physiol., 157:393-401.
Hille, B., 1984, "Ionic channels of excitable membranes", Sinauer Associates, Inc., Sunderland, Mass.

Jackson, H. and Usherwood, P.N.R., 1988, Spider toxins as tools for dissecting elements of excitatory amino acid transmission, Trends in Neurosci., 11:278-282.

Keiji, M., Mizoguchi, Y. and Osa, T., 1986, Effects of polyamines on the mechanical and electrical properties of the isolated circular muscle of rat uterus, Jpn. J. Physiol., 35:903-916.

Kerry, C.J., Kitts, K.S., Ramsey, R.L., Sansom, M.S.P. and Usherwood, P.N.R., 1987a, Single channel kinetics of a glutamate receptor, Biophys. J., 51:137-144.

Kerry, C.J., Ramsey, R.L., Sansom, M.S.P., Usherwood, P.N.R. and Washio, H., 1987b, Single-channel studies of the action of (+)-tubocurarine on locust muscle glutamate receptors, J. exp. Biol., 127:121-134.

Kerry, C.J., Ramsey, R.L., Sansom, M.S.P. and Usherwood, P.N.R., 1988a, Glutamate receptor channel kinetics. The effect of glutamate concentration, Biophys. J., 53:39-52.

Kerry, C.J., Ramsey, R.L., Sansom, M.S.P. and Usherwood, P.N.R., 1988b, Single channel studies of non-competitive antagonisms of a quisqualate-sensitive glutamate receptor by argiotoxin$_{636}$ - a fraction isolated from orb-web spider venom, Brain Res., 459:312-327.

Lea, T.J. and Usherwood, P.N.R., 1973, The site of action of ibotenic acid and the identification of two populations of glutamate receptors on insect muscle-fibres, Comp. Gen. Pharmacol., 4:351-363.

Long, C., King, E.J. and Sperry, W.M. (Eds.), 1961, Biochemists' Handbook, vol. 46. E. and F.N. Spon Ltd., London.

Magazanik, L.G., Antonov, S.M., Fedorova, I.M., Volkova, T.M. and Grishin, E.V., 1986, Effects on the venom of the spider *Argiope lobata* and its low molecular component-argiopin on the postsynaptic membrane, Biol. Memb., 12:1204-1209.

Nadler, S.G. and Takahashi, M.T., 1985, Putrescine transport in human platelets, Biochem. Biophys. Acta, 812:345-352.

Neher, E. and Stevens, C.F., 1977, Conductance fluctuations and ionic pores in membranes, Ann. Rev. Biophys. Bioeng., 6:345-381.

Nothig-Laslo, V., Weygand-Durasevic, I., Zivkovig, T. and Kucan, V., 1981, Binding of spermine to tRNA stabilizes the conformation of the anticodon loop and creates strong binding sites for divalent cations, Eur. J. Biochem., 117:263-267.

Patlak, J.B., Gration, K.A.F. and Usherwood, P.N.R., 1979, Single glutamate-activated channels in locust muscle, Nature (Lond.), 278:643-645.

Pauling, L., 1960, "Nature of the Chemical Bond and Structure of Molecules and Crystals", 3rd edn. Cornell University Press, Ithaca, N.Y.

Rinehart, C.A. Jr., and Chen, K.Y., 1984, Characterisation of the polyamine transport system in mouse neuroblastoma cells, J. Biol. Chem., 259:4750-4756.

Robinson, N.L., 1980, Neuromuscular blockade by polyamines at the bodywall muscles of the third instar larvae of the sheepfly *Lucilia sericcata*, in "Insect Neurobiology and Pesticide Action (Neurotox 79)", Soc. Chem. Ind. Lond., pp. 237-238.

Takeda, Y., Samejima, K., Nagano, K., Watanabe, M., Sugeta, H. and Kyoguku, Y., 1983, Determination of protonation sites in thermospermine and in some other polyamines by ^{15}N and ^{13}C nuclear magnetic resonance spectroscopy., Eur. J. Biochem., 130:383-389.

Tam, S.-C. and Williams, R.J.P., 1984, Electrostatic interactions between organic ions, J. Chem. Soc. Faraday Trans. I., 80:2255-2267.

Teshima, T., Wakamiya, T., Aramaki, Y., Nakajima, T., Kawai, N. and Shiba, T., 1987, Synthesis of a new neurotoxin NSTx-3 of Papua New Guinean spider, Tetrahedron Letters, 28:3509-3510.

Usherwood, P.N.R., 1969, Glutamate sensitivity of denervated insect muscle fibres, Nature (Lond.), 223:411-413.

Usherwood, P.N.R., 1987, Interactions of spider toxins with arthropod and mammalian glutamate receptors, in "Neurotoxins and their Pharmacological Implications", P. Jenner (Ed.), Raven Press, New York, pp. 133-151.

Usherwood, P.N.R. and Duce, I.R., 1985, Antagonism of glutamate receptor channel complexes by spider venom polypeptides, Neurotoxicology, 6(2), 239-250.

Usherwood, P.N.R. and Grundfest, H., 1965, Peripheral inhibition in skeletal muscles of insects, J. Neurophysiol., 28:497-518.

Usherwood, P.N.R. and Machili, P., 1968, Pharmacological properties of excitatory neuromuscular synapses in the locust, J. exp. Biol., 49:341-361.

Usherwood, P.N.R., Duce, I.R. and Boden, P., 1984, Slowly reversible block of glutamate receptor channels by venoms of *Argiope trifasciata* and *Araneus gemma*, J. Physiol. (Paris), 79:169-171.

Yamamoto, D. and Washio, H., 1979, Curare has a voltage-dependent blocking action on the glutamate synapse, Nature (Lond.), 281:372-373.

Yung, M.W. and Green, J., 1986, The binding of polyamines to phospholipid bilayers, Biochem. Pharm., 35:4037-4042.

MOLECULAR INTERACTIONS OF ORGANOPHOSPHATES (OPs), OXIMES AND CARBAMATES AT NICOTINIC RECEPTORS

Edson X. Albuquerque[1,2], Manickavasagom Alkondon[1],
Sharad S. Deshpande[1], Vanga K. Reddy[1] and Yasco Aracava[1,2]

[1]Department of Pharmacology and Experimental Therapeutics
University of Maryland School of Medicine
Baltimore, Maryland

[2]Molecular Pharmacology Training Program
Institute of Biophysics "Carlos Chagas Filho"
Federal University of Rio de Janeiro
Rio de Janeiro, Brazil

INTRODUCTION

The nicotinic acetylcholine receptor–ion channel macromolecule (AChR), which is densely distributed at the endplate region of skeletal muscle and in *Torpedo* electric organ, is the best characterized among agonist–gated ion channels (Karlin, 1980; Spivak and Albuquerque, 1982; Noda et al., 1983; Changeux et al., 1984; Sakmann et al., 1985). This molecule comprises the neurotransmitter recognition site and the ion channel and is formed by five polypeptide subunits α, β, γ, and δ with a stoichiometry of 2:1:1:1. The subunits of the ion channel traverse the membrane and protrude about 50 Å towards the extracellular side and 15 Å towards the interior side of the cell membrane (Ross et al., 1977; Klymkowsky et al., 1980).

The discovery of α–bungarotoxin (α–BGT) (Lee, 1972) and histrionicotoxin (Albuquerque et al., 1974, 1988b) in the early 1970's, marked the beginning of an era of great progress in the understanding of the structure and function of the AChR. In the case of α–BGT, the high specificity and rather irreversible binding to the transmitter recognition site led to isolation of the AChR, determination of its polypeptide subunits and aminoacid composition, and reconstitution into artificial membranes (Heidmann and Changeux, 1978). More recently, successful genetic encoding of this macromolecule was achieved (Noda et al., 1983; Sakmann et al., 1985).

The study of the pharmacology of the AChR disclosed its susceptibility to blockade by a number of drugs, many of them with widespread clinical use and with important toxic actions. The ACh–activated ion channels can be altered either by a competitive blockade or by a variety of noncompetitive mechanisms (Spivak and Albuquerque, 1982; Aracava et al., 1987; Albuquerque et al., 1988b). α–BGT and *d*–tubocurarine are the best representatives of the group of agents that block the ACh–recognition site (Lee, 1972; Lapa et al., 1974). The existence of other sites and their role in AChR function was revealed by another group of alkaloids produced by Colombian frogs *Dendrobates histrionicus* and named histrionicotoxins (HTX) (for recent review see Albuquerque et al., 1988b). The partial or total synthesis of derivatives of HTX and their radiolabelled analogs provided important tools for biochemical characterization of sites responsible for noncompetitive blockade. Many drugs were also reported to alter AChR activation by noncom–petitive mechanisms, namely, open channel blockade, closed channel blockade and desensitization. These chemicals include, among others, local anesthetics (Neher and Steinbach, 1978; Aracava et al., 1984), the hallucinogenic agent phencyclidine (PCP) (Aguayo and Albuquerque, 1986), acetylcholinesterase (AChE) inhibitors (Aracava et al., 1987), and oxime AChE reactivators (Alkondon et al., 1988).

The intensive studies that have been carried out in the last decade using binding and electrophysiological techniques have provided important quantitative data on the sequence of events that leads to the opening of the ion channel (Magleby and Stevens, 1972; Katz and Miledi, 1973; Anderson and Stevens, 1973; Kuba et al., 1974; Albuquerque et al., 1974; Adler et al., 1978). Most of the data support the idea that the nicotinic AChR, upon binding of agonist molecule, undergoes a conformational change which leads to an opening of its ion channel. According to the following sequence of reactions, AChR isomerization is more likely to occur when two agonist molecules (2A) bind to their sites (R) on the α subunit:

$$2A + R \underset{k_{-1}}{\overset{k_1}{\rightleftharpoons}} A_2R \underset{k_{-2}\text{ or }\alpha}{\overset{k_2\text{ or }\beta}{\rightleftharpoons}} A_2R^* \underset{\overset{k_{-3}}{D}}{\overset{\overset{D}{k_3}}{\rightleftharpoons}} A_2R^*D$$

This scheme also depicts sequential blockade of the open channels (A_2R^*) by a number of drugs, named open channel blockers, leading to a nonconductive state (A_2R^*D), as discussed below (Ruff, 1977; Adler et al., 1978; Neher and Steinbach, 1978; Ikeda et al., 1984). In addition, it is known that prolonged AChR activation or exposure to high concentrations of the ACh results in enhanced agonist affinity for its binding sites and acceleration of the conformational change of the AChR towards a desensitized state (Katz and Thesleff, 1957; Heidmann and Changeux, 1980; Spivak and Albuquerque, 1982).

Our understanding of the microscopic kinetics of the activation and blockade of the AChR was greatly enhanced with the development of patch–clamp technique in 1976 (Neher and Sakmann, 1976) and its further improvement in the early 1980's (Hamill et al., 1981). This technique allows one to record the opening and closing of the ion channels resulting from the conformational changes or gating of the single AChR macromolecules. Also, this recording mode has revealed some additional characteristics of channel activation, such as the occurrence of very brief closures during the open state of the channels generating bursting–type channel openings (Colquhoun and Sakmann, 1981; Swanson et al., 1986).

It has been shown that lifetime of the open state, frequency and length of gaps within a burst, and burst duration are influenced by a number of factors such as concentration and chemical structure of the agonist, transmembrane voltage, temperature, etc. The stereochemical requirements for the agonist to open the ion channel gate and the influence of agonist structure on other kinetic properties of the channel could be analyzed in more detail by virtue of the discovery of more rigid agonists such as (+)–anatoxin–a and its analogs (Spivak et al., 1980; Spivak and Albuquerque, 1982; Swanson et al., 1986).

The analysis of single channel currents also provided definitive evidence for the direct interactions of classic AChE inhibitors with site(s) at the AChR macromolecule. Furthermore, these interactions were implicated in the ability of certain carbamates to protect against organophosphate (OP) compounds. Our studies demonstrated that carbamates (Albuquerque et al., 1985, 1987, 1988a; Aracava et al., 1987; Kawabuchi et al., 1988, 1989) and oximes (Alkondon et al., 1988), chemicals used as antidotes for OP intoxication, alter AChR function by mechanisms not related to AChE inhibition or reactivation. The results showed that all of these agents including the OP compounds altered the kinetics of the AChR by acting as weak agonists or noncompetitive blockers. In addition, our studies disclosed that the targets of OPs and carbamates were not restricted to cholinergic synaptic elements; glutamate–activated postsynaptic receptors were similarly affected by these compounds (Idriss et al., 1986).

In this study we examined the interactions of the carbamate physostigmine, the oximes pyridine–2–aldoxime (2–PAM) and 1–(2–hydroxyiminomethyl–1–pyridino)–3–(4–carbamoyl–1–pyridino)–2–oxapropane (HI–6), and chemically related 1,1'–oxybis(methylene)bis 4–(1,1–dimethylethyl)pyridinium (SAD–128) with different sites on the AChR macromolecule. Our studies demonstrated that (+) physostigmine in spite of having no significant anticholinesterase action could effectively prevent the lethal actions of OP compounds. This property may be the result of direct interactions of this carbamate with sites on the nicotinic AChR. In addition, the present study provided evidence for the important role of interactions of 2–PAM and HI–6 with sites on the AChR in the antidotal efficacy of the oximes against OPs. This hypothesis was strongly strengthened by the findings that the compound named SAD–128, an HI–6 analog without the oxime group and therefore without significant AChE reactivating potency, was a powerful open–channel blocker at the nicotinic AChR. SAD–128 was able to revert the toxic effects in animals exposed to soman (Oldiges and Schoene, 1970; Oldiges, 1976; Clement, 1981. Our data not only provided the molecular basis for the antidotal properties of carbamates and oximes against OPs but more importantly widened our perspective for the design of new and more effective drugs for both protection and reversion of the toxic effects of irreversible AChE inhibitors.

RESULTS AND DISCUSSION

Carbamates and Related Compounds

Prophylactic effect of carbamates against OPs. Electrophysiological, toxicological and morphological studies have provided strong evidence that most of the clinically used reversible AChE inhibitors, namely neostigmine, pyridostigmine, edrophonium and (–) physostigmine, interfere with neuromuscular transmission. The agents not only prevent ACh hydrolysis but also directly interact with site(s) located on the AChR macromolecule (Meshul et al., 1985; Deshpande et al., 1986; Kawabuchi et al., 1986, 1988, 1989; Aracava et al., 1987; Albuquerque et al., 1988a). The morphology of the neuromuscular junction and the function of the postsynaptic AChRs were altered differently depending upon the carbamate used. Each of these compounds interacts with multiple targets at nicotinic synapses producing a unique spectrum of alterations in the kinetics of the AChR activation process. In addition, we believe that because of these interactions, some of the carbamates are more effective than others as antidotes against poisoning by a particular OP (Albuquerque et al., 1985, 1987; Deshpande et al., 1986; Kawabuchi et al., 1988, 1989). Many lines of evidence have also emerged from our studies that have strengthened this hypothesis:

i) (+) Physostigmine, a synthetic isomer of the natural (−) form, in spite of having no significant anti−AChE activity prevented OP−induced myopathic lesions and protected animals against lethal doses of irreversible AChE inhibitors (Albuquerque et al., 1987; see also Fig. 1 in Kawabuchi et al., 1988).

ii) All of these carbamates, including (+) physostigmine and the non−carbamate edrophonium, though with different affinity and potency, can activate or competitively block AChR through interactions with ACh−recognition site, and alter the kinetics of AChR through noncompetitive site(s) and produce different types of ion channel blockade such as reversible open channel blockade, closed channel blockade, desensitization, etc. Usually, the final effect results from a combination of two or more of these actions (Aracava et al., 1987; Albuquerque et al., 1988a).

iii) The pharmacological and toxicological effects differ among the reversible AChE inhibitors. The physostigmine enantiomers and particularly the (+) isomer induced the least damage to the neuromuscular synaptic elements (Meshul et al., 1985; Kawabuchi et al., 1988, 1989).

iv) Carbamates have distinct antidotal potencies and selective actions against OPs. Accordingly, among the carbamates (+) and (−) physostigmine provided the best protection to animals exposed to lethal doses of OP compounds (Albuquerque et al., 1985, 1987; Deshpande et al., 1986; Kawabuchi et al., 1988, 1989). *In vitro* experiments carried out for light microscopic and ultrastructural analyses confirmed these findings. Thus, whereas sublethal doses (0.08 mg/kg) of sarin, an irreversible AChE inhibitor, produced a severe and extensive loss of band pattern, and induced vacuolation, supercontracture of the subjunctional regions, and phagocyte infiltration, the muscles of rats pretreated with (−) and (+) physostigmine and injected with lethal doses of sarin (0.13 mg/kg) showed marked reduction in the severity and extent of neuromuscular synapse destruction (Fig. 1; see also Kawabuchi et al., 1988, 1989).

v) (−) Physostigmine's prophylactic potency was greatly enhanced when co−administered with drugs such as mecamylamine or chlorisondamine that had no anti−AChE activity but exhibited definite blocking actions on the nicotinic AChR (Albuquerque et al., 1985; Deshpande et al., 1986).

Moreover, the carbamate−OP antagonism was further complicated because OP actions, similar to carbamates, were not restricted to inhibition of AChE activity. All OP compounds tested interacted with sites at the AChR macromolecule producing multiple, simultaneous alterations of the kinetics of AChR ion channel activation (Rao et al., 1986, 1987).

Interactions of carbamates with the nicotinic AChR. In this study we compared the actions of physostigmine's optical isomers on the single−channel currents induced by nicotinic AChR activation. Single−channel currents were recorded from the perijunctional regions of fibers isolated from interosseal and lumbricalis muscles of adult frogs *Rana pipiens* using patch−clamp techniques. The details of the isolation of the single muscle fibers and recording procedure were described elsewhere (Hamill et al., 1981; Allen et al., 1984).

Agonist property. Both of the physostigmine enantiomers acted as weak agonists at muscle nicotinic AChRs. (−) Physostigmine was about 10−fold more potent than its optical isomer, disclosing a lower degree of stereospecificity for the agonist recognition site as compared to 40−fold potency ratio for the AChE

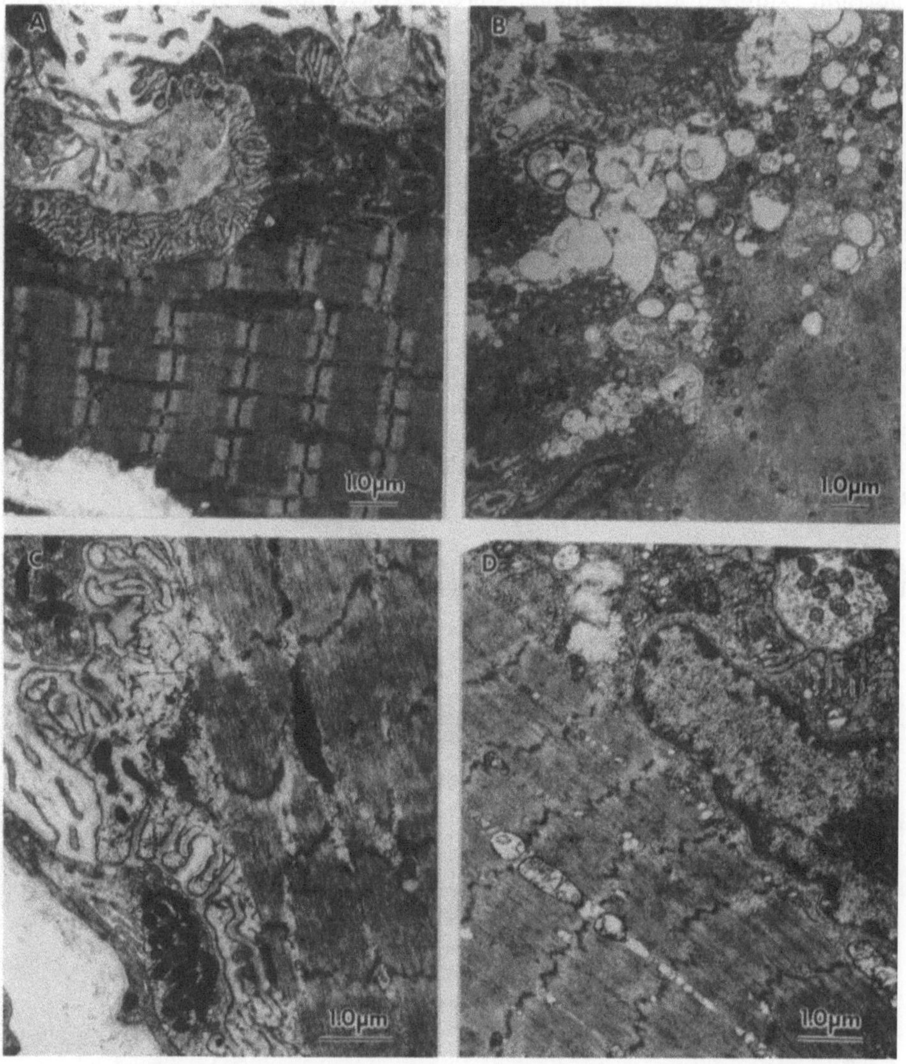

Figure 1. A: Motor endplate and synaptic sections from a soleus muscle which was removed from a rat after it was subjected to a dose of (+) physostigmine (0.3 mg/kg). B: Motor endplate from a rat injected subcutaneously with a sublethal dose of sarin (0.08 mg/kg). The soleus muscle was removed 1 hr after injection and longitudinally cut sections disclosed the intact motor nerve terminal. Note that the sarcoplasm as seen is enlarged and filled with vacuoles of mitochondrial origin. The myofibrils were totally disorganized with clear loss of the original sarcoplasmatic bands. C: Motor endplate after treatment with (−) physostigmine (0.1 mg/kg) for 1 hr. D: Motor endplate from a rat previously treated with (+) physostigmine (0.3 mg/kg) prior to a lethal injection of sarin (0.13 mg/kg). There was a marked decrease in myopathic lesions with a small number of vacuoles of mitochondrial origin. Z lines showed some slight irregularities and dislocation. It is obvious that (+) physostigmine offers significant protection against the marked damage induced by irreversible organophosphate poisoning (Kawabuchi et al., 1987, 1988, 1989).

inhibitory site (Albuquerque et al., 1988a). The enantiomers of anatoxin–a, for comparison, exhibited a much higher stereospecificity for the ACh recognition site (Swanson et al., 1986). However, the kinetics of the ion channels activated by physostigmine enantiomers were markedly distinct. (–) Physostigmine (>0.5 μM) activated currents that showed a high frequency of flickers during the open state of the channel (Shaw et al., 1985) (Fig. 2). These flickers were too brief to be adequately recorded considering our filter bandwidth and digitization rate. This contributed to a broader noise level during the channel open state and most likely accounts for the apparent decrease in single–channel conductance observed in the presence of (–) physostigmine. In contrast to (–) physostigmine, brief, square, well separated pulses with few flickers were recorded in the presence of 10 μM (+) isomer (Albuquerque et al., 1988a) (Fig. 2). The mean open times were shorter than those induced by ACh, e.g. 5.2 vs. 13 msec, at –140 mV holding potential. However, the decrease in the mean open time with increasing concentrations of (+) physostigmine and the gradual change in its sensitivity to membrane voltage suggested that at this concentration range this isomer may be acting as an open channel blocker. This pattern was also exhibited by the (–) enantiomer. Therefore, the actual characteristics of both the (+) and (–) physostigmine–activated currents could not be determined.

Blocking actions of (+) and (–) physostigmine. On the nerve–elicited endplate currents (EPCs), the studies with physostigmine enantiomers confirmed the

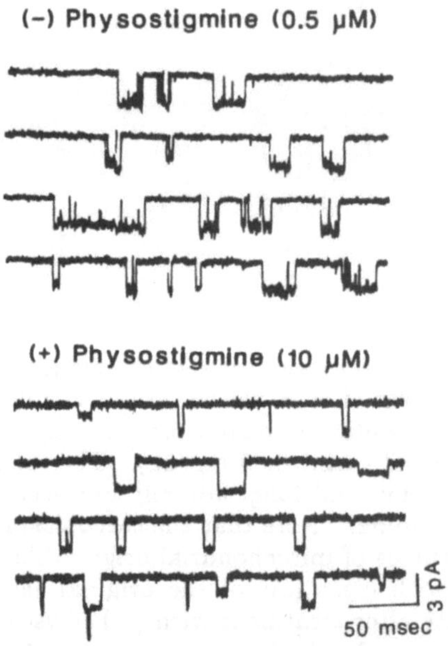

Figure 2. Samples of single–channel currents (cell–attached configuration) activated by isomers of physostigmine. Recorded from frog interosseal muscle at 10°C. Data were filtered at 3 kHz.

previous report of the lack of stereospecificity of noncompetitive blockade of ion channel sites (Spivak et al., 1983). With (+) physostigmine, analysis of EPCs eliminated the possibility of any significant anti–AChE activity owing to the absence of the potentiation of peak amplitude and prolongation of EPC decay typical of AChE inhibitors including the (–) isomer (Fig. 3; see also Fig. 1 to 5 of Shaw et al., 1985 and Fig. 4 of Albuquerque et al., 1988a). Also, the data showed that like the (–) isomer, (+) physostigmine, at concentrations higher than 2 μM, produced significant reduction of both peak amplitude and EPC decay time constant (τ_{EPC}). This suggests a noncompetitive blockade of the AChR in the open state in a manner described by the sequential model presented earlier (Shaw et al., 1985).

Single channel currents, however, enabled us to distinguish differences in the alterations induced by these enantiomers on the microkinetics of the ACh–activated currents. As in recordings obtained with physostigmine's enantiomers alone, (–) physostigmine produced channel blockade characterized by very fast blocking and unblocking reactions. Events in the presence of the (+) isomer appeared as well separated brief square–wave–like pulses. In the latter case, bursts could not be discerned, denoting a very slow unblocking rate. The blocking actions of (+) physostigmine reflected a decreased mean channel open time (τ_o) that was both

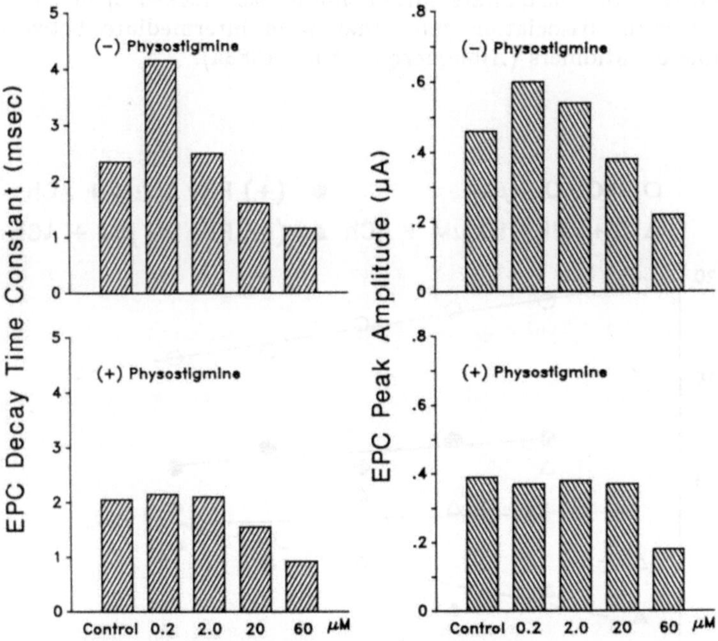

Figure 3. Effect of isomers of physostigmine on the endplate current (EPC) recorded from frog sciatic nerve–sartorius muscle preparation at 21°C. Values given were obtained at –100 mV holding potential. Note that in contrast to (–) physostigmine due to the lack of AChE blockade by (+) physostigmine neither increase in peak amplitude nor lengthening of decay time constant was observed.

concentration–and voltage–dependent (Fig. 4). The blockade increased linearly with (+) physostigmine concentration (1–50 μM) and exponentially with hyper–polarization. In addition, as the concentrations of (+) physostigmine increased, the semilogarithmic plots of τ_o vs. membrane holding potential disclosed a progressive loss of the voltage dependence. This dependence is typical of control ACh–activated currents such that at high concentrations (>20 μM) an inversion of the slope sign of these plots was observed.

The sequential model introduced earlier (Steinbach, 1968; Adler et al., 1978) was used to analyze (+) physostigmine actions. According to this model, in the presence of the blocker, the reciprocal of the mean open times ($1/\tau_o$) is governed by the rate constants k_{-2} or α and k_3 and is linearly dependent on the concentration ([D]) of the blocker. It can be represented by the following equation: $1/\tau_o = (k_{-2}(V) + k_3(V) \times [D])$. The reversal in the slope of the plot of τ_o vs. membrane potential (V) can be attributed to the strong voltage dependence of k_3 which is opposite to that of k_{-2}.

The lack of clearly defined bursts in the presence of (+) physostigmine precluded the determination of both blocked and burst times. This type of long–lasting blockade was also described for other drugs like the local anesthetics bupivacaine (Aracava et al., 1984) and QX314 (Neher and Steinbach, 1978) and the OP compound VX (Rao et al., 1987). The carbamates neostigmine and pyrido–stigmine and the non–carbamate edrophonium also blocked open nicotinic AChR channels but with dissociation rates that were intermediate between the two physostigmine enantiomers (Albuquerque et al., 1988a).

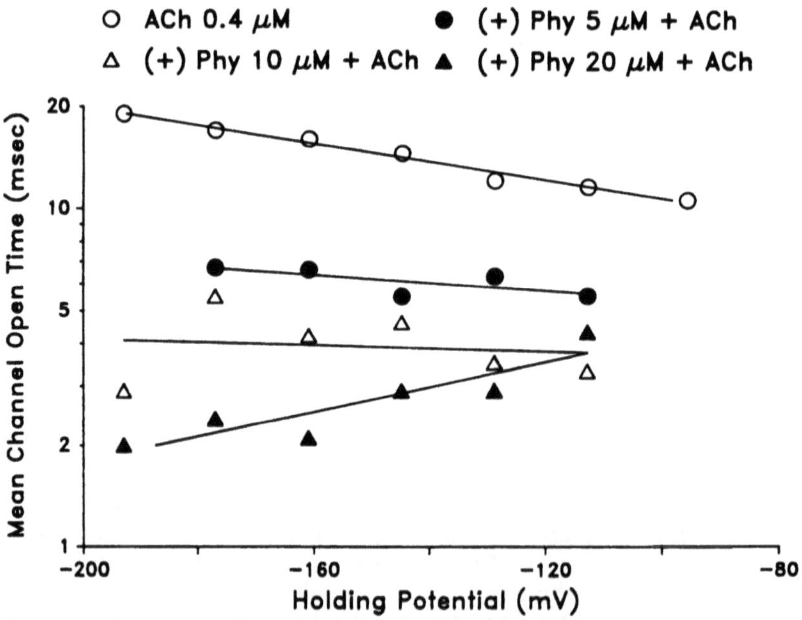

Figure 4. Relationship between mean channel open time and holding potential of channels activated by ACh in the absence and in the presence of different concentrations of (+) physostigmine. Note the marked decrease in mean channel lifetime as a function of (+) physostigmine concentration.

Table 1. Effects of 2–PAM and HI–6 on the recovery of function in muscles[a] paralyzed by OPs

OP used Dose (μM)	Experimental Condition	Twitch Tension	Tetanus Tension 50 Hz	Tetanus–Sustaining Ability	AChE Activity
None	Control	100	100	100	100
Soman (0.2)	15–min exposure	51	13	12	4
	3–hr wash	108	56	6	7
	2–PAM[b]	93	54	0	18
	HI–6[b]	108	64	100	21
Tabun (0.4)	15–min exposure	59	15	0	6
	3–hr wash	67	42	5	21
	2–PAM	92	74	100	6
	HI–6	136	53	2	21

[a]Rat phrenic nerve–diaphragm muscle preparation was used, and the results are expressed as % of control values.

[b]Muscles were treated with 2–PAM (0.1 mM) or HI–6 (0.1 mM) for 1 hr after 15–min exposure to OP and subsequent removal of its excess.

Oximes and Related Compounds

Antidotal potency: specificity against OPs. Studies carried out with 2–PAM and HI–6, mono– and bispyridinium oximes, respectively, disclosed that in general, HI–6 was more potent than 2–PAM. However, against tabun and soman, a very specific antidotal interaction occurred which was independent of the AChE–reactivation potency (Table 1). Thus, against tabun, in spite of insignificant reactivation of the enzyme (less than 5%), 2–PAM produced complete recovery of twitch tension and tetanus sustaining ability blocked by the OP. In contrast, HI–6, although reactivating AChE to a higher level (20%) than 2–PAM was unable to provide any improvement of tetanus sustaining ability. On the other hand, against soman, HI–6 was effective in recovering muscle function, although it reactivated the same 20% of the AChE activity. Against VX and sarin poisoning, in spite of better reactivation of AChE activity by HI–6 (100% vs. 50–70% for 2–PAM), both oximes were equally effective in recovering muscle function.

More recent studies carried out with SAD–128, a bispyridinium compound closely related to HI–6, reinforced the hypothesis of a mechanism unrelated to AChE reactivation underlying the antidotal actions of the classical oximes. SAD–128, although devoid of an oxime moiety which confers the AChE reactivating effect, provided effective protection of animals exposed to lethal doses of soman (Oldiges and Schoene, 1970; Oldiges, 1976; Clement, 1981). However, SAD–128 produced alterations in the kinetics of the ion channels activated by the neurotransmitter that were quite similar to those produced by 2–PAM and HI–6, and even more potent. These alterations resulted from the direct interactions of these compounds with sites located on the ion channel component of the nicotinic

AChR (Alkondon et al., 1988; Alkondon and Albuquerque, 1988). When the actions were studied in detail at the single–channel current level, all the compounds showed definite actions on the nicotinic AChR, enhancing its activation, blocking the open ion channels and/or accelerating its recovery from the desensitized state. The differential contribution of all these actions accounted for the relative potency and the selectivity of the compounds in relation to a particular OP.

Activation and blockade of the postsynaptic nicotinic AChR. 2–PAM and HI–6 did not affect presynaptic elements, membrane electrical properties or the contractile apparatus. Thus, neither of these compounds affected resting membrane potential, action potential generation or muscle twitches elicited by direct stimulation. At high micromolar or even millimolar concentrations, they failed to alter significantly the neurotransmitter release process as determined by analysis of quantal content, quantal size and frequency of spontaneously occurring miniature endplate potentials (Alkondon et al., 1988). Therefore, most of the effects were restricted to motor endplate AChRs.

Increase in AChR activation. This effect was particularly evident with 2–PAM and resulted from an increase in activation of the post–synaptic AChR (Fig. 5; see also Fig. 16 of Alkondon et al., 1988) since, as mentioned before, this oxime and others did not affect presynaptic processes (Alkondon et al., 1988). Also, AChE inhibition was not sufficient to account for this facilitation owing to the fact that enzyme activity was only affected at much higher doses of this oxime (Alkondon et al., 1988). At the macroscopic level this effect resulted in potentiation of muscle twitch tension and increased peak amplitude of the EPCs at holding potentials ranging from −50 to +50 mV. As described below, at more negative potentials blocking actions became prevalent such that the facilitatory effects were not evident.

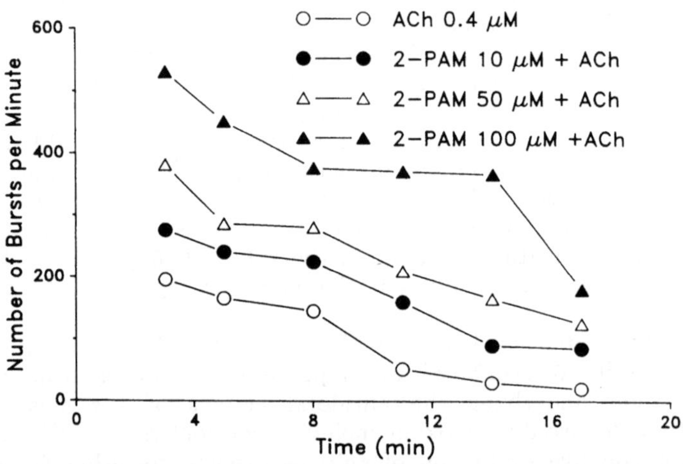

Figure 5. Effect of increasing concentrations of 2–PAM on the frequency of channel activation produced by ACh. This increase in channel opening probability in the presence of 2–PAM could be of significant value in revitalizing the function of OP–poisoned endplates.

Single–channel recordings provided the ultimate evidence for the direct interactions of these compounds with the AChR sites. Neither 2–PAM, HI–6, nor SAD–128 produced channel openings by itself, i.e. when applied in the patch micropipette alone, without ACh. However, in the presence of ACh (0.4 μM), 2–PAM (Fig. 5) and HI–6 produced a marked concentration–dependent increase in the frequency of bursts. Under control conditions, ACh (0.4 μM) activated channel openings which appeared as square–wave pulses with very few flickers during the open state and no clear bursting activity. At this concentration, desensitization appeared very slowly as evidenced by the gradual decline of the frequency of openings over the 40– to 60–min recording period. Upon the addition of 2–PAM at concentrations of 1 to 50 μM along with 0.4 μM ACh, the frequency curve was shifted to a higher level while maintaining the same slope of the declining phase (Fig. 5). Although dependent upon oxime concentration, this increase in the frequency was neither voltage– nor time–dependent. With HI–6 this facilitatory effect was significantly less marked, and it was not seen with SAD–128.

The increased channel activation could result from a primary action of 2–PAM increasing the affinity of ACh for its binding site and/or the isomerization rate constant (β) facilitating the ion channel opening. Another possible mechanism for 2–PAM's action is that this oxime could enhance channel activation by counteracting the already existing agonist–mediated receptor desensitization. This explanation seems particularly tempting, considering that OPs block neuromuscular transmission by enhancing AChR desensitization via ACh accumulation and by direct actions (Alkondon et al., 1988). Indeed, OPs have been reported to enhance AChR desensitization through direct interactions with the nicotinic AChR molecule (Eldefrawi et al., 1988).

Assuming that there is no synthesis or incorporation into the muscle membrane of new nicotinic AChRs during patch–clamp recording, one could argue that in the presence of 2–PAM more receptors become available for ACh–activation. This greater availability of activatable AChRs could result from the shift of the existing AChRs from the desensitized state. It is known that the neurotransmitter and other nicotinic agonists, at equilibrium, shift the AChRs from a low agonist–affinity state to a high agonist–affinity state(s) responsible for the development of desensitization (Changeux et al., 1984). Biochemical and electrophysiological techniques have disclosed at least two phases of desensitization (Heidmann and Changeux, 1980; Feltz and Trautmann, 1982). The onset of fast desensitization occurring on a millisecond time–scale would usually be missed under control patch–clamp recording conditions. Therefore, our recordings obtained with ACh alone may only depict the activation of those receptors that escaped the fast desensitization induced by the agonist. Under these conditions, the increased channel activation could result from 2–PAM's ability to prevent AChR isomerization towards a fast desensitized state. Slow desensitization, however, appeared to be refractory to 2–PAM's facilitatory actions since at all concentrations of this oxime parallel decline of channel activation was observed following the initial increase in frequency of openings.

<u>Blockade of AChR ion channels.</u> The analyses of the kinetics of the macroscopic EPC decays and single channel currents disclosed noncompetitive blockade of the AChR function through direct interactions of 2–PAM, HI–6 and SAD–128 with site(s) on the AChR ion channels. The ion–channel blockade was more evident with HI–6 and SAD–128. On the macroscopic EPCs, plots of τ_{EPC} vs. membrane holding potential revealed ion channel blockade only at hyperpolarized potentials (from –150 to –80 mV). Denoting a very strong voltage–dependent process, the decrease in the decay time constant was accompanied by an inversion

of the slope sign of these plots as the concentration of these drugs was increased. In the presence of HI–6 (1 μM to 2 mM) the acceleration of the EPC decay occurred without changing the single exponential function observed under control conditions. In contrast, with SAD–128, double exponential decays could be discerned at all concentrations (10–100 μM) tested at membrane potentials between −150 and −100 mV.

For better interpretation of these alterations, the microkinetics of the elementary currents were analyzed. 2–PAM (10–200 μM), HI–6 (1–50 μM) (Fig. 6) and SAD–128 (1–40 μM) (Fig. 7) when added to fixed concentrations of ACh (0.4 μM for 2–PAM and HI–6, and 0.1–0.2 μM for SAD–128), induced openings with marked increase in the frequency of flickers during the open state as compared to control ACh–induced currents. This flickering was interpreted as resulting from successive blocking and unblocking reactions before the ion channel was closed towards its resting state. The bursts with SAD–128 were much longer than those observed in the presence of 2–PAM or HI–6 because of the much longer blocked states. With HI–6 and especially with 2–PAM the high frequency of these flickers made the noise level during the open state broader than that observed during the closed state or in the absence of channel activity. In addition, as the frequency of these flickers increased with higher concentrations of these oximes (>100 μM), the inadequate recording and digitization of the very fast events induced an apparent decrease in the single channel conductance.

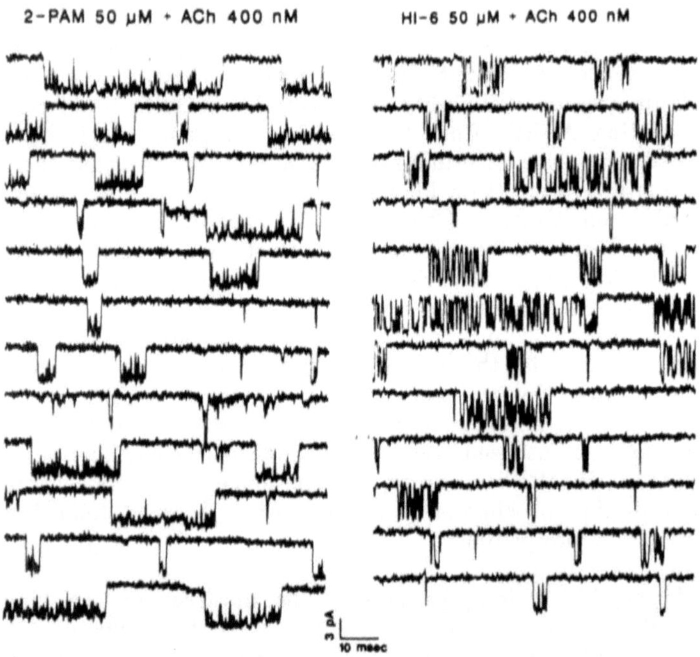

Figure 6. Samples of ACh–activated channel currents recorded from frog interosseal muscle in the presence of 2–PAM (50 μM) (left) or HI–6 (50 μM) (right) included in the patch pipette together with ACh (400 nM). Holding potential, −165 mV.

Table 2. Comparison of the channel–blocking rates[a] for different pyridinium drugs

Holding Potential (mV)	$k_3 \times 10^{-6}$ sec^{-1} M^{-1}		
	2–PAM	HI–6	SAD–128
−100	2.7	8.7	104
−120	4.0	14.5	130
−140	5.9	24.0	148
−160	8.7	39.6	170

[a]The blocking rates were obtained from single channel studies with frog muscle fibers.

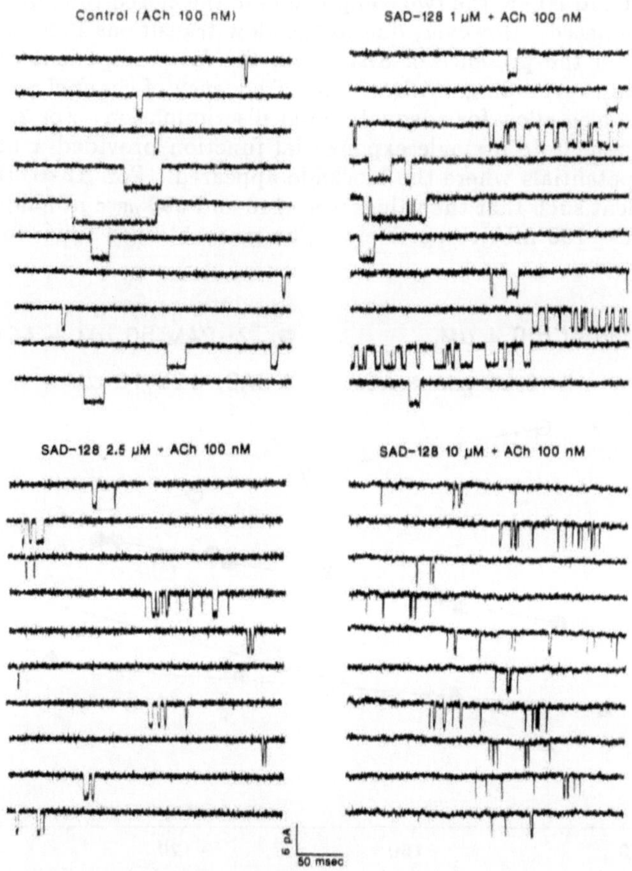

Figure 7. Samples of ACh–activated single–channel currents recorded from frog interosseal muscle in the absence and presence of SAD–128 inside the patch pipette solution. Holding potential, −140 mV.

The analysis of the open channel kinetics showed that both oximes 2–PAM and HI–6 and SAD–128 produced a concentration– and voltage–dependent reduction of the mean open times (Fig. 8). At low concentrations, this effect was only apparent at greatly hyperpolarized potentials. As the concentration of these agents increased, the effect became apparent at less negative potentials. The voltage dependence of the mean open times followed the predictions of the sequential model used to describe the actions of many ion channel blockers as presented before. As discussed earlier, the opposite voltage dependence of the rate constants k_{-2} and k_3 resulted in the blocking pattern exhibited by these drugs. Table 2 shows the k_3 ($sec^{-1}.M^{-1}$) values and its voltage sensitivity for 2–PAM, HI–6 and SAD–128. The k_3 values changed an e–fold per 52 mV and 40 mV for 2–PAM and HI–6. SAD–128 blocking actions were less voltage–dependent such that the k_3 for this drug changed an e–fold per 150 mV. Also, many other blockers such as QX–222 and (–) physostigmine produced much less voltage–dependent reduction of the mean open times.

Analysis of the distribution of the closed times showed that in the presence of these drugs they were best fitted by the sum of two exponentials. The fast component represented the numerous fast flickers or blocked state induced by 2–PAM and HI–6 and, on a much slower time scale, by SAD–128. With both oximes, 2–PAM and HI–6, the two components in the closed time histograms could be easily discriminated. However, due to the slow transitions between the blocked and open states in the presence of SAD–128, the fast component could be ade–quately separated in recordings with very low frequency of channel openings. These conditions would also allow for adequate burst discrimination. For 2–PAM, the fit of the fast component to a single exponential function provided a mean of about 130 μsec at all potentials where the blockade appeared. For HI–6, this value was voltage–dependent such that the values were 140 and 390 μsec at holding potentials of –120 mV and –180 mV, respectively. The mean blocked times for SAD–128

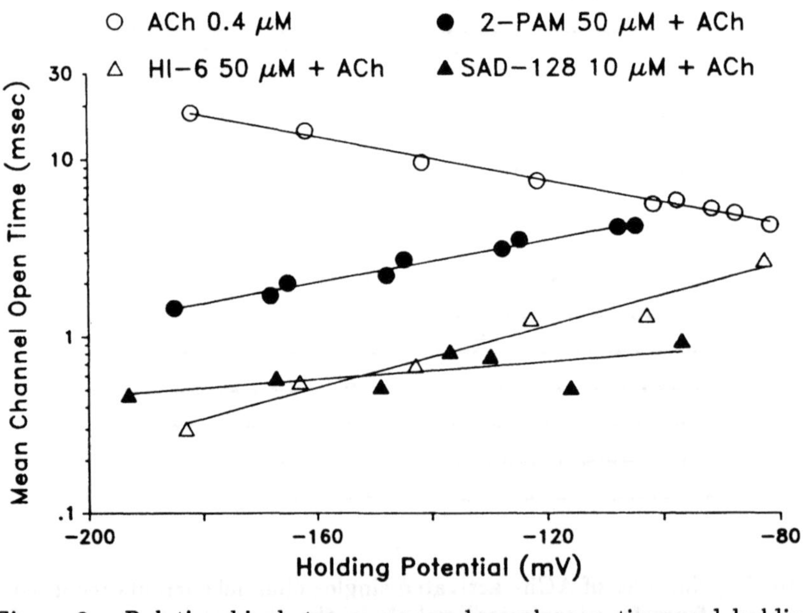

Figure 8. Relationship between mean channel open time and holding potential of channels activated by acetylcholine in the absence and in the presence of 2–PAM, HI–6 or SAD–128.

Table 3. Comparison of the channel–unblocking rates[a] for different pyridinium drugs

Holding Potential (mV)	$k_{-3} \times 10^{-3}$ sec^{-1}		
	2–PAM	HI–6	SAD–128
–100	7.8	9.9	0.70
–120	7.8	7.1	0.38
–140	7.8	5.0	0.21
–160	7.8	3.6	0.12

[a]All the unblocking rates were obtained from single channel studies with frog muscle fibers.

were also voltage–dependent but 10– to 20–fold more prolonged compared to HI–6. According to the sequential model, the mean blocked times depend solely on the rate constant for the unblocking reaction (k_{-3}). The values and the voltage dependence of k_{-3} determined from the reciprocals of the mean blocked time are shown in Table 3. The k_{-3} values changed an e–fold per 58 and 32 mV for HI–6 and SAD–128, respectively, whereas the dissociation rate constant for 2–PAM was not significantly influenced by the voltage.

The dissociation constant (K_D) values obtained for 2–PAM and HI–6 were 1.27 and 0.204 mM at –140 mV holding potential, respectively. In comparison to HI–6, SAD–128's K_D value was almost 100– to 150–fold lower, around 1.5 μM at –140 mV holding potential. The high K_D values for the oximes indicated that they bind to a low–affinity site. Using the Boltzmann distribution to describe the voltage dependence of K_D values, the location of the binding site can be estimated. For both 2–PAM and HI–6 the binding site is roughly half way across the membrane (Alkondon and Albuquerque, 1988). Similar values were determined for SAD–128 (Alkondon and Albuquerque, 1988) and for other blockers such as neostigmine and edrophonium binding sites (Aracava et al., 1987), suggesting that they bind to the same site with different affinities.

Some additional features of the blockade produced by these drugs can be discussed in light of the predictions of the sequential model. The following points argue in favor of the sequential blocking model: i) a linear decrease in the mean open times with concentration of the blocker, in the case of 2–PAM up to 100 μM and HI–6 up to 50 μM; ii) an increase in the mean blocked time with hyperpolarization in the case of HI–6 and SAD–128, but not of 2–PAM; and iii) the blocked times independent of drug concentration for HI–6 up to 50 μM and for SAD–128 up to 40 μM. However, some deviations from the predictions of the sequential model have been observed with the oximes and with SAD–128 that can be enumerated as follows: i) The model predicts that the total time that the channel

Table 4. Interaction between pyridinium drugs and acetylthiocholine

Concentration of Pyridinium Drug (μM)	Rate of Acetylthiocholine Breakdown (μM/min)		
	2–PAM	HI–6	SAD–128
50	1.1	0.4	0
100	2.0	0.9	0
200	4.8	1.9	0
500	9.7	5.2	0

spends in the open state is unaltered by the blocker. ii) The model also predicts that both the number of flickers or openings per burst and the mean burst time should increase with drug concentration. The analysis showed that the total open time in a burst and the duration of the bursts were decreased in a voltage–dependent manner as the concentrations of these drugs increased. At lower concentrations of the oximes, the number of openings per burst increased with concentration. However, at higher concentrations of oxime, particularly with HI–6, and at large negative holding potentials, a reduction in this number was observed. With a typical blocker like QX–222 (up to 40 μM) the mean burst time along with the number of openings per burst increased with concentration. The total open time per burst was thus maintained equal to that determined in the absence of the drug (Neher and Steinbach, 1978). iii) With SAD–128 the linear relationship between the reciprocal of τ_o and its concentration as predicted by the model was observed only at a low concentration of the blocker (up to 10 μM). Above this, a departure from the linearity became evident.

AChE–like actions of the oximes. It has been reported that hydroxylamine is able to hydrolyze acetylthiocholine, in a manner similar to the enzyme AChE. Therefore, the existence of a similar reaction between either 2–PAM or HI–6 and acetylthiocholine used as substrate was investigated. The data in Table 4 indicated significant hydrolysis particularly with 2–PAM that was about 2–2.5 times more potent than HI–6. This reaction could also be predicted to occur between these oximes and the neurotransmitter ACh. Although such a hydrolysis plays no role under normal conditions with the AChE fully functioning, it may be of great relevance under conditions of irreversible phosphorylation of the enzyme, reducing the activity of excess ACh in the cholinergic synaptic cleft. The occurrence of this reaction *in vivo* could partly account for the antidotal efficacy of oximes against OP. Weak anti–AChE activity observed with high concentrations of 2–PAM may not be of any importance during OP–poisoning.

Molecular mechanisms of the antidotal efficacy. The oximes and related compounds, 2–PAM, HI–6 and SAD–128, produce multiple alterations of the AChR function through mechanisms unrelated to reactivation of the phosphorylated AChE. Although interacting with the same sites, the final action of each of these compounds represents the result of distinct contributions of multiple interactions with the nicotinic AChR and of the chemical reaction between the oximes and the neurotransmitter. This differential contribution of the various interactions makes each compound specifically or particularly potent against a given OP (Reddy et al., 1987).

As mentioned before, in the increase of AChR activation, 2–PAM was much more potent than HI–6, whereas this effect was not observed with SAD–128. The increase in the channel opening probability, which may result from the ability of oximes to arrest fast desensitization, could become relevant under conditions of OP poisoning, when AChR desensitization may be the dominant process produced not only by excess ACh but also by direct interactions of OP with the nicotinic AChR (Rao et al., 1986). Thus, especially 2–PAM, through this mechanism could counteract the effect of OPs and restore neuromuscular transmission.

In addition, all three compounds produced reversible channel blockade. Comparatively, SAD–128 produced more stable blockade at much lower doses. The availability of activatable AChR through the first mechanism described above, followed by a reversible channel blockade may release a significant number of AChR from the desensitized state and thereby reestablish the synaptic function.

And, finally, an AChE–like action, particularly with 2–PAM and HI–6, may play a significant role in the antidotal efficacy of these agents. This effect, although irrelevant under normal conditions, may greatly contribute to diminish ACh concentration at the cholinergic synapses in OP–poisoned animals.

CONCLUSIONS

Our studies provide insights into the molecular mechanisms underlying the antidotal properties of the carbamates, oximes and non–oxime related compounds against lethal effects of irreversible AChE inhibitors. The data disclosed that carbamylation or reactivation of phosphorylated AChE is not the primary mechanism responsible for the antidotal properties of these agents against OPs. (+) Physostigmine's results from ultrastructural and *in vivo* toxicological studies provided the ultimate evidence for this theory. Moreover, the electrophysiological data showed that carbamates' protecting potency was strongly related to specific interactions with the molecular targets at the postsynaptic nicotinic AChR. Regarding the actions of oximes, studies on SAD–128 showed definite correlation between the antidotal efficacy of these compounds and their actions at AChR macromolecule. Furthermore, our studies suggested that the direct interactions of OP compounds with nicotinic AChR targets (Rao et al., 1987) should be taken into account in the investigations of the carbamate–OP and oxime–OP antagonisms.

ACKNOWLEDGEMENTS

This work was supported in part by U.S. Army Medical Research and Development Command Contract DAMD17–88–C–8119 and U.S.P.S.H. Grant NS–25296. We thank Ms. M. Zelle and Mrs. B. Marrow for computer and technical assistance.

REFERENCES

Adler, M., Albuquerque, E.X. and Lebeda, F.J., 1978, Kinetic analysis of endplate currents altered by atropine and scopolamine, Mol. Pharmacol., 14:514–529.

Aguayo, L.G. and Albuquerque, E.X., 1986, The voltage– and time–dependent effects of phencyclidines on the endplate currents arise from open and closed channel blockade, Proc. Natl. Acad. Sci. USA, 83:3523–3527.

Albuquerque, E.X., Aracava, Y., Cintra, W.M., Brossi, A., Schönenberger, B. and Deshpande, S.S., 1988a, Structure–activity relationship of reversible cholinesterase inhibitors: activation, channel blockade and stereospecificity of nicotinic acetylcholine receptor–ion channel complex, Brazilian J. Med. Biol. Res., 21:1173–1196.

Albuquerque, E.X., Aracava, U., Idriss, M., Schönenberger, B., Brossi, A. and Deshpande, S.S., 1987, Activation and blockade of the nicotinic and glutamatergic synapses by reversible and irreversible cholinesterase inhibitors, in: "Neurobiology of Acetylcholine," N.J. Dun and R.L. Perlman, eds., pp. 301–328, Plenum Publ. Corp., New York, NY.

Albuquerque, E.X., Daly, J.W. and Warnick, J.E., 1988b, Macromolecular sites for specific neurotoxins and drugs on chemosensitive synapses and electrical excitation in biological membranes, in: "Ion Channels," T. Narahashi, ed., Vol. I, pp. 95–162, Plenum Publ. Corp., New York, NY.

Albuquerque, E.X., Deshpande, S.S., Kawabuchi, M., Aracava, Y., Idriss, M., Rickett, D.L. and Boyne, A.F., 1985, Multiple actions of anticholinesterase agents on chemosensitive synapses: Molecular basis for prophylaxis and treatment of organophosphate poisoning, Fundam. Appl. Toxicol., 5:S182–S203.

Albuquerque, E.X., Kuba, K., and Daly, J., 1974, Effect of histrionicotoxin on the ionic conductance modulator of the cholinergic receptor: A quantitative analysis of the endplate current, J. Pharmacol. Exp. Ther., 189:513–524.

Alkondon, M. and Albuquerque, E.X., 1988, Non–oxime bispyridinium compound SAD–128 alters the kinetics of ACh–activated channels, Neurosci. Abs., 14:640.

Alkondon, M. and Rao, K.S. and Albuquerque, E.X., 1988, Acetylcholinesterase reactivators modify the functional properties of the nicotinic acetylcholine receptor ion channel, J. Pharmacol. Exp. Ther., 245:543–556.

Allen, C.N., Akaike, A. and Albuquerque, E.X., 1984, The frog interosseal muscle fiber as a new model for patch clamp studies of chemosensitive and voltage–sensitive ion channels, J. Physiol. (Paris), 79:338–343.

Anderson, C.R. and Stevens, C.F., 1973, Voltage clamp analysis of acetylcholine produced end–plate current fluctuations at frog neuromuscular junction, J. Physiol. (Lond.), 236:655–691.

Aracava, Y., Deshpande, S.S., Rickett, D.L., Brossi, A., Schönenberger, B. and Albuquerque, E.X., 1987, The molecular basis of anticholinesterase actions on nicotinic and glutamatergic synapses, in: "Myasthenia Gravis: Biology and Treatment," D.B. Drachman, ed., Ann. N.Y. Acad. Sci., 505:226–255.

Aracava, Y., Ikeda, S.R., Daly, J.W., Brookes, N., and Albuquerque, E.X., 1984, Interactions of bupivacaine with ionic channels of the nicotinic receptor, Analysis of single channel currents, Mol. Pharmacol., 26:304–313.

Changeux, J.–P., Devillers–Thiéry, A. and Chemouilli, P., 1984, Acetylcholine receptor: an allosteric protein, Science, 225:1335–1345.

Clement, J.G., 1981, Toxicology and pharmacology of bispyridinium oximes–insight into the mechanism of action vs soman poisoning *in vivo*, Fundam. Appl. Toxicol., 1:193–202.

Colquhoun, D. and Sakmann, B., 1981, Fluctuations in the microsecond time range of the current through single acetylcholine receptor ion channels, Nature (Lond.), 294:464–466.

Deshpande, S.S., Viana, G.B., Kauffman, F.C., Rickett, D.L. and Albuquerque, E.X., 1986, Effectiveness of physostigmine as a pretreatment drug for protection of rats from organophosphate poisoning, Fundam. Appl. Toxicol., 6:566–577.

Eldefrawi, M.E., Schweizer, G., Bakry, N.M. and Valdes, J.J., 1988, Desensitization of the nicotinic acetylcholine receptor by diisopropylfluorophosphate, J. Biochem. Toxicol., 3:21–32.

Feltz, A., and Trautmann, A., 1982, Desensitization at the frog neuromuscular junction: A biphasic process, J. Physiol. (Lond.), 322:257–272.

Hamill, O.P., Marty, A., Neher, E., Sakmann, B. and Sigworth, F.J., 1981, Improved patch–clamp techniques for high–resolution current recording from cells and cell–free membrane patches, Pflügers Arch., 391:85–100.

Heidmann, T. and Changeux, J.–P., 1978, Structural and functional properties of the acetylcholine receptor protein in its purified and membrane bound states, Ann. Rev. Biochem., 47:317–357.

Heidmann, T. and Changeux, J.–P., 1980, Interaction of fluorescent agonist with the membrane–bound acetylcholine receptor from *Torpedo marmorata* in the millisecond time range: Resolution of an "intermediate" conformational transition and evidence for positive cooperative effects, Biochem. Biophys. Res. Commun., 97:889–896.

Idriss, M.K., Aguayo, L.G., Rickett, D.L. and Albuquerque, E.X., 1986, Organophosphate and carbamate compounds have pre– and postjunctional effects at the insect glutamatergic synapse, J. Pharmacol. Exp. Ther., 239:279–285.

Ikeda, S.R., Aronstam, R.S., Daly, J.W., Aracava, Y. and Albuquerque, E.X., 1984, Interactions of bupivacaine with ionic channels of the nicotinic receptor. Electrophysiological and biochemical studies, Mol. Pharmacol., 26:293–303.

Karlin, A., 1980, Molecular properties of nicotinic acetylcholine receptor, in: "The Cell Surface and Neuronal Function," C.W. Cotman, G. Poste and G.L. Nicolson, eds., pp. 191–260, Elsevier North Holland Biomedical Press, Amsterdam.

Katz, B. and Miledi, R., 1973, The characteristics of 'endplate noise' produced by different depolarizing drugs, J. Physiol. (Lond.), 230:707–717.

Katz, B. and Thesleff, S., 1957, A study of the 'desensitization' produced by acetylcholine at the motor endplate, J. Physiol. (Lond.), 138:63–80.

Kawabuchi, M., Boyne, A.F., Deshpande, S.S. and Albuquerque, E.X., 1986, Comparison of the endplate myopathy induced by two different carbamates in rat soleus muscle, Neurosci. Abs., 12:740.

Kawabuchi, M., Boyne, A.F., Deshpande, S.S., Cintra, W.M., Brossi, A. and Albuquerque, E.X., 1988, Enantiomer (+)physostigmine prevents organophosphate–induced subjunctional damage at the neuromuscular synapse by a mechanism not related to cholinesterase carbamylation, Synapse, 2:139–147.

Kawabuchi, M., Boyne, A.F., Deshpande, S.S., and Albuquerque, E.X., 1989, The reversible carbamate, (−) physostigmine, reduces the size of synaptic endplate lesions induced by sarin, an irreversible organophosphate, Toxicol. & Appl. Pharmacol., 97:98–106.

Klymkowsky, M., Heuser, J.E., and Stroud, R.M., 1980, Protease effects on the structure of acetylcholine receptor membranes from *Torpedo californica*, J. Cell Biol., 85:823–838.

Kuba, K., Albuquerque, E.X., Daly, J., and Barnard, E.A., 1974, A study of the irreversible cholinesterase inhibitor, diisopropylfluorophosphate on time course of endplate currents in frog sartorius muscle, J. Pharmacol. Exp. Ther., 193:232–245.

Lapa, A.J., Albuquerque, E.X. and Daly, J., 1974, An electrophysiological study of the effects of *d*–tubocurarine, atropine, and α–bungarotoxin on the cholinergic receptor in innervated and chronically denervated mammalian skeletal muscles, Exp. Neurol., 43:375–398.

Lee, C.Y., 1972, Chemistry and pharmacology of polypeptide toxins in snake

venoms, Ann. Rev. Pharmacol., 12:265–286.
Magleby, K.L. and Stevens, C.F., 1972, A quantitative description of end–plate currents, J. Physiol. (Lond.), 233:173–197.
Meshul, C.K., Boyne, A.F., Deshpande, S.S. and Albuquerque, E.X., 1985, Comparison of the ultrastructural myopathy induced by anticholinesterase agents at the end plates of rat soleus and extensor muscle, Exp. Neurol., 89:96–114.
Neher, E. and Sakmann, B., 1976, Single channel currents recorded from membrane of denervated frog muscle fibers, Nature (Lond.), 260:799–802.
Neher, E. and Steinbach, J.H., 1978, Local anesthetics transiently block currents through single acetylcholine receptor channels, J. Physiol. (Lond.), 277:153–176.
Noda, M., Furutani, Y., Takahashi, H., Toyosato, M., Tanabe, T., Shimizu, S., Kikyotani, S., Kayano, T., Hirose, T., Inayama, S., Miyata, T. and Numa, S., 1983, Cloning and sequence analysis of calf cDNA and human genomic DNA encoding α–subunit precursor of muscle acetylcholine receptor, Nature (Lond.), 305:818–823.
Oldiges, H., 1976, Comparative studies of the protective effects of pyridinium compounds against organophosphate poisoning, in: "Medical Protection Against Chemical Warfare Agents," J. Stares, ed., pp. 101–108, SIPRI Books, Almqvist and Wiksells, Stockholm.
Oldiges, H., and Schoene, K., 1970, Pyridinium and imidazolium salts as antidotes for soman and paraoxon poisoning in mice, Arch. Toxicol., 26:293–305.
Reddy, F.K., Deshpande, S.S. and Albuquerque, E.X., 1987, Bispyridinium oxime HI–6 reverses organophosphate (OP)–induced neuromuscular depression in rat skeletal muscle, Fed. Proc., 46:862.
Rao, K.S., Aracava, Y., Rickett, D.L. and Albuquerque, E.X., 1987, Noncompetitive blockade of the nicotinic acetylcholine receptor–ion channel complex by an irreversible cholinesterase inhibitor, J. Pharmacol. Exp. Ther., 240:337–344.
Rao, K.S., Alkondon, M., Aracava, Y. and Albuquerque, E.X., 1986, A comparative study of organophosphorus compounds on frog neuromuscular transmission, Neurosci. Abs., 12:739.
Ross, M.J., Klymkowsky, M.W., Agard, D.A., and Stroud, R.M., 1977, Structural studies of a membrane–bound acetylcholine receptor from *Torpedo californica*, J. Mol. Biol., 116:645–659.
Ruff, R.L., 1977, A quantitative analysis of local anaesthetic alteration of miniature end–plate current fluctuations, J. Physiol. (Lond.), 264:89–124.
Sakmann, B., Methfessel, C., Mishina, M., Takahashi, T., Takai, T., Kurasaki, M., Fukuda, K. and Numa, S., 1985, Role of acetylcholine receptor subunits in gating of the channel, J. Physiol. (Lond.), 318:538–543.
Sakmann, B., Patlak, J., and Neher, E., 1980, Single acetylcholine–activated channels show burst–kinetics in presence of desensitizing concentrations of agonists, Nature (Lond.), 286:71–73.
Shaw, K.–P., Aracava, Y., Akaike, A., Daly, J.W., Rickett, D.L. and Albuquerque, E.X., 1985, The reversible cholinesterase inhibitor physostigmine has channel–blocking and agonist effects on the acetylcholine receptor–ion channel complex, Mol. Pharmacol., 28:527–538.
Spivak, C.E. and Albuquerque, E.X., 1982, Dynamic properties of the nicotinic acetylcholine receptor ionic channel complex: activation and blockade. in: "Progress in Cholinergic Biology: Model Cholinergic Synapses," I. Hanin and A.M. Goldberg, eds., pp. 323–357, Raven Press, New York, NY.
Spivak, C.E., Maleque, M.A., Takahashi, K., Brossi, A. and Albuquerque, E.X., 1983, The ionic channel of the nicotinic acetylcholine receptor is unable to differentiate between the optical antipodes of perhydrohistrionicotoxin, FEBS Lett., 163:189–198.

Spivak, C.E., Witkop, B., and Albuquerque, E.X., 1980, Anatoxin–A: A novel, potent agonist at the nicotinic receptor, Mol. Pharmacol., 18:384–394.
Steinbach, A.B., 1968, A kinetic model for the action of xylocaine on receptors for acetylcholine, J. Gen. Physiol., 52:162–180.
Swanson, K.L., Allen, C.N., Aronstam, R.S., Rapoport, H. and Albuquerque, E.X., 1986, Molecular mechanisms of the potent and stereospecific nicotinic receptor agonist (+)–Anatoxin–a, Mol. Pharmacol., 29:250–257.

THE ROLE OF ION CHANNELS IN INSECTICIDE ACTION

Toshio Narahashi

Department of Pharmacology
Northwestern University Medical School
303 East Chicago Avenue
Chicago, IL 60611

ABSTRACT

It has been well established that the sodium channel of the nerve membrane is the major target site of both type I and type II pyrethroids. Changes in the gating kinetics of the sodium channel including the prolongation of open time and the shift of activation voltage toward hyperpolarization are responsible for the hyperactivity of the nervous system. The different symptoms of poisoning in mammals caused by the two types of pyrethroids can be accounted for in terms of different efficacies of the actions of these pyrethroids on various parameters of the sodium channel gating. The pyrethroid molecules bind to the gating machinery of the sodium channel at a site different from those of other sodium channel agents including tetrodotoxin, grayanotoxin, and local anesthetics. The inactive isomers of pyrethroids appear to bind to the sodium channel sites without exerting effects. The high potency of pyrethroids is explicable by a profound amplification of toxic action from channel modification to repetitive discharges. The voltage-activated calcium channels do not appear to play any significant role in pyrethroid toxicity. The GABA receptor-channel complex has been suspected to be a target site for type II pyrethroids, but our recent patch clamp experiments have clearly ruled out that possibility. The well-known temperature dependence of the insecticidal action of pyrethroid and DDT is due primarily to the high temperature dependence of the sodium channel modification caused by the insecticides.

INTRODUCTION

It has been well established that certain ion channels are the major target site of insecticides. DDT and pyrethroids have been studied most extensively for their interactions with sodium channels. This chapter gives highlights of our studies of the mechanism of action of insecticides on the nervous system which have been conducted during the past 39 years, with special emphasis of recent work that dealt with ion channels. Despite the drastic difference in chemical structure, DDT and pyrethroids act on sodium channels in a very similar, if not identical, manner. Therefore, the both insecticides will be discussed together where appropriate. Several reviews on the mechanisms of action of DDT and pyrethroids have been published (Narahashi, 1971, 1976, 1981, 1985, 1987; Ruigt, 1984; Wouters and van den Bercken, 1978; Woolley, 1981).

HISTORICAL BACKGROUND

The earliest study of the effect of insecticides on the nervous system was performed by Lowenstein (1942) who discovered that impulse discharges recorded from insect nerve were increased by pyrethrum extract. In 1946, Roeder and Weiant found that sensory neurons of cockroach legs were stimulated by DDT to evoke repetitive discharges. DDT was also found to cause repetitive discharges in isolated arthropod nerve fiber preparations (Welsh and Gordon, 1947; Yamasaki and Ishii (Narahashi), 1952a). However, it was not until 1952 that a clue to the mechanism of production of repetitive discharges by DDT was obtained. While studying the effect of DDT on the synaptic transmission of the cockroach ganglion, Yamasaki and Ishii (Narahashi) (1952b) found that postsynaptic after-discharges were greatly prolonged and the extracellularly recorded action potentials from individual nerve fibers were markedly prolonged in duration (Fig. 1). This finding was later confirmed by the intracellular recording technique which disclosed a marked increase and prolongation of the action potential by DDT giving rise to repetitive after-discharges (Yamasaki and Narahashi, 1957; Narahashi and Yamasaki, 1960). Thus it became clear that the depolarizing after-potential was elevated to the threshold level for action potential generation. A similar increase in depolarizing after-potential was observed in the presence of allethrin (Narahashi, 1962a,b; van den Bercken et al., 1973) and

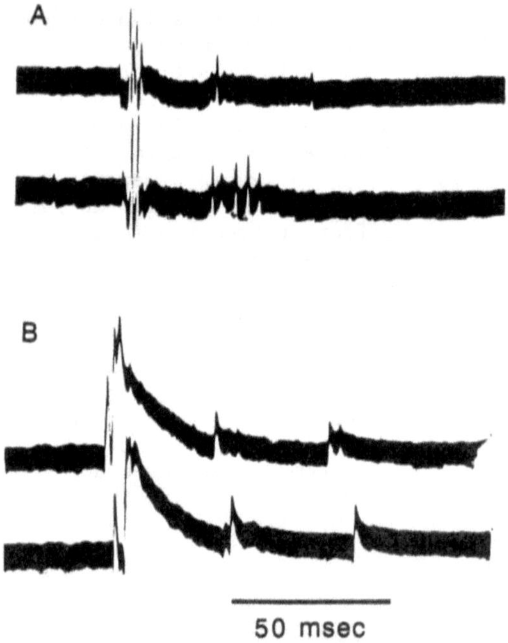

Fig. 1. Postsynaptic responses recorded extracellularly from the abdominal nerve cord of the cockroach as evoked by a presynaptic stimulus applied to the cercal nerve. (A) Control; the initial large postsynaptic action potentials are followed by after-discharges of smaller action potentials originated from individual nerve fibers. (B) 40 min after application of 1×10^{-5}g/ml DDT; the initial large action potentials originated from individual fibers exhibits a prolonged after-potential. From Yamasaki and Ishii (Narahashi) (1952b).

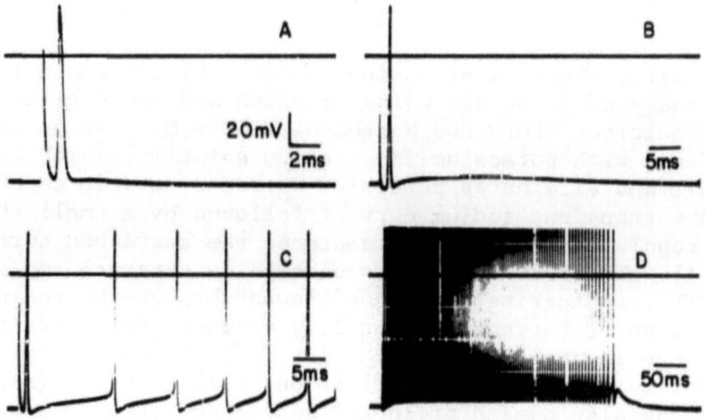

Fig. 2. Repetitive discharges induced by a single stimulus in a crayfish giant axon exposed to 10 μM (+)-trans tetramethrin. Intracellular recording at 22°C. (A) Control. (B) 5 min after application of tetramethrin. (C) and (D) 10 min after application. From Lund and Narahashi (1981a).

tetramethrin (Lund and Narahashi, 1981a) (Fig. 2). The next question was how the depolarizing after-potential would be increased by DDT and pyrethroids. It was clear that the mechanism was related to modulations of gating kinetics of ion channels, but voltage clamp experiments that are required for channel gating analyses were not easily done in the early 1960s.

The improvements of voltage clamp techniques, especially those using double sucrose-gap chambers, have made it possible to study the effects of DDT on the ion channels of lobster giant axons (Narahashi and Haas, 1967, 1968). Squid giant axons are far better than lobster giant axons for voltage clamp experiments using axial wire electrodes, but at that time DDT was found ineffective on squid axons when test solutions were prepared from DDT stock solutions in ethanol. It was later found that DDT was effective on squid axons if it was first dissolved in dimethylsulfoxide (DMSO) instead of ethanol (Lund and Narahashi, 1981c). The nodes of Ranvier of frogs were also used for voltage clamp studies of DDT action (Hille, 1968; Vijverberg et al., 1982a). The pyrethroid allethrin was effective on both squid and crustacean giant axons, so voltage clamp experiments were performed with squid axons (Narahashi and Anderson, 1967). Both DDT and allethrin prolonged the sodium current and suppressed the peak sodium current and steady-state potassium current. However, the most remarkable change was the prolongation of the sodium current. Unless the nerve chamber, which was used for a pyrethroid or DDT experiment, was thoroughly washed with appropriate solvents or detergents, the next nerve preparation would exhibit prolonged sodium currents even before exposure to the insecticides. It has now become clear that the prolonged sodium current would elevate and prolong the depolarizing after-potential. Temperature had a profound effect on the allethrin-induced modulation of the sodium current (Wang et al., 1972; Vijverberg et al., 1982b). Temperature dependence of pyrethroid and DDT effects will be discussed in a later section.

CHANGES IN SODIUM CHANNEL GATING KINETICS

Changes in gating kinetics of sodium channels by tetramethrin and allethrin were analyzed in detail using crayfish and squid giant axons under voltage clamp conditions (Lund and Narahashi, 1981a,b). In an axon internally perfused with potassium-free cesium solution, which blocks the potassium channel and eliminates potassium currents, a step depolarizing pulse generated a transient sodium current followed by a small sustained current. Upon repolarization of the membrane, the sustained current decayed quickly (Fig. 3A). These currents were blocked by external application of tetrodotoxin (TTX), indicating that they flowed through the sodium channels. Internal application of tetramethrin greatly enhanced the sustained current without much change in the transient current, and caused a large and sustained current to flow upon repolarization (tail current) (Fig. 3A). Several pyrethroids also caused prolonged sodium currents to flow during and after a depolarizing pulse in crayfish axons (Lund and Narahashi, 1983; Salgado and Narahashi, 1989; Salgado et al., 1989); in squid axons (Brown and Narahashi, 1987); in frog nodes of Ranvier (Hille, 1968; Vijverberg et al., 1982a,b); in cockroach axons (Pichon, 1969); and in neuroblastoma cells (Ogata et al., 1988; Ruigt et al., 1984). Experiments with prolonged depolarizing pulses revealed that the large sustained currents in tetramethrin slowly increased, attained a maximum, and decayed slowly, and that the tail current also decayed slowly (Fig. 3B and C). The tetramethrin-induced large slow current and tail current were blocked by TTX indicating that these currents flowed through the sodium channels. These results suggest that the transient sodium current unaltered by tetramethrin represents the activity of the normal sodium channels and that the slow current and tail current increased by tetramethrin represent the activity of the modified sodium channels. The tetramethrin-modified sodium channels could be activated at membrane potentials more negative than those at which the normal channels were activated (Lund and Narahashi, 1981a). The prolonged sodium current at large negative potentials in the

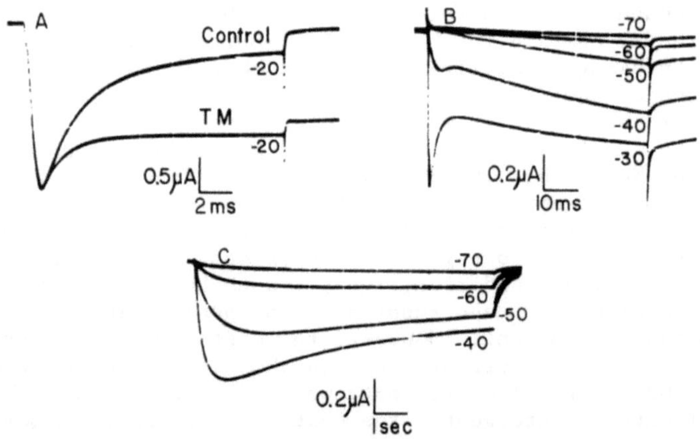

Fig. 3. Prolongation of sodium currents associated with step depolarizations to the levels indicated in a crayfish giant axon before and after application of 20 μM (+)-trans tetramethrin. (A) Control current and current after application of tetramethrin. (B) Currents associated with longer step depolarizations to various levels in tetramethrin. (C) Similar currents to those in B, but with even longer depolarizing pulses. From Lund and Narahashi (1981a).

pyrethroid-poisoned axon can account for membrane depolarization which has been observed to various extents depending on the kind of pyrethroids and preparations.

Pyrethroids are conveniently divided into two large groups: Type I pyrethroids lack a cyano group and are represented by tetramethrin, allethrin, phenothrin and permethrin; type II pyrethroids contain a cyano group at the α position, and are represented by deltamethrin, cyphenothrin, cypermethrin and fenvalerate. Although type I and type II pyrethroids cause somewhat different symptoms of poisoning in mammals, they both modify the sodium channel gating kinetics similarly. The only major difference is that in the presence of type II pyrethroids the sodium current is prolonged more drastically and the activation potential is shifted more in the hyperpolarizing direction (Narahashi, 1985). The more efficacious actions of type II pyrethroids explain a greater degree of membrane depolarization than that caused by type I pyrethroids.

MODIFICATION OF SINGLE SODIUM CHANNELS

The changes in sodium current caused by pyrethroids represent the algebraic sum of those occurring in a large number of sodium channels present in the preparation. The activity of individual sodium channels can be recorded by the patch clamp technique originally developed by Neher and Sakmann (1976) and improved later by Hamill et al. (1981). We used cultured neuroblastoma cells, N1E-115 line, as the experimental material. This cell line has several advantages for single channel recording: it is endowed with a variety of channels including sodium channels; the cell is naked and not surrounded by connective tissues or other cells making it suitable for patch clamp recording; and it is easy to maintain in cell culture.

Tetramethrin has been found to modify the single sodium channel activity drastically (Fig. 4). Whereas the normal sodium channels open for a short period of time during a depolarizing step, the sodium channels exposed to tetramethrin are kept open for much longer (Fig. 4A and B). The current amplitude is not changed by tetramethrin (Fig. 4C and D). The open time distribution in tetramethrin is expressed by two exponential functions: one has a time constant of 1.8 msec which is similar to that of control (1.7 msec), and the other has a time constant of 16.6 msec (Fig. 4E and F). This indicates that the individual sodium channels are modified by tetramethrin in an all-or-none manner, and that in the presence of tetramethrin there are two populations of sodium channels, one being unmodified and the other being modified.

Type II pyrethroids such as deltamethrin and fenvalerate have also been demonstrated to modify the gating kinetics of single sodium channels in neuroblastoma cells (Chinn and Narahashi, 1986; Holloway et al., 1984). Three important features have been disclosed. First, the open time was greatly prolonged by type II pyrethroids, the degree of which was more pronounced than the prolongation caused by type I pyrethroids (Fig. 5). The open time histogram of single sodium channel currents in the presence of deltamethrin shows a single exponential function with a time constant of 1.1 sec, which is almost three orders of magnitude longer than that of control (Fig. 4E) and almost two orders of magnitude longer than that of tetramethrin-poisoned channels (Fig. 4F). Second, the deltamethrin-modified sodium channels could open with a long delay after the onset of depolarizing pulse (see record B3 of Fig. 5). Normal sodium channels open with only a brief delay during a depolarizing pulse (Fig. 5A). Third, the deltamethrin-modified sodium channels could open after termination of a depolarizing pulse (Fig. 6). Some modified channels open during a 3 sec depolarization from -100 mV to -30 mV, and some others open after

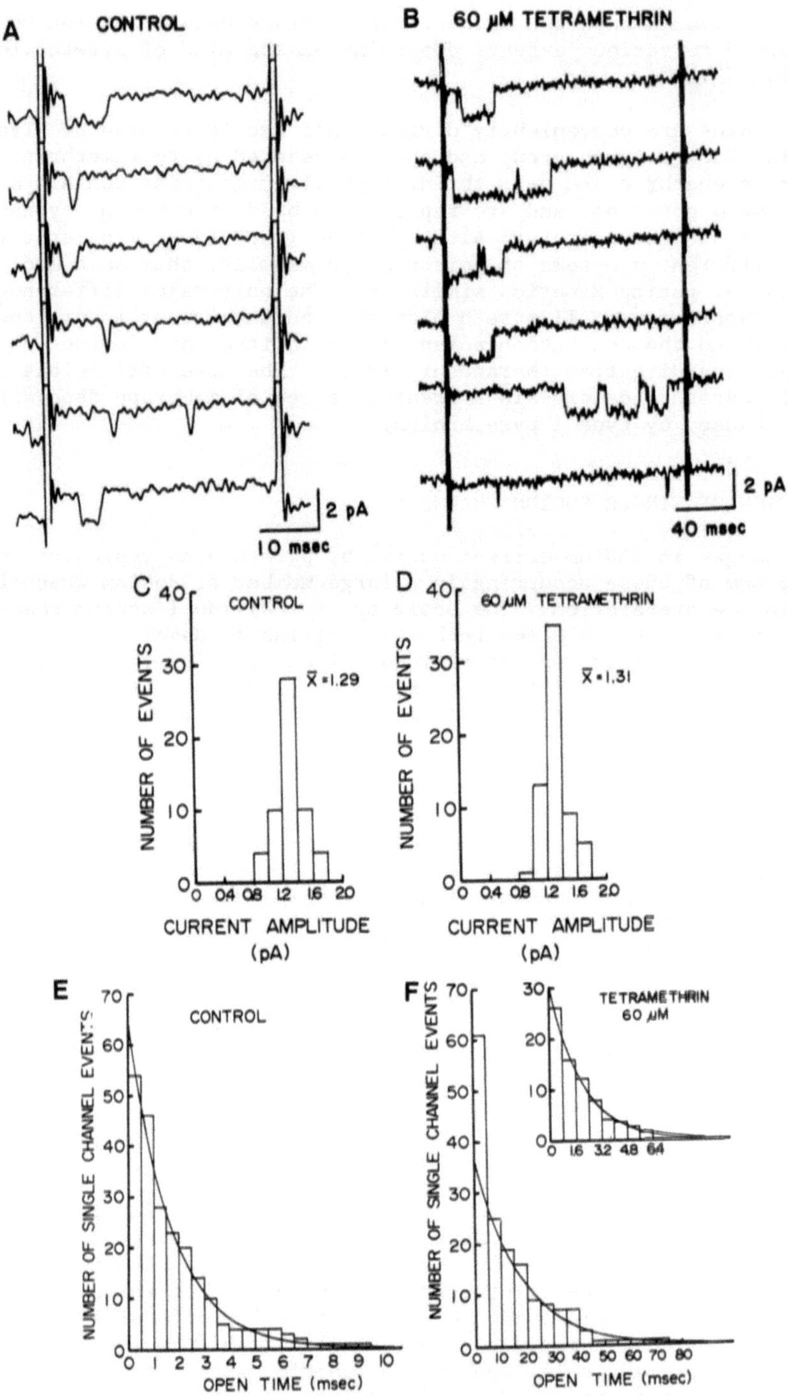

Fig. 4. Effects of 60 μM (+)-trans tetramethrin on single sodium channels in an inside-out membrane patch excised from a neuroblastoma cell (N1E-115 line). (A) Sample records of sodium channel currents (downward deflections) associated with step depolarizations from -90 mV to -50 mV. (B) As in A, but after application of tetramethrin to the internal surface of the membrane. (C) Current amplitude histogram in the control. (D) As in C, but after application of tetramethrin. (E) Channel open time distribution in the control. (F) As in E, but after application of tetramethrin; inset shows the distribution of short open times. From Yamamoto et al. (1983).

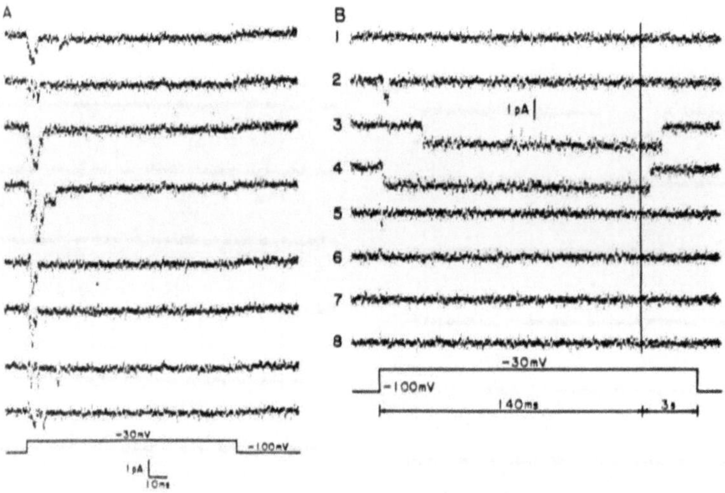

Fig. 5. Effects of deltamethrin on single sodium channel currents of a neuroblastoma cell (N1E-115). (A) Currents from a cell before drug treatment in response to 140 ms depolarizing steps from a holding potential of -100 to -30 mV with a 3 s interpulse interval. Records were taken at a rate of 100 μs per point. (B) Currents after exposure to 10 μM deltamethrin. The membrane patch was depolarized for 3140 ms from a holding potential of -100 to -30 mV. The interpulse interval was 3 s. The time scale changes during the voltage step as indicated in the Figure. During the first 140 ms, data records were taken at a rate of 100 μs per point and after the vertical line records were taken at a rate of 10 ms per point. From Chinn and Narahashi (1986).

repolarization to -100 mV. Openings at -100 mV do not occur without a depolarizing pulse, indicating that a depolarizing stimulus is necessary for the modified channels to open at -100 mV (Fig. 6). These three changes in gating kinetics of deltamethrin-modified single sodium channels would cause a prolonged sodium current during and after a depolarizing pulse in a whole cell or axon. These results are interpreted as indicating that type II pyrethroids stabilize a variety of channel states by reducing the transition rates between them.

CHANGES IN GATING CURRENT

The decrease in transition rates between the closed and open states of individual sodium channels by type II pyrethroids as described in the preceding section predicts that the gating current would be reduced. The gating current results from asymmetric charge movements within the nerve membrane, and flows in the outward direction upon depolarization (ON gating current) before the sodium current begins to flow. The same gating current flows in the opposite direction upon repolarization of the membrane (OFF gating current). Experiments were performed with internally perfused and voltage clamped crayfish giant axons (Salgado and Narahashi, 1989). Both ON and OFF gating currents were suppressed by application of fenvalerate, a type II pyrethroid. It was concluded that fenvalerate traps the gating charges of sodium channels in the open state. The results are compatible with those of single sodium channels.

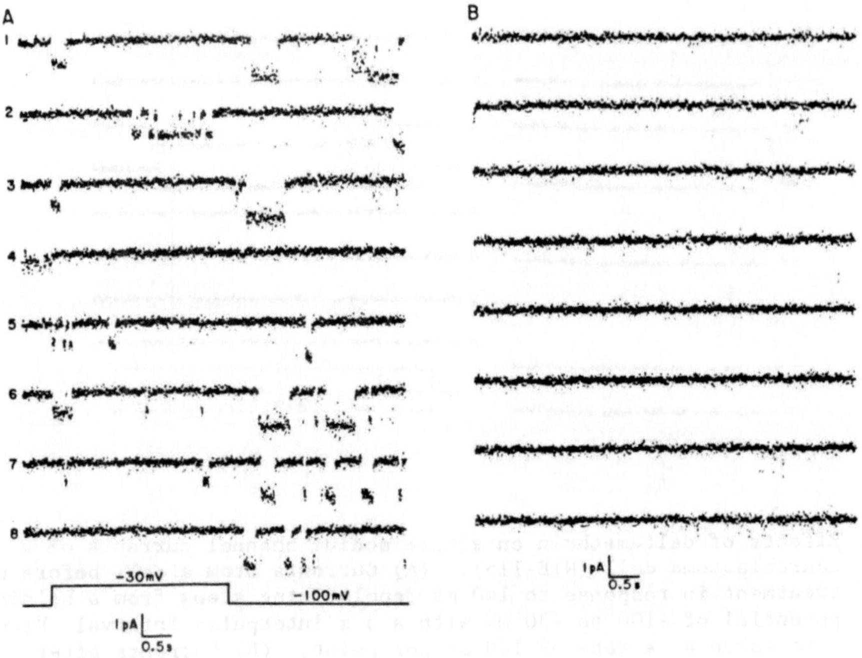

Fig. 6. Sodium channel openings in a neuroblastoma cell (N1E-115) after a voltage pulse in the presence of 10 μM deltamethrin. (A) A series of consecutive current traces taken from a membrane patch depolarized from a holding potential of -100 to -30 mV for 3 s and then repolarized for 6 s (only a 3 s period after each voltage step is shown). (B) A series of consecutive current traces taken after the stimulus pulse was turned off. Note the lack of spontaneous channel activity at -100 mV. Data in A and B were taken at a rate of 10 ms per point. From Chinn and Narahashi (1986).

CLOSED OR OPEN CHANNEL MODIFICATION

There are at least two possibilities of pyrethroid modification of sodium channels: The pyrethroid molecules bind to the sodium channels either in the closed configuration or in the open configuration. Opening and closing of the sodium channels in the presence of pyrethroid may be represented by a simplified diagram shown in Fig. 7. The normal unmodified sodium channel is in the closed state (C) at the resting potential, and opens upon depolarization (O). The normal open channel is inactivated (I) during sustained depolarization. The pyrethroid molecule may bind to the closed channel to produce the modified closed channel (C^*) which in turn opens upon depolarization to become the modified open channel (O^*). Alternatively, the pyrethroid molecule binds to the normal open channel to produce the modified open channel. The modified open channel is hardly inactivated during sustained depolarization. Several different types of experiments have shown that both closed and open channel modifications occur depending on the kind of pyrethroids. Some of the experiments are briefly described below.

Since the pyrethroid-modified open sodium channel exhibits kinetics much slower than those of the normal open channel, the activity of the former can be measured from the slow component of tail current associated with step repolarization of the membrane (Lund and Narahashi, 1981a,b). Therefore, the rate at which the sodium channel reaches the modified open state in the

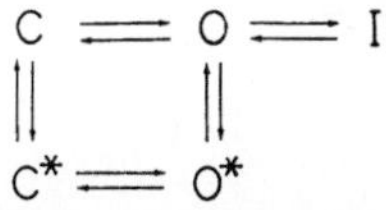

Fig. 7. Scheme of opening and closing of normal and pyrethroid-modified sodium channels. C, O and I refer to normal closed, open and inactivated channels, respectively. C^* and O^* refer to pyrethroid-modified closed and open channels, respectively.

presence of pyrethroid can be measured from the time course of change in the amplitude of slow tail currents associated with step repolarizations following various durations of depolarization. Such experiments with tetramethrin-treated squid giant axons showed that the time course of the development of the modified open channel was expressed by two exponential functions (Lund and Narahashi, 1981b). Thus the modified open state may be reached by two separate routes, one via open channel modification and the other via closed channel modification.

If modification occurred solely in the closed state of the sodium channel, we would expect that the rate at which the modified open channel develops during depolarization would be independent of the pyrethroid concentration. Experiments with internally perfused squid axons exposed to several pyrethroids yielded different results depending on the kind of pyrethroids (de Weille et al., 1988). Figure 8 shows the time course of the development of slow tail current amplitude during depolarization in the presence of various concentrations of phenothrin (type I) (panel A) and cyphenothrin (type II) (panel B). The curves such as those shown in Fig. 8A and B are sometimes fit to a single exponential function and sometimes to a dual exponential function. The latter example is shown in Fig. 8C. The logarithm of the time constant is plotted against the logarithm of the pyrethroid concentration (Hill plot). Both of the two time constants for the phenothrin-treated axon are independent of the concentration. As for cyphenothrin, one time constant is slightly dose-dependent with a slope of 0.12, whereas the other time constant is dose-independent. It can be concluded that a large fraction of sodium channels is modified in the closed state with a small fraction of channels being modified in the open state. The latter fraction may be limited by channel inactivation.

In order to determine the role of channel inactivation in open channel modification, similar experiments were performed after the inactivation had been removed by internal perfusion of pronase (de Weill et al., 1988). The pyrethroid molecule would have more chances to interact with the open channel under this condition. The Hill plots of the data obtained from axons exposed to either phenothrin or cyphenothrin have a slope of 0.25 (Fig. 9). Thus the dose dependence of the time constant increased somewhat after removal of inactivation which resulted in an increased likelihood of open channel modification.

OPEN CHANNEL PROPERTIES

The results of experiments summarized in foregoing sections clearly show that pyrethroids drastically modify the gating kinetics of sodium channels thereby causing various symptoms of poisoning in animals. A question remains as to whether there is any change in the properties of open sodium channels as a result of intoxication with pyrethroids. Several

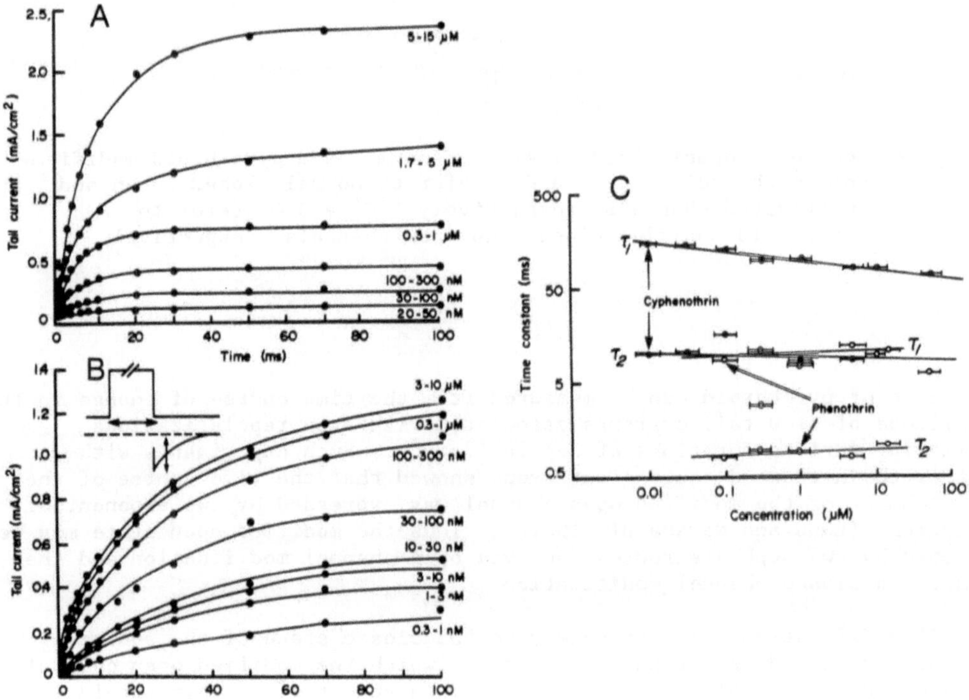

Fig. 8. (A) and (B) The amplitude of the phenothrin-(A) or cyphenothrin-(B) induced sodium tail current in squid giant axons as a function of the duration of depolarizing pulse at several concentrations of the pyrethroids. Inset shows the pulse protocol used and applies to both A and B. (C) The time constants as obtained by dual exponential fits are plotted against the concentrations of phenothrin and cyphenothrin on double exponential scales (Hill plots). From de Weille et al. (1988).

properties associated with open channels were studied using internally perfused, voltage clamped squid giant axons (Yamamoto et al., 1986). Open sodium channels are known to be blocked, in a manner dependent on membrane potential, by certain monovalent and divalent cations such as sodium, lithium, ammonium, guanidine, formamidine, calcium, and magnesium (Yamamoto et al., 1984, 1985). This voltage-dependent blocking action remained unchanged after exposure to tetramethrin (Fig. 10). The cation block of open sodium channel is interpreted as being due to binding of the permeating cation to a site inside the channel, thus reflecting the profile of energy barriers that exist in the channel (Yamamoto et al., 1984). It was therefore concluded that tetramethrin did not interfere with or bind to these intrachannel sites (Yamamoto et al., 1984, 1986).

The ionic permeability ratios of open sodium channels can be calculated from the shifts of the reversal potential for cationic current (Fig. 10). The permeability ratios before and after application of tetramethrin as obtained in two different cation concentrations are given in Table 1. Tetramethrin had virtually no effect on the permeability ratios. Since the relative permeability of sodium channels to various cations is known to be controlled by the selectivity filter located near the external orifice of the channel (Hille, 1971), it was concluded that tetramethrin did not alter its property (Yamamoto et al., 1986).

SITE OF ACTION OF PYRETHROIDS IN THE SODIUM CHANNEL

The results of experiments summarized in preceding sections indicate that pyrethroids do not act on intrachannel sites, leaving the open channel properties unaltered. It is also shown that pyrethroids drastically modify the gating kinetics of the sodium channel. Since the pyrethroid molecules are highly hydrophobic, they are expected to be dissolved in the lipid phase of the nerve membrane thereby reaching the gating machinery. One way of determining the exact site of action of pyrethroids in the sodium channel is to examine the possible interactions or competition between pyrethroids and other chemicals known to bind to specific sites in the sodium channel.

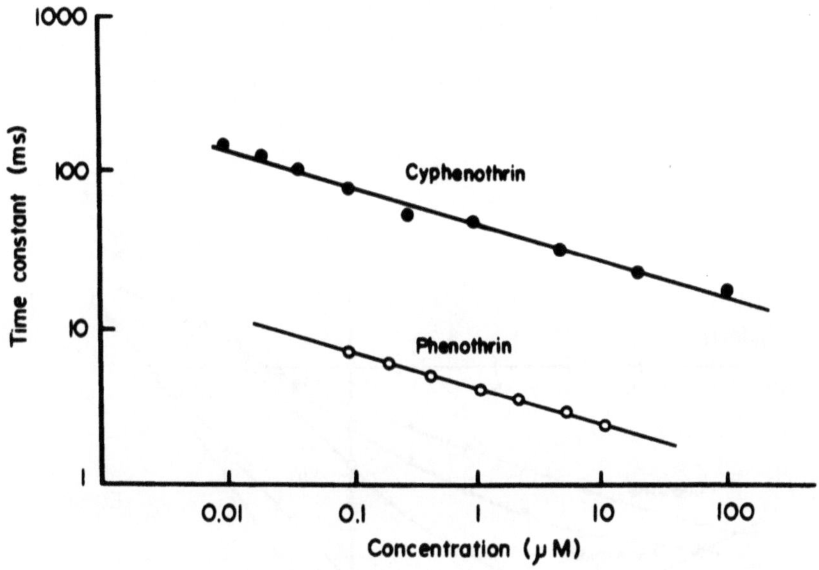

Fig. 9. Tail current time constant-concentration relationships obtained in squid axon experiments similar to those of Fig. 8 but after removal of sodium channel inactivation by pronase treatment. The time constants were obtained by single exponential fits. From de Weille et al. (1988).

n-Octylguanidine has been demonstrated to block the open sodium channel from inside the membrane (Kirsch et al., 1980) in a manner similar to that of local anesthetics (Strichartz, 1973; Courtney, 1975, 1980; Courtney et al., 1978; Hille, 1977; Khodorov et al., 1976; Yeh, 1978, 1980), pancuronium (Yeh and Narahashi, 1977), 9-aminoacridine (Yeh, 1979), and strychnine (Shapiro, 1977). The major route of entry to the sodium channel is via the open gates located at the inner orifice. The dose-response relationships of octylguanidine block of peak sodium current before and after treatment with phenothrin are shown in Fig. 11. The apparent dissociation constant for octylguanidine block is estimated to be 80 μM. This value was greatly reduced when the block was tested by the tenth depolarizing pulse in a train

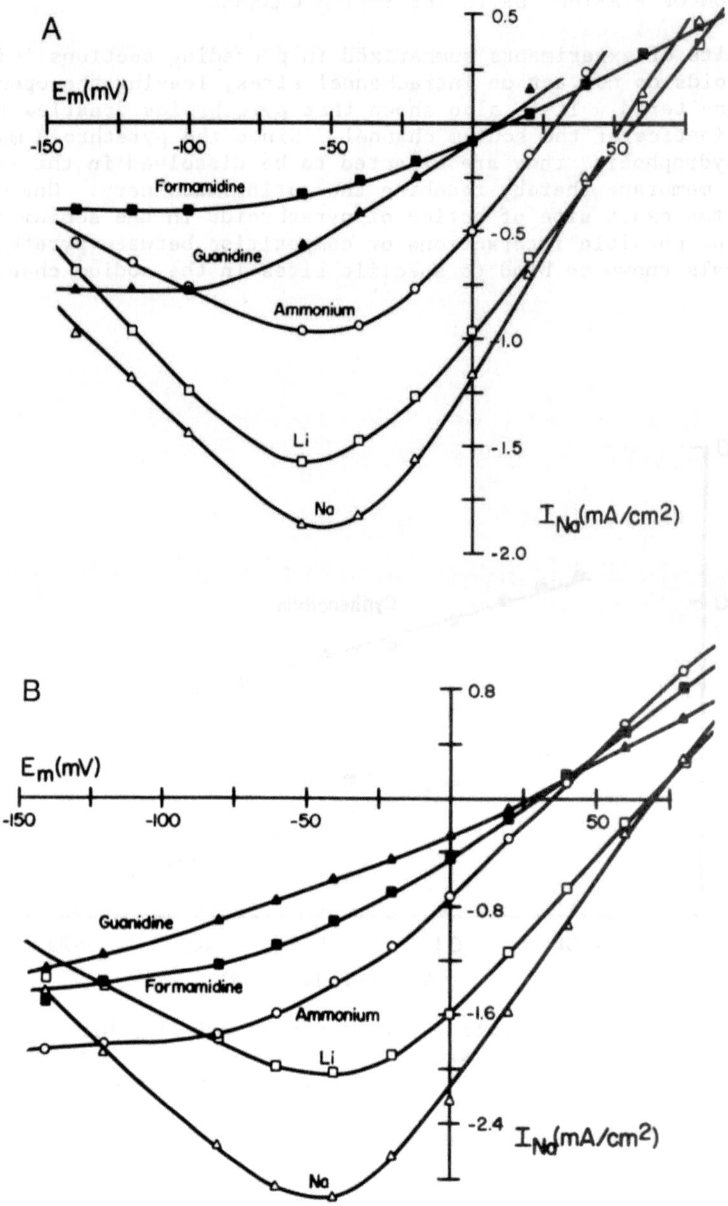

Fig. 10. Instantaneous I-V relationships in external solutions containing various permeant cations (300 mM) in a normal squid axon (A) and in an axon internally perfused with 50 μM tetramethrin (B). Permeant cations are Na (Δ), Li (□), NH_4 (O), formamidine (■) and guanidine (▲). The amplitude of the instantaneous tail current in the normal axon was measured upon step repolarization of the membrane to various levels following a 1 ms depolarizing pulse to 0 mV. The tail current amplitude in the tetramethrin-treated axon was measured upon step repolarization to various levels following an 80 ms depolarizing pulse to +20 mV. Different axons were used for each cation in the control, whereas one axon was used for all cations in tetramethrin. From Yamamoto et al. (1986).

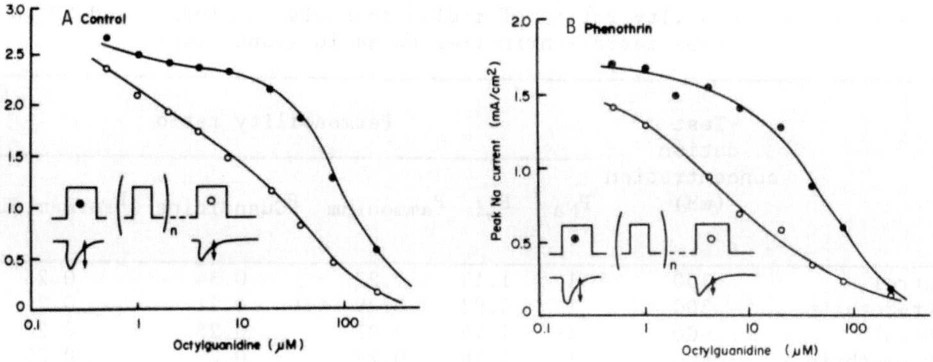

Fig. 11. The dose-response curves for octylguanidine block of the peak sodium current of a squid giant axon during the first (O) and the tenth (O) depolarizing pulse in a train of stimuli before (A) and after (B) treatment with 10 μM phenothrin. From de Weille et al. (1986).

of stimuli, indicating use-dependent block. Phenothrin did not change the dose-response curves for octylguanidine as examined either at the first or at the tenth pulse, suggesting independent effects and sites of action of phenothrin and octylguanidine (de Weille et al., 1988).

Fig. 8 shows the time course of development of pyrethroid-modified open channels. In the presence of octylguanidine, however, the amplitude of the phenothrin-induced tail current first increased and then decreased as the duration of depolarization was increased (Fig. 12), suggesting that pyrethroid-modified channels were progressively blocked by octylguanidine during depolarization (de Weille et al., 1988). Measurements in the declining phases of the curves in Fig. 12 were fit to a single exponential function, and the time constants were independent of the pyrethroid concentration at values close to 20 msec. This observation is again in keeping with the independence of the action of pyrethroids and octylguanidine. Thus pyrethroids appear to bind to a site different from the intrachannel binding site of octylguanidine and presumably of local anesthetics.

Batrachotoxin (BTX) and grayanotoxin (GTX) depolarize the nerve membrane (Narahashi et al., 1971; Albuquerque et al., 1973; Seyama and Narahashi, 1973; Narahashi and Seyama, 1974) by inhibiting the sodium inactivation mechanism and by shifting the sodium channel activation potential in the direction of hyperpolarization (Khodorov, 1978, 1979, 1985; Khodorov and Revenko, 1979; Khodorov et al., 1976; Seyama and Narahashi, 1981; Tanguy et al., 1984). It appears that the BTX molecule binds to an intrachannel site to which the inactivation gate normally binds causing inactivation (Tanguy et al, 1984). Thus BTX and GTX serve as useful tools to identify the site of action of pyrethroids.

A family of sodium currents was first recorded from an internally perfused squid axon exposed to tetramethrin (Fig. 13A) (Takeda and Narahashi, 1988). The membrane was step depolarized from a holding potential of -80 mV to various levels ranging from -60 mV to +60 mV in 20 mV steps. No sodium current flowed at -60 mV, and a small current appeared when the membrane was depolarized to -40 mV. The same axon was then treated with tetramethrin plus grayanotoxin I (GTX I), and a family of sodium

Table 1. Permeability ratios of sodium channels in control and 50 μM (+)-trans tetramethrin-treated squid giant axons

	Test cation concentration (mM)	Permeability ratio				
		P_{Na}	P_{Li}	$P_{ammonium}$	$P_{Guandidine}$	$P_{Formamidine}$
Control	300	1	1.13	0.27	0.34	0.23
Tetramethrin	300	1	0.93	0.29	0.21	0.21
Control	600	1	1.19	0.21	0.28	0.20
Tetramethrin	600	1	1.18	0.29	0.29	0.25

From Yamamoto et al. (1986).

currents was recorded as the membrane was step depolarized from a holding potential of -140 mV to various levels ranging from -120 mV to +20 mV in 20 mV steps (Fig. 13B). A small current was generated at a large negative potential of -100 mV at which no current flowed in the absence of GTX I. Fig. 13C shows the sodium currents recorded from another axon when the membrane was depolarized from a holding potential of -140 mV to -60 mV. No current flowed in the absence of either of the chemicals (record a), a large, non-inactivating current was generated after exposure to GTX I (record b), and an even larger, non-inactivating current was generated and the tail current was augmented after exposure to GTX I plus tetramethrin (record c). It is clear that GTX I and tetramethrin act independently

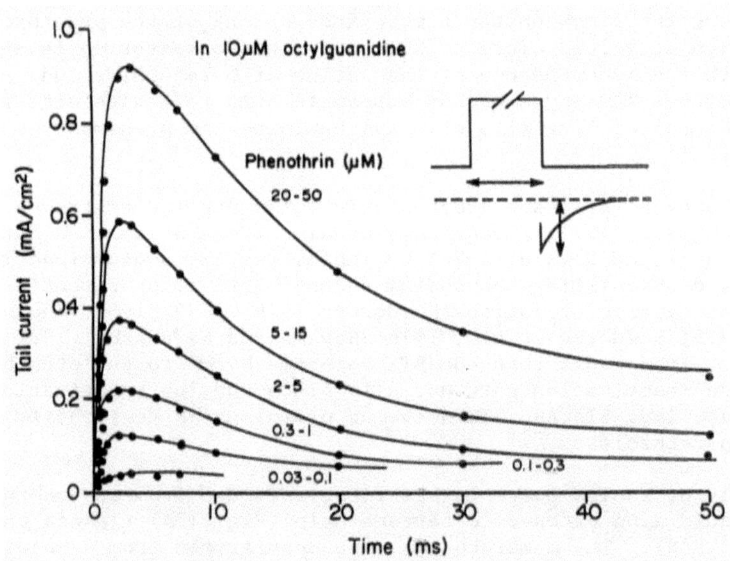

Fig. 12. The amplitude of the phenothrin-induced tail current of a squid giant axon in the presence of 10 μM octylguanidine after perfusion with several concentrations of the pyrethroid as a function of the duration of depolarizing pulse as shown in the inset. From de Weille et al. (1986).

(Takeda and Narahashi, 1988). Similar results were obtained with a combination of BTX and tetramethrin (Tanguy and Narahashi, unpublished observation). Therefore, it is concluded that pyrethroids bind to the inactivation gating machinery at a site different from the BTX and GTX site which is located inside the channel and which is the binding site of the inactivation gate itself.

TTX may be used to determine the site of action of pyrethroids in the sodium channel by blocking the sodium channel with an extremely high selectivity and potency (Narahashi, 1974, 1988; Narahashi et al., 1964). TTX blocks the sodium channel solely from the external membrane surface (Narahashi et al., 1967). Based on the hypothesis originally proposed by Kao

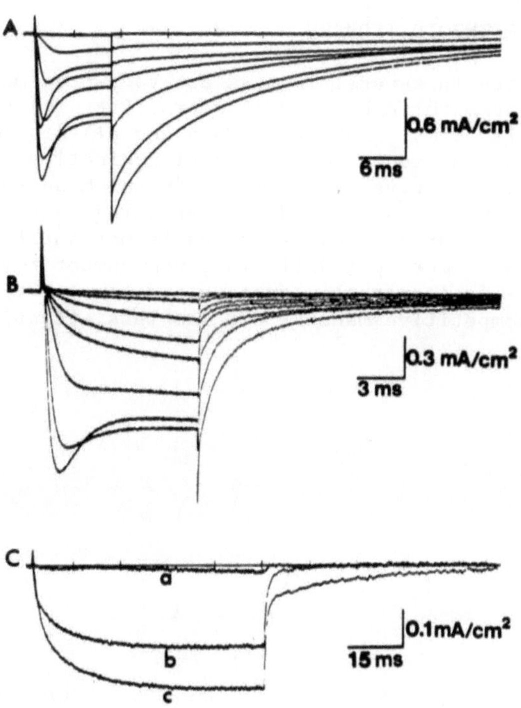

Fig. 13. Simultaneous effects of GTX I and (+)-tetramethrin on sodium current of squid giant axons. (A) Membrane currents for an axon treated with 10 μM (+)-tetramethrin. Test pulse duration 10 ms; holding potential -80 mV; test pulses from -60 to 60 mV in 20 mV steps. Note the prolonged sodium current time course during the depolarizing step and the pronounced sodium tail current following step repolarization. (B) Same axon following treatment with both (+)-tetramethrin and 50 μM GTX I. The characteristic tetramethrin-induced tail current and the GTX-induced slow sustained sodium current are both present. Test pulse duration 10 ms; holding potential -140 mV; test pulses from -120 to 20 mV in 20 mV steps. (C) Membrane currents from another axon for control 50 ms depolarizations to -60 mV from a holding potential of -140 mV (a), and successively following perfusion of 50 μM GTX I (b) and both GTX I and 10 μM (+)-tetramethrin (c). From Takeda and Narahashi (1988).

and Nishiyama (1965), Hille (1975) developed an idea that called for binding of the TTX molecule to the selectivity filter of the sodium channel thereby occluding the passage of sodium ions. As expected from this hypothesis, the gating current of sodium channel was not affected by TTX (Armstrong and Bezanilla, 1974; Keynes and Rojas, 1974; Gilly and Armstrong, 1982; Neumcke et al., 1976). Recently, Kao and his associates proposed a hypothesis that called for binding of the TTX molecule to an external membrane site outside the sodium channel orifice (Kao, 1983; Kao and Walker, 1982; Kao et al., 1983). Our experiments with squid giant axons showed that TTX suppressed the tetramethrin-induced tail current in a non-competitive manner (Fig. 14) (Lund and Narahashi, 1982). The dose-response curve for the tetramethrin block of sodium tail current was brought down along the ordinate without a shift along the concentration axis. Thus it is concluded that tetramethrin binds to a site other than the TTX binding site which is located near the external orifice of the sodium channel.

SITE OF ACTION OF PYRETHROID ISOMERS

A pyrethroid exists in several isomers due to two or more chiral centers in its molecule. Tetramethrin has four isomers, i.e. (+)-trans, (+)-cis, (-)-trans, and (-)-cis forms. Both (+)-trans and (+)-cis forms were effective in modifying the sodium channel gating kinetics, whereas (-)-trans and (-)-cis forms were inactive. Experiments in which an active form and an inactive form were combined have yielded interesting results shown in Fig. 15 (Lund and Narahashi, 1982). The (-)-trans isomer antagonized both (+)-trans and (+)-cis isomers primarily in a non-competitive manner (Fig. 15A and B). The (-)-cis isomer also antagonized the (+)-trans isomer primarily in a non-competitive manner, but was less effective than the

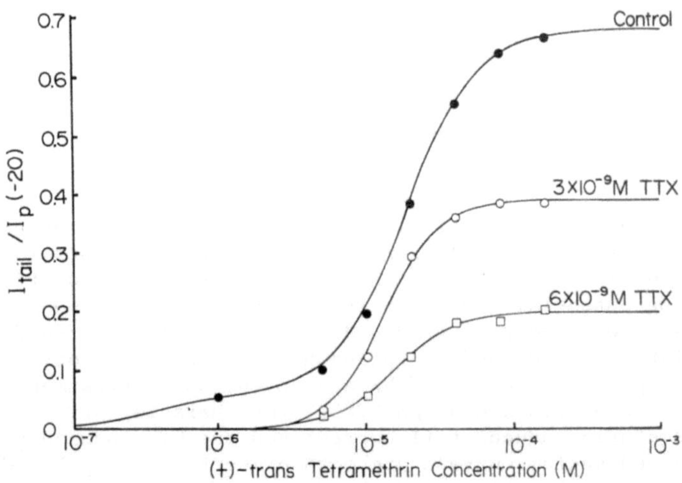

Fig. 14. The dose dependence of the slowly decaying tail current (I_{tail}) amplitude in a squid giant axon after 15 msec conditioning pulses. The current amplitude is normalized to the peak current recorded at -20 mV (I_p (-20)) before application of (+)-trans tetramethrin. TTX blocks the tetramethrin-induced tail current in a non-competitive manner. From Lund and Narahashi (1982).

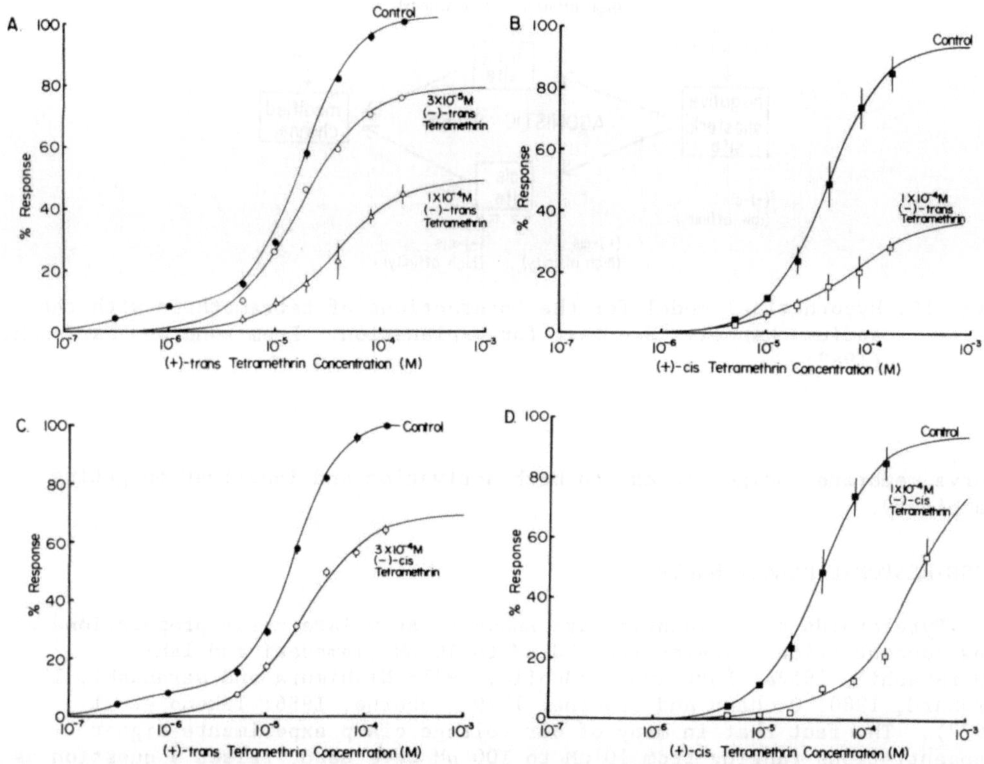

Fig. 15. Dose-response relationships obtained as in Fig. 14 from squid giant axons and plotted as percent of the maximum response. (A) Antagonism of (+)-trans tetramethrin by 3 x 10^{-5}M and 1 x 10^{-4}M (-)-trans tetramethrin. (B) Antagonism of (+)-cis tetramethrin by 1 x 10^{-4}M (-)-trans tetramethrin. (C) Antagonism of (+)-trans tetramethrin by 3 x 10^{-4}M (-)-cis tetramethrin. (D) Antagonism of (+)-cis tetramethrin by 1 x 10^{-4}M (-)-cis tetramethrin. Values shown are mean ± S.E.M. where n = 3. From Lund and Narahashi (1982).

(-)-trans isomer (Fig. 15C). In contrast, the (-)-cis isomer antagonized the (+)-cis isomer primarily in a competitive manner (Fig. 15D).

These results can be interpreted using a scheme for the tetramethrin binding sites shown in Fig. 16. There are two agonistic sites, a trans site and a cis site, to which the (+)-trans and (+)-cis isomers bind, respectively, with a high affinity. The (-)-trans isomer binds primarily to a negative allosteric site with a high affinity thereby causing a non-competitive antagonism against the (+)-trans isomer. The (-)-cis isomer binds to the cis site with a high affinity, thereby antagonizing the (+)-cis isomer in a competitive manner. These agonistic and negative allosteric sites are located somewhere in or around the sodium channels. TTX binds to an entirely different site as discussed in a preceding section. Since sites within the sodium channel and near the external orifice of the sodium channel have been excluded as pyrethroid binding sites as described above, and also since pyrethroid molecules are highly hydrophobic, it seems very likely that the pyrethroid molecules are dissolved in the lipid phase of the

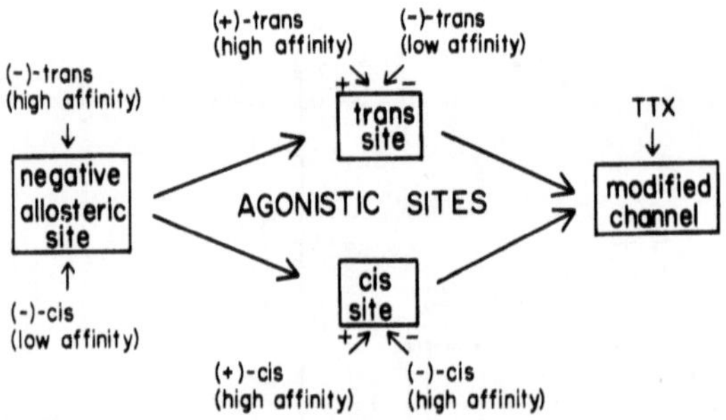

Fig. 16. Hypothetical model for the interactions of tetramethrin with the sodium channel. See text for explanation. From Lund and Narahashi (1982).

nerve membrane and get access to both activation and inactivation gating machinery.

DOSE-RESPONSE RELATIONSHIP

Pyrethroids and pyrethrins are known to stimulate nerve preparations at low concentrations ranging from 50 pM to 10 nM (Yamasaki and Ishii (Narahashi), 1952a; Burt and Goodchild, 1971; Nishimura and Narahashi, 1978; Orchard, 1980; Orchard and Osborne, 1979, Osborne, 1986; Takeno et al., 1977). The fact that in many of our voltage clamp experiments higher concentrations ranging from 10 nM to 100 μM were used, raised a question as to whether the pyrethroid-induced modification of the sodium channel gating kinetics was related to the symptoms of poisoning in animals.

The answer to this question lies in the threshold phenomenon involved in pyrethroid intoxication. As the depolarizing after-potential is elevated as a result of pyrethroid-induced prolongation of sodium current, repetitive after-discharges are induced as soon as the after-potential reaches the excitation threshold. The percentage of sodium channels that need to be modified by tetramethrin to increase the depolarizing after-potential to the threshold level is calculated to be less than 0.1% (Lund and Narahashi, 1982). The concentration of tetramethrin that would cause this degree of channel modification of sodium channel gating kinetics is sufficient to account for repetitive discharges in nerves observed at low concentrations. The same "toxicological amplication" can also account for the actions of type II pyrethroids. When the membrane depolarization caused by type II pyrethroids reaches the threshold for an increase in discharges from sensory neurons or in transmitter release from the nerve terminal, the animal would develop symptoms of poisoning characterized by hypersensitivity, choreoathetosis and tremors. Thus ED_{50} values could provide misleading information in order to interpret the animal's symptoms of poisoning in terms of changes in nervous function.

ROLE OF GABA RECEPTOR-CHANNEL COMPLEX

It was proposed that the GABA receptor-channel complex was an important target site of type II pyrethroids. This hypothesis was based on the observation that the binding of [^{35}S]*t*-butylbicyclophosphorothionate (TBPS), a ligand for the channel site of GABA receptors, and that of Ro5-4864, a diazepam analog and benzodiazepine receptor ligand, were inhibited by type II pyrethroids but not type I pyrethroids (Crofton et al., 1987; Gammon and Sander, 1985; Lawrence and Casida, 1983; Lawrence et al., 1985; Lummis et al., 1987). Chloride uptake by mouse brain vesicles as stimulated by GABA was inhibited by deltamethrin (Bloomquist and Soderlund, 1985). GABA-induced decrease in input resistance of crayfish muscle was also antagonized by type II pyrethroids (Gammon and Casida, 1983). Diazepam protected cockroaches from poisoning with type II pyrethroids but not type I pyrethroids (Gammon et al., 1982). However, the effective concentrations for these actions were high.

In order to settle the controversy of whether or not the GABA receptor-channel complex represents an important site of action of type II pyrethroids, we have performed whole cell patch clamp experiments to record chloride currents as induced by GABA application (Ogata et al., 1988). Rat dorsal root ganglion neurons in primary culture were used as material. Bath application of GABA caused an inward current which was desensitized to a lower, sustained level (Fig. 17). This current was blocked by bicuculline, and its reversal potential was shifted by changing the external chloride concentration in a manner predicted by the Nernst equation for chloride ions. Therefore, the current is carried by chloride ions. Both GABA-induced chloride current and voltage-activated sodium current were recorded alternately from the same neuron, and the effect of deltamethrin was examined. While the sodium current underwent characteristic changes such as marked prolongation, the GABA-induced current remained completely unaffected (Fig. 17). Thus it is concluded that the GABA receptor-channel complex, at least in the dorsal root ganglion neurons of the rat, is not the target site

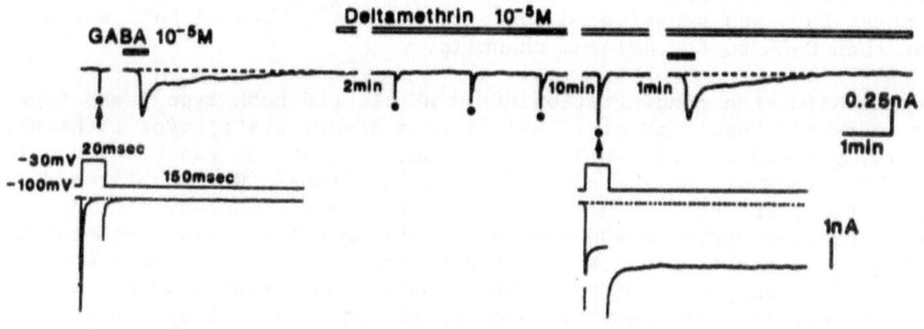

Fig. 17. Effects of deltamethrin on the sodium current and the GABA-induced inward current in a rat dorsal root ganglion neuron in primary culture. The membrane was held at -100 mV. In the upper tracings, the membrane current was registered on a pen-recorder. At points indicated by dots, 20 msec depolarizing pulses to -30 mV were applied to generate sodium currents. At the two points indicated by upward arrows, the sodium currents were recorded on an oscilloscope with an expanded time scale. GABA (1×10^{-5}M) and deltamethrin (1×10^{-5}M) were applied during periods indicated by black and white horizontal bars, respectively. While the GABA-induced current remained unchanged by deltamethrin, the sodium current underwent drastic changes with prolongation of currents during and after the depolarizing pulse. From Ogata et al. (1988).

of deltamethrin. The effects of type II pyrethroids on this complex observed by other investigators appear to play a negligible role in producing the symptoms of poisoning in animals which can be explained by the modulation of the sodium channel gating kinetics. The GABA hypothesis is not supported by some other studies which include the specific binding of [^{3}H]-dihydropicrotoxinin, a ligand for the GABA-activated channel (Matsumura and Tanaka, 1984) and high concentrations of type II pyrethroids required for affecting the GABA system (Chalmers et al., 1985).

ROLE OF CALCIUM CHANNELS

There has been some speculation as to whether calcium channels, which are known to play a variety of important roles in nerve and muscle function, are the target site of pyrethroids. Permethrin at a concentration as low as 50 pM increased the electrical activity of neurosecretory cells of the stick insect (Orchard and Osborne, 1979), which generate action potentials by inward calcium currents (Osborne, 1980). Thus it was proposed that pyrethroids act on calcium channels to exert their toxic effects (Osborne, 1980; Gammon and Sander, 1985).

In order to verify the validity of this hypothesis, we have performed patch clamp experiments to examine the effect of pyrethroids on the voltage-activated calcium channels. In neuroblastoma cells (N1E-115 line), there are two types of calcium channels (Narahashi et al., 1987; Yoshii et al., 1988). One is activated at relatively large negative potentials (around -40 mV), and the channel current is inactivated during a step depolarizing pulse (type I calcium channel). The other can be activated only when the membrane is depolarized to -20 mV or more positive potentials, and the current is hardly inactivated (type II calcium channel). In many of our experiments, Ba^{2+} was substituted for external Ca^{2+}, because Ba^{2+} was more permeant than Ca^{2+} to the calcium channels.

Tetramethrin at a concentration of 50 μM blocked both type I and type II calcium currents (Yoshii et al., 1985). At a steady state, type I channel current was blocked by 75%, while type II channel current was blocked only by 30%. The block of the two types of calcium channels was complicated comprising two components. One was a time-dependent component of block which was enhanced during a 400 msec depolarizing pulse; this component was reversible after washing the cell with drug-free media. The other was a time-independent component of block which persisted during a 40 min washing period. In contrast, deltamethrin and fenvalerate, type II pyrethroids, were without effect on either type of calcium channels. The two components of calcium channel block, a time-dependent reversible block and a time-independent irreversible block, suggest two separate sites of activation of tetramethrin in calcium channels. The observed tetramethrin block of calcium channels is not compatible with the idea that pyrethroids stimulate central neurons via actions on calcium channels.

This seemingly controversial situation may be interpreted in the following manner: Even though calcium channels are directly responsible for generation of action potentials in the secretory cells of stick insect (Osborne, 1980), the observed stimulation of the cells by permethrin (Orchard and Osborne, 1979) does not necessarily indicate actions on calcium channels, because usually both sodium and calcium channels are present depending on the region of the cell. Whereas the neurosecretory cell may generate action potentials through the activation of calcium channels, the axon generates and propagates action potentials via sodium channels, and neurotransmitter is released from nerve terminals through activity of both sodium and calcium channels. Thus impulse discharges from the neurosecretory cells reflect activity of not only the cells themselves but also presynaptic neurons contained in the preparation. The observed impulse discharges from the neurosecretory cells may be derived from repetitive discharges of presynaptic neurons and their nerve terminals through prolonged opening of sodium channels. In our study of calcium channels, tetramethrin but not deltamethrin and fenvalerate blocked calcium channels (Yoshii et al., 1985). These differential actions between type I and type II pyrethroids may be partially responsible for different symptoms of poisoning in animals. The exact role of the observed block of calcium channels in pyrethroid intoxication remains to be elucidated.

COMPARISON OF DDT AND PYRETHROIDS

Despite the drastic difference in chemical structure, DDT and pyrethroids alter the activity of sodium channels in a very similar manner. In both cases, the symptoms of poisoning in animals can be attributed to the modification of sodium channel gating kinetics. As described below, DDT particularly resembles type I pyrethroids.

Repetitive after-discharges are induced in axons by both DDT and pyrethroids through an increase in depolarizing after-potential (Yamasaki and Ishii (Narahashi), 1952a; Narahashi and Yamasaki, 1960a,b; Narahashi, 1962a,b; Narahashi and Haas, 1967, 1968; Narahashi and Anderson, 1967; Wu et al., 1980; Lund and Narahashi, 1981a,b,c). In the axons exposed to either DDT or type I pyrethroids, sodium current is prolonged during a step depolarization and after termination of the depolarization (Lund and Narahashi, 1981a,b,c; Narahashi and Haas, 1967, 1968; Narahashi and Anderson, 1967; Wu et al., 1980; Wang et al., 1972). When a series of compounds including DDT, DDT derivatives and type I and type II pyrethroids were compared for their effects on sodium current, it became clear that all of these compounds had essentially the same action to increase and prolong the current, both during and after a depolarizing step (Lund and Narahashi, 1983). The time constants of tail sodium current associated with step repolarization from a depolarized level in the axons poisoned with the various chemicals could be arranged in a continuous spectrum ranging from the shortest time constant in DDT (9.5 msec at -100 mV), to an intermediate value in tetramethrin (620 msec at -100 mV), and to an unmeasurably long time constant in deltamethrin (Table 2). Thus DDT and its derivatives (plifenate and EDO) cause tail currents with relatively short time constants, type I pyrethroids and a DDT-pyrethroid hybrid (GH401) with intermediate time constants, and type II pyrethroid with long time constants. In conclusion, DDT, type I pyrethroids and type II pyrethroids modify the sodium channel gating kinetics in a qualitatively similar manner.

TEMPERATURE EFFECTS

It has been known for a long time that temperature has a profound effect on the insecticidal activity of DDT and pyrethroids (see Narahashi, 1971). For both insecticides, the insecticidal activity increases when the temperature is lowered. The mechanism underlying the negative temperature dependence lies in the temperature-dependent sensitivity of the nerve to these insecticides. Other factors including insecticide penetration to the cuticle and detoxication in the body can be dismissed as the major contributing events.

The rate at which DDT penetrated the insect cuticle increased by elevating the temperature (Barker, 1957; Vinson and Kearns, 1952).

Table 2. Time constants (msec) of tail currents associated with step repolarizations of the membrane to the levels indicated in crayfish giant axons treated with various compounds

Compound	Conc. (M)	Tail current time constant (msec) at -160 mV	-120 mV	-100 mV
DDT	1×10^{-4}	3.0	6.1	9.5
Plifenate	3×10^{-6}	9.4	14.8	17.0
EDO	1×10^{-4}	16	44	86
Tetramethrin	2×10^{-5}	30	225	620
Phenothrin	3×10^{-5}	200	750	1340
GH401	1×10^{-4}	700	1450	2220
Cyphenothrin	1×10^{-6}	(min)	(min)	(∞)
Fenvalerate	1×10^{-6}	(min)	(min)	(∞)
Deltamethrin	1×10^{-6}	(min)	(min)	(∞)

From Lund and Narahashi (1983).

Therefore, the penetration factor does not contribute to the negative temperature dependence of insecticidal activity, but rather acts to lessen the temperature dependence. Enzymatic detoxication of DDT (Narahashi, 1971) was augmented by increasing the temperature, thereby contributing to the negative temperature dependence of insecticidal activity (see Narahashi, 1971). However, the detoxication factor is not sufficient to account for the negative temperature dependence. One line of evidence is that a larger amount of undetoxified DDT was detected in insects that survived at high temperature than in the dead insects at low temperature (Vinson and Kearns, 1952). Thus the insects could tolerate a larger amount of DDT at a high temperature than at a low temperature. Another line of evidence comes from the observation that the symptoms of DDT poisoning, when appropriate doses were applied, were reversible upon changing the temperature. The DDT-poisoned cockroaches exhibited the symptoms of poisoning at 15°C, but the symptoms disappeared when the temperature was raised to 30°C (Yamasaki and Ishii (Narahashi), 1954; Narahashi, 1971). This process could be repeated several times by changing the temperature between the low and high levels. In order to explain this phenomenon by detoxication, we must assume the reversible changes between DDT and its metabolites such as DDE with respect to temperature, but no trace of DDT was discovered in insects injected with DDE or DDA (Sternburg and Kearns, 1950). Thus it was concluded that the detoxication factor alone could not explain the negative temperature dependence of insecticide action.

The nervous system was found to be the critical factor for the negative temperature dependence of the insecticidal action of DDT (Yamasaki and Ishii (Narahashi), 1954). The potency of DDT to initiate trains of impulses in the sensory neurons was vastly different depending on the temperature, being much higher at 16°C than at 30°C with a Q_{10} of 0.181 (Yamasaki and Ishii (Narahashi), 1954; Fig. 38 of Narahashi, 1971). This Q_{10} value is in the same order of magnitude as that for developing symptoms of poisoning by DDT in cockroaches (Vinson and Kearns, 1952; Yamasaki and Ishii (Narahashi), 1954; Table I of Narahashi, 1971). Furthermore, trains of impulses appeared and disappeared depending on the temperature; after an appropriate dose of DDT, trains appeared upon lowering the temperature and disappeared upon raising it, and this procedure could be repeated (Fig. 18). Therefore, it is clear that the temperature dependence of the action of DDT to evoke repetitive discharges in nerves is responsible for the negative temperature dependence of producing symptoms of poisoning in insects (see also van den Bercken and Akkermans, 1971).

Pyrethroids also exhibit higher insecticidal activity at a lower temperature than at a higher temperature (see Narahashi, 1971). Similar to the case of DDT, a change in nerve sensitivity to pyrethroids with respect to temperature is the key factor for the negative temperature dependence of insecticidal action. However, the mechanism underlying the temperature-dependent nerve sensitivity to pyrethroids is more complicated than in the case of DDT. First, there was an optimal temperature range at which repetitive after-discharges were evoked in allethrin-poisoned squid axons (Starkus and Narahashi, 1978). This usually occurred between 22°C and 28°C, and lowering or raising the temperature beyond this range caused repetitive after-discharge to stop. The initiation of repetitive discharges could be accounted for by the fact that at the optimal temperatures the amplitudes of the depolarizing after-potential exceeded the threshold membrane potential level for the generation of action potentials. At temperatures higher than the optimal range, the threshold membrane potential became increasingly less negative while the amplitude of the depolarizing after-potential drastically decreased. In the cockroach giant axons poisoned with allethrin, there also was a critical temperature (around 26°C) beyond which repetitive after-discharges were produced (Narahashi, 1962a). However, repetitive discharges did not disappear at 33°C, the maximum temperature tested in the study. It is not known whether repetitive discharges stop at temperatures higher than 33°C in the cockroach axon, yet this is likely to happen as cockroaches are acclimated to higher temperatures than squid.

Pyrethroids also depolarize the nerve membrane which in turn would cause a conduction block in axons (Narahashi, 1971; Wang et al., 1972; Narahashi and Anderson, 1967), an increase in sensory discharges (van den Bercken et al., 1973; Vijverberg et al., 1982c), and an increase in transmitter release from nerve terminals (Eells et al., 1987). The action potentials recorded from the cockroach nerve cord were blocked by allethrin when the temperature was lowered below around 15°C and restored when the temperature was elevated beyond that level (Narahashi, 1971). This procedure could be repeated a few times in the presence of the same concentration of allethrin. The negative temperature dependence of allethrin block can be explained partially by an increase in sodium conductance block by lowering the temperature as observed in squid axons (Wang et al., 1972) and partially by an increase in membrane depolarization by lowering the temperature as observed in tetramethrin- and fenvalerate-poisoned crayfish giant axons (Salgado et al., 1989). These negative temperature coefficients appear to be responsible for the increase in the action of pyrethroids to depolarize the membrane and therefore to disturb sensory discharges and synaptic transmission in general. However, more systematic studies are needed to construct the whole picture of pyrethroid actions on various membrane and sodium channel parameters that can explain the negative temperature dependence of insecticidal action.

SUMMARY AND CONCLUSIONS

Both type I and type II pyrethroids modify the gating kinetics of the sodium channel. The rates at which the sodium channels open and close are markedly reduced causing prolonged opening of individual channels. At the whole cell level, these changes in gating kinetics cause a prolonged sodium current to flow both during and after a step depolarizing pulse. Thus the action potential is followed by an increased and prolonged depolarizing after-potential from which repetitive after-discharges may be produced. The pyrethroid-modified sodium channels can open at membrane potentials more negative than those at which normal channels open. This shift in the activation potential and the prolonged opening of sodium channels cause a depolarization of the nerve membrane. Type II pyrethroids are generally more efficacious than type I pyrethroids in exerting these effects. Thus the modification of the sodium channel gating kinetics would cause functional changes in various regions of the nervous system. These include repetitive discharges in nerve fibers and nerve terminals; an increase in discharges from sensory neurons due to membrane depolarization; an increase in transmitter release from nerve terminals due to membrane depolarization; and the severe disturbances of synaptic transmission. The differences in the degree to which each of these changes is brought about by type I and type II pyrethroids may account for somewhat different symptoms of poisoning in animals between the two types of pyrethroids.

Despite drastic modifications of sodium channel gating kinetics by pyrethroids, the properties of the open sodium channel, including ionic selectivity and cation block, remain unaffected. It appears that the pyrethroid molecule gets access to the gating machinery of the sodium channel via the lipid phase of the membrane. The pyrethroid binding site is different from the sites for batrachotoxin, grayanotoxin, TTX, octylguanidine and local anesthetics. Three different sites in the sodium channel have been identified for the active (+)-isomers and the inactive (-)-isomers of tetramethrin, i.e. a trans site, a cis site, and a negative allosteric site. The inactive isomers can bind to some of these sites thereby antagonizing the effect of the active isomers.

Only a very small fraction of the sodium channel population needs to be modified by pyrethroids to elevate the depolarizing after-potential to the threshold level for the generation of repetitive discharges. For tetramethrin this fraction has been estimated to be in the order of 0.1%. The concentration of tetramethrin to cause this degree of channel modification is calculated to be 1-10 nM in agreement with the observed value. This kind of toxicological amplification through a threshold phenomenon plays an important role in high potency of the pyrethroids.

Calcium channels are blocked by tetramethrin, but not by deltamethrin or fenvalerate. The block comprises a time-dependent reversible component and a time-independent irreversible component. The toxicological significance of the pyrethroid block of calcium channels remains to be seen. It is unlikely that the potent stimulating action of pyrethroids on certain insect neurons, thought to generate action potentials by calcium channels, is due to modulation of the calcium channels.

GABA receptor-channel complexes are not affected by deltamethrin. While the sodium channel of a rat dorsal root ganglion neuron undergoes drastic changes in gating kinetics after application of deltamethrin, the GABA-induced chloride current remains completely unaffected. It seems unlikely that the inhibition of the GABA system reported in the literature plays a very significant role in the symptoms of poisoning in animals.

DDT and its derivatives, despite the fact that the chemical structures are considerably different from those of pyrethroids, modify the sodium channel gating kinetics in a manner very similar to that of pyrethroids, especially type I pyrethroids.

The well-documented negative temperature dependence of the insecticidal action of DDT and pyrethroids can be accounted for by the higher nerve sensitivity to the insecticides at lower temperature than at higher temperatures. This temperature dependence of action is related to the high temperature dependence of channel modifications caused by these insecticides.

ACKNOWLEDGEMENTS

Our studies quoted in this article were supported by NIH grant NS14143. Thanks are due to Janet Henderson and Vicky James-Houff for secretarial assistance.

REFERENCES

Albuquerque, E.X., Seyama, I., and Narahashi, T., 1973, Characterization of batrachotoxin-induced depolarization of the squid giant axons, J. Pharmacol. Exp. Ther., 184:308-314.

Armstrong, C.M., and Bezanilla F., 1974, Charge movement associated with the opening and closing of the activation gates of the Na channels, J. Gen. Physiol., 63:533-552.

Barker, R.J., 1957, Some effects of temperature on adult house flies treated with DDT, J. Econ. Entom., 50:446-450.

Bloomquist, J.R., and Soderlund, D.M., 1985, Neurotoxic insecticides inhibit GABA-dependent chloride uptake by mouse brain vesicles, Biochem. Biophys. Res. Comm., 133:37-43.

Brown, L.D., and Narahashi, T., 1987, Activity of tralomethrin to modify the nerve membrane sodium channel, Toxicol. Appl. Pharmacol., 89:305-313.

Burt, P.E., and Goodchild, R.E., 1971, The site of action of pyrethrin I in the nervous system of the cockroach *Periplaneta americana*, Entomol. Exp. Appl., 14:179-189.

Chalmers, A.E., Miller, T.A., and Olsen, R.W., 1985, A pharmacological investigation of invertebrate GABA receptors, Neurotox '85, Neuropharmacology and Pesticide Action, Univ. Bath, Abstr. p. 41-42.

Chinn, K., and Narahashi, T., 1986, Stabilization of sodium channel states by deltamethrin in mouse neuroblastoma cells, J. Physiol. (London), 380:191-207.

Courtney, K.R., 1975, Mechanism of frequency-dependent inhibition of sodium currents in frog myelinated nerve by the lidocaine derivative GEA 968, J. Pharmacol. Exp. Ther., 195:225-236.

Courtney, K.R., 1980, Structure-activity relations for frequency-dependent sodium channel block in nerve by local anesthetics, J. Pharmacol. Exp. Ther., 213:114-119.

Courtney, K.R., Kendig, J.J., and Cohen, E.N., 1978, The rates of interaction of local anesthetics with sodium channels in nerve, J. Pharmacol. Exp. Ther., 207, 594-604.

Crofton, K.M., Reiter, L.W., and Mailman, R.B., 1987, Pyrethroid insecticides and radioligand displacement from the GABA receptor chloride ionophore complex, Toxicol. Letters, 35:183-190.

de Weille, J.R., Vijverberg, H.P.M., and Narahashi, T., 1988, Interactions of pyrethroids and octylguanidine with sodium channels of squid giant axons, Brain Res., 445:1-11.

Eells, J.T., Watabe, S., Ogata, N., and Narahashi, T., 1987, The effects of pyrethroid insecticides on synaptic transmission in slices of guinea pig olfactory cortex, NATO Advanced Study Institute Series, Vol. H13, Toxicology of Pesticides: Experimental, Clinical and Regulatory Aspects, L.G. Costa et al., ed., Springer-Verlag, Berlin, Heidelberg, pp. 267-271.

Gammon, D., and Casida, J.E., 1983, Pyrethroids of the most potent class antagonize GABA action at the crayfish neuromuscular junction, Neurosci. Letters, 40:163-168.

Gammon, D.W., and Sander, G., 1985, Two mechanisms of pyrethroid action: Electrophysiological and pharmacological evidence, NeuroToxicology 6:63-86.

Gammon, D.W., Lawrence, L.J., and Casida, J.E., 1982, Pyrethroid toxicology: Protective effects of diazepam and phenobarbital in the mouse and the cockroach, Toxicol. Appl. Pharmacol., 66:290-296.

Gilly, W.F., and Armstrong, C.M., 1982, Slowing of sodium channel opening kinetics in squid axon by extracellular zinc, J. Gen. Physiol., 79:935-964.

Hamill, O.P., Marty, A., Neher, E., Sakmann, B., and Sigworth, F.J., 1981, Improved patch-clamp techniques for high-resolution current recording from cells and cell-free membrane patches, Pflügers Arch., 391:85-100.

Hille, B., 1968, Pharmacological modification of the sodium channels of frog nerves, J. Gen. Physiol., 51:199-219.

Hille, B., 1971, The permeability of the sodium channel to organic cations in myelinated nerve, J. Gen. Physiol., 58:599-619.

Hille, B., 1975, The receptor for tetrodotoxin and saxitoxin, A structural hypothesis, Biophys.J., 15:615-619.

Hille, B., 1977, Local anesthetics: Hydrophilic and hydrophobic pathways for the drug-receptor reaction, J. Gen. Physiol., 69:497-515.

Holloway, S.F., Salgado, V.L., Wu, C.H., and Narahashi, T., 1984, Maintained opening of single Na channels by fenvalerate, 14th Ann. Mtg. Soc. Neurosci. Abstr. 10:864.

Kao, C.Y., 1983, New perspectives on the interactions of tetrodotoxin and saxitoxin with excitable membranes, Toxicon Suppl., 3:211-219.

Kao, C.Y., and Nishiyama, A., 1965, Actions of saxitoxin on peripheral neuromuscular systems, J. Physiol. (London), 180:50-66.

Kao, C.Y., and Walker, S.E., 1982, Active groups of saxitoxin and tetrodotoxin as deduced from actions of saxitoxin analogues on frog muscle and squid axon, J. Physiol. (London), 323:619-637.

Kao, P.N., James-Kracke, M.R., and Kao, C.Y., 1983, The active guanidinium group of saxitoxin and neosaxitoxin identified by the effects of pH on their activities on squid axon, Pflügers Arch., 398:199-203.

Keynes, R.D., and Rojas, E., 1974, Kinetics and steady state properties of the charged system controlling sodium conductance in the squid giant axon, J. Physiol. (London), 239:393-434.

Khodorov, B.I., 1978, Chemicals as tools to study nerve fiber sodium channels: Effects of batrachotoxin and some local anesthetics, in: "Membrane Transport Processes," D.C. Tosteson, A.O. Yu and R. Latorre, ed., Vol. 2, Raven Press, New York, pp. 153-174.

Khodorov, B., 1979, Some aspects of the pharmacology of sodium channels in nerve membrane, Process of inactivation, Biochem. Pharmacol., 28:1451-1459.

Khodorov, B.I., 1985, Batrachotoxin as a tool to study voltage-sensitive sodium channels of excitable membranes, Progr. Biophys. Mol. Biol., 45:57-148.

Khodorov, B.I., Shishkova, L., Peganov, E., and Revenko, S., 1976, Inhibition of sodium currents in frog Ranvier node treated with local anesthetics, Role of slow sodium inactivation, Biochim. Biophys. Acta, 433:409-435.

Kirsch, G.E., Yeh, J.Z., Farley, J.M., and Narahashi, T., 1980, Interaction of n-alkylguanidines with the sodium channels of squid axon membrane, J. Gen. Physiol., 76:315-335.

Lawrence, L.J., and Casida, J.E., 1983, Stereospecific action of pyrethroid insecticides on the γ-aminobutyric acid receptor-ionophore complex, Science, 221:1399-1401.

Lawrence, L.J., Gee, K.W., and Yamamura, H.I., Interactions of pyrethroid insecticides with chloride ionophore-associated binding sites, NeuroToxicology, 6(2):87-98.

Lowenstein, O., 1942, A method of physiological assay of pyrethrum extracts, Nature (London), 150:760-762.
Lummis, S.C.R., Chow, S.C., Holan, G., and Johnston, G.A.R., 1987, γ-Aminobutyric acid receptor ionophore complexes: Differential effects of deltamethrin, dichlorodiphenyltrichloroethane, and some novel insecticides in a rat brain membrane preparation, J. Neurochem., 48:689-694.
Lund, A.E., and Narahashi, T., 1981a, Modification of sodium channel kinetics by the insecticide tetramethrin in crayfish giant axons, Neurotoxicology, 2:213-229.
Lund, A.E., and Narahashi, T., 1981b, Kinetics of sodium channel modification by the insecticide tetramethrin in squid axon membranes, J. Pharmacol. Exp. Ther., 219:464-473.
Lund, A.E., and Narahashi, T., 1981c, Interaction of DDT with sodium channels in squid giant axon membranes, Neuroscience, 6, 2253-2258.
Lund, A.E., and Narahashi, T., 1982, Dose-dependent interaction of the pyrethroid isomers with sodium channels of squid axon membranes, Neurotoxicology, 3:11-24.
Lund, A.E., and Narahashi, T., 1983, Kinetics of sodium channel modification as the basis for the variation in the nerve membrane effects of pyrethroids and DDT analogs, Pestic. Biochem. Physiol., 20:203-216.
Matsumura, F., and Tanaka, K., 1984, Molecular basis of neuroexcitatory actions of cyclodiene-type insecticides, in: "Cellular and Molecular Neurotoxicology," T. Narahashi, ed., Raven Press, New York, pp. 225-240.
Narahashi, T., 1962a, Effect of the insecticide allethrin on membrane potentials of cockroach giant axons, J. Cell. Comp. Physiol., 59:61-65.
Narahashi, T., 1962b, Nature of the negative after-potential increased by the insecticide allethrin in cockroach giant axons, J. Cell. Comp. Physiol., 59:67-76.
Narahashi, T., 1971, Effects of insecticides on excitable tissues, in: "Advances in Insect Physiology," Vol. 8, J.W.L. Beament, J.E. Treherne and V.B. Wiggleworth, ed., Academic Press, London and New York, pp. 1-93.
Narahashi, T., 1974, Chemicals as tools in the study of excitable membranes, Physiol. Rev., 54:813-889.
Narahashi, T., 1976, Effects of insecticides on nervous conduction and synaptic transmission, in: "Insecticide Biochemistry and Physiology," C.F. Wilkinson, ed., Plenum Press, New York, pp. 327-352.
Narahashi, T., 1981, Mode of action of chlorinated hydrocarbon pesticides on the nervous system, in: Halogenated Hydrocarbons: Health and Ecological Effects," M.A.Q. Khan, ed., Pergamon Press, Elmsford, NY, pp. 222-242.
Narahashi, T., 1985, Nerve membrane ionic channels as the primary target of pyrethroids, NeuroToxicology, 6(2):3-22.
Narahashi, T., 1987, Neuronal target sites of insecticies, in: "Sites of Action for Neurotoxic Pesticides," American Chemical Society Symposium Series, No. 356, R.M. Hollingworth and M.B. Green, ed., ACS, Washington, DC, pp. 226-250.
Narahashi, T., 1988, Mechanism of tetrodotoxin and saxitoxin action, in: Handbook of Natural Toxins," Vol. 3, Marine Toxins and Venoms, A.T. Tu, ed., Marcell Dekker, New York, pp. 185-210.
Narahashi, T., and Anderson, N.C., 1967, Mechanism of excitation block by the insecticide allethrin applied externally and internally to squid giant axons, Toxicol. Appl. Pharmacol., 10:529-547.
Narahashi, T., and Haas, H.G. 1967, DDT: Interaction with nerve membrane conductance changes, Science, 157:1438-1440.
Narahashi, T., and Haas, H.G., 1968, Interaction of DDT with the components of lobster nerve membrane conductance, J. Gen. Physiol., 51:177-198.

Narahashi, T., and Seyama, I., 1974, Mechanism of nerve membrane depolarization caused by grayanotoxin I., J. Physiol. (London), 242:471-487.

Narahashi, T., and Yamasaki, T., 1960a, Mechanism of increase in negative after-potential by dicophanum (DDT) in the giant axons of the cockroach, J. Physiol. (London), 152:122-140.

Narahashi, T., and Yamasaki, T., 1960b, Behaviors of membrane potential in the cockroach giant axons poisoned by DDT, J. Cell. Comp. Physiol., 55:131-142.

Narahashi, T., Moore, J.W., and Scott, W.R., 1964, Tetrodotoxin blockage of sodium conductance increase in lobster giant axons, J. Gen. Physiol., 47:965-974.

Narahashi, T., Anderson, N.C., and Moore, J.W., 1967, Comparison of tetrodotoxin and procaine in internally perfused squid giant axons, J. Gen. Physiol., 50:1413-1428.

Narahashi, T., Albuquerque, E.X., and Deguchi, T., 1971, Effects of batrachotoxin on membrane potential and conductance of squid giant axons, J. Gen. Physiol., 58:54-70.

Narahashi, T., Tsunoo, A., and Yoshii, M., 1987, Characterization of two types of calcium channels in mouse neuroblastoma cells, J. Physiol. (London), 383:231-249.

Neher, E., and Sakmann, B., 1976, Single-channel currents recorded from membrane of denervated frog muscle fibres, Nature (London), 260:779-802.

Neumcke, B., Nonner, W., and Stämpfli, W., 1976, Asymmetrical displacement current and its relation with the activation of sodium current in the membrane of frog myelinated nerve, Pflügers Arch., 363:193-203.

Nishimura, K., and Narahashi, T., 1978, Structure-activity relationships of pyrethroids based on direct action on nerve, Pestic. Biochem. Physiol., 8:53-64.

Ogata, N., Vogel, S.M., and Narahashi, T., 1988, Lindane but not deltamethrin blocks a component of GABA-activated chloride channels, FASEB J., 2:2895-2900.

Orchard, I., 1980, The effects of pyrethroids on the electrical activity of neurosecretory cells from the brain of *Rhodnius prolixus*, Pestic. Biochem. Physiol., 13:220-226.

Orchard, I., and Osborne, M.P., 1979, The action of insecticides on neurosecretory neurons in the stick insect, *Carausius morosus*, Pestic. Biochem. Physiol., 10:197-202.

Osborne, M.P., 1980, The insect synapse: structure functional aspects in relation to insecticidal action, in: "Insect Neurobiology and Pesticide Action (Neurotox 79)," Society of Chemical Industry, London, pp. 29-39.

Osborne, M.P., 1986, Insect neurosecretory cells - structural and physiological effects induced by insecticides and related compounds, in: "Neuropharmacology and Pesticide Action, M.G. Ford, G.G. Lunt, R.C. Reay and P.N.R. Usherwood, ed., Ellis Horwood Ltd., Chichester, England, pp. 203-243.

Pichon, Y., 1969, Aspects Électriques et Ioniques du Fonctionnement Nerveux chez les Insectes, Cas Particulier de la Chaine Nerveuse Abdominale d'une Blatte *Periplaneta americana* L, Thèse, University of Rennes, France.

Roeder, K.D., and Weiant, E.A., 1946, The site of action of DDT in the cockroach, Science, 103:304-306.

Ruigt, G.S.F., 1984, Pyrethroids, in: "Comprehensive Insect Physiology, Biochemistry and Pharmacology," G.A. Kerkut and L.I. Gilbert, ed., Vol. 12, Chapter 7, Pergamon Press, Oxford, pp. 183-263.

Ruigt, G.S.F., Neijt, H.C., de Weille, J.R., van der Zalm, J.M., and van den Bercken, J., 1984, Interaction of pyrethroid insecticides with sodium

channels in internally perfused, cultured mouse neuroblastoma cells, in, "An Electrophysiological Investigation into the Mode of Action of Pyrethroid Insecticides," Thesis, G.S.F. Ruigt, University of Utrecht, The Netherland, pp. 131-148.
Salgado, V.L., and Narahashi, T., 1989, Block of sodium channel gating current by the pyrethroid fenvalerate in crayfish giant axons, Biophys. J., in press.
Salgado, V.L., Herman, M.D., and Narahashi, T., 1989, Interactions of the pyrethroid fenvalerate with nerve membrane sodium channels: Temperature dependence and mechanism of depolarization, NeuroToxicology, in press.
Seyama, I., and Narahashi, T., 1973, Increase in sodium permeability of squid axon membranes by α-dihydrograyanotoxin II, J. Pharmacol. Exp. Ther., 184:299-307.
Seyama, I., and Narahashi, T., 1981, Modulation of sodium channels of squid nerve membranes by grayanotoxin I, J. Pharmacol. Exp. Ther., 219:614-624.
Shapiro, B.I., 1977, Effects of strychnine on the sodium conductance of the frog node of Ranvier, J. Gen. Physiol., 69:915-926.
Starkus, J.G., and Narahashi, T., 1978, Temperature dependence of allethrin-induced repetitive discharges in nerves, Pestic. Biochem. Physiol., 9:225-230.
Sternburg, J., and Kearns, C.W., 1950, Degradation of DDT by resistant and susceptible strains of house flies, Ann. Entom. Soc. Amer., 43:444-458.
Strichartz, G., 1973, The inhibition of sodium currents in myelinated nerve by quaternary derivatives of lidocaine, J. Gen. Physiol., 62:37-57.
Takeda, K., and Narahashi, T., 1988, Chemical modification of sodium channel inactivation: separate sites for the action of grayanotoxin and tetramethrin, Brain Res., 448:308-312.
Takeno, K., Nishimura, K., Parmentier, J., and Narahashi, T., 1977, Insecticide screening with isolated nerve preparation for structure-activity relationships, Pestic. Biochem. Physiol., 7:486-499.
Tanguy, J., Yeh, J.Z., and Narahashi, T., 1984, Interaction of batrachotoxin with sodium channels in squid axons, Biophys. J., 45:184a.
van den Bercken, J., and Akkermans, L.M.A., 1971, Negative temperature coefficient of the action of DDT in a sense organ, Europ. J. Pharmacol., 16:241-244.
van den Bercken, J., Akkermans, L.M.A., and van der Zalm, J.M., 1973, DDT-like action of allethrin in the sensory nervous system of *Xenopus laevis*, Europ. J. Pharmacol., 21:95-106.
Vijverberg, H.P.M., van der Zalm, J.M., and van den Bercken, J., 1982a, Similar mode of action of pyrethroids and DDT on sodium channel gating in myelinated nerves, Nature (London), 295:601-603.
Vijverberg, H.P.M., van der Zalm, J.M., van Kleef, R.G.D.M., and van den Bercken, J., 1982b, Temperature and structure-dependent interaction of pyrethroids with the sodium channels in frog node of Ranvier, Biochim. Biophys. Acta, 728:73-82.
Vijverberg, H.P.M., Ruigt, G.S.F., and van den Bercken, J., 1982c, Structure-related effects of pyrethroid insecticides on the lateral-line sense organ and on peripheral nerves of the clawed frog, *Xenopus laevis*, Pestic. Biochem. Physiol., 18:315-324.
Vinson, E.B., and Kearns, C.W., 1952, Temperature and the action of DDT on the American cockroach, J. Econ. Entom., 45:484-496.
Wang, C.M., Narahashi, T., and Scuka, M., 1972, Mechanism of negative temperature coefficient of nerve blocking action of allethrin, J. Pharmacol. Exp. Ther., 182:442-453.
Welsh, J.H., and Gordon, H.T., 1947, The mode of action of certain insecticides on the arthropod nerve axon, J. Cell. Comp. Physiol., 30:147-172.

Woolley, D.E., 1981, The neurotoxicity of DDT and possible mechanisms of action, in: Mechanisms of Neurotoxic Substances, K.N. Prasal and A. Vernadakis, ed., Raven Press, New York, pp. 95-141.
Wouters, W., and van den Bercken, J., 1978, Action of pyrethroids, Gen. Pharmacol., 9:387-398.
Wu, C.H., Oxford, G.S., Narahashi, T., and Holan, G., 1980, Interaction of a DDT analog with the sodium channel of lobster axon, J. Pharmacol. Exp. Ther., 212:287-293.
Yamamoto, D., Quandt, F.N., and Narahashi, T., 1983, Modification of single sodium channels by the insecticide tetramethrin, Brain Res., 274:344-349.
Yamamoto, D., Yeh, J.Z., and Narahashi, T., 1984, Voltage-dependent calcium block of normal and tetramethrin-modified single sodium channels, Biophys. J., 45:337-344.
Yamamoto, D., Yeh, J.Z., and Narahashi, T., 1985, Interactions of permeant cations with sodium channels of squid axon membranes, Biophys. J., 48:361-368.
Yamamoto, D., Yeh, J.Z., and Narahashi, T., 1986, Ion permeation and selectivity of squid axon sodium channels modified by tetramethrin, Brain Res., 372:193-197.
Yamasaki, T., and Ishii (Narahashi), T., 1952a, Studies on the mechanism of action of insecticides (IV), The effects of insecticides on the nerve conduction of insect, Oyo-Kontyu (J. Nippon Soc. Appl. Entomol.), 7:157-164.
Yamasaki, T., and Ishii (Narahashi), T., 1952b, Studies on the mechanism of action of insecticides (V), The effects of DDT on the synaptic transmission in the cockroach, Oyo-Kontyu (J. Nippon Soc. Appl. Entomol.), 8:111-118.
Yamasaki, T., and Ishii (Narahashi), T., 1954, Studies on the mechanism of action of insecticides (VIII), Effects of temperature on the nerve susceptibility to DDT in the cockroach, Botyu-Kagaku, 19:39-46, English translation, 1957, in: Japanese Contributions to the Study of the Insecticide-Resistance Problem, Publ. by Kyoto Univ. for W.H.O., pp. 155-162.
Yamasaki, T., and Narahashi, T., 1957, Intracellular microelectrode recording of resting and action potentials from the insect axon and the effects of DDT on the action potential, Studies on the mechanism of action of insecticides (XIV), Botyu-Kogaku, 22:305-313.
Yeh, J.Z., 1978, Sodium inactivation mechanism modulates QX-314 block of sodium channels in squid axons, Biophys. J., 24:569-574.
Yeh, J.Z., 1979, Dynamics of 9-aminoacridine block of sodium channels in squid axons, J. Gen. Physiol., 73:1-21.
Yeh, J.Z., 1980, Blockage of sodium channels by stereoisomers of local anesthetics, in: "Molecular Mechanisms of Anesthesia," B.R. Fink, ed., Raven Press, New York, pp. 35-44.
Yeh, J.Z., and Narahashi, T., 1977, Kinetic analysis of pancuronium interaction with sodium channels in squid axon membranes, J. Gen. Physiol., 69:293-323.
Yoshii, M., Tsunoo, A., and Narahashi, T., 1985, Effects of pyrethroids and veratridine on two types of Ca channels in neuroblastoma cells, Soc. Neurosci. Abstr., 11:518.
Yoshii, M., Tsunoo, A., and Narahashi, T., 1988, Gating and permeation properties of two types of calcium channels in neuroblastoma cells, Biophys. J., 54:885-895.

PHARMACOLOGICAL CHARACTERIZATION OF INSECTICIDE-BINDING DOMAINS OF THE VOLTAGE-SENSITIVE SODIUM CHANNEL

David M. Soderlund, Jeffrey R. Bloomquist,[1] Gregory T. Payne, and James A. Ottea

Department of Entomology
New York State Agricultural Experiment Station
Cornell University
Geneva, NY 14456

[1] Department of Entomology
Virginia Polytechnic Institute and State University
Blacksburg, VA 24061

ABSTRACT

The voltage-sensitive sodium channel mediates the transient sodium permeability of the cell membrane that is associated with action potentials in vertebrate and invertebrate nerves and vertebrate skeletal muscle. Since this channel is gated by changes in cell membrane potential rather than by the binding of a ligand, it is not considered to be a "receptor" as that term is understood by neurobiologists. Nevertheless, the voltage-sensitive sodium channel has been shown to possess binding domains for a variety of potent neurotoxins, and it is implicated as the principal molecular target for the action of pyrethroid insecticides and DDT and its analogs. More recent studies have suggested that synthetic analogs of naturally-occurring *N*-alkylamide insecticides may also act at the sodium channel. In this paper, we summarize current information on the number and properties of neurotoxin binding domains on the voltage-sensitive sodium channel and review recent studies from this laboratory that implicate the involvement of these and other binding domains in the action of insecticides.

INTRODUCTION

Neurotoxin Recognition Sites on the Sodium Channel

Four principal neurotoxin recognition sites associated with the sodium channel have been characterized in both functional assays and in radioligand binding experiments (Catterall, 1984). Site 1 binds the water-soluble toxins tetrodotoxin (TTX) and saxitoxin, which interact at or near the extracellular opening of the ion pore and block ion transport through the channel. Site 2 binds a structurally heterogeneous group of lipophilic toxins that alter both the opening (activation) and closing (inactivation) of sodium channels. Site 2 neurotoxins [e.g. veratridine (VTD), aconitine (ACN), batrachotoxin (BTX), and the grayanotoxins] shift the voltage dependency of sodium channel activation toward more negative membrane potentials, thereby increasing the probability that channels will open at normal membrane resting potentials. These toxins also slow or completely block sodium channel inactivation, thereby prolonging the open state of the channel. [^{3}H]Batrachotoxinin A 20-α-benzoate ([^{3}H]BTX-B), a close structural analog of BTX, has been developed as a radioligand to study binding interactions at Site 2 (Catterall et al., 1981). Site 3 binds a group of polypeptide toxins, called α-toxins, isolated from scorpion venoms or sea anemone nematocysts. These toxins selectively prolong sodium channel inactivation without affecting the rate or voltage dependency of channel opening (Strichartz et al., 1987). Binding and ion flux studies have documented allosteric interactions between Sites 2 and 3, by which α-polypeptide toxins are able to increase both the potency and efficacy of activators binding at Site 2 (Catterall, 1984). Site 4 binds a second group of polypeptide scorpion toxins, called β-toxins, which selectively enhance sodium channel activation and do not interact allosterically with Site 2 neurotoxins (Bablito et al., 1986).

Recently, an additional group of lipophilic neurotoxins also has been shown to interact with the sodium channel at a binding domain distinct from Sites 1-4. The brevetoxins, which are isolated from marine dinoflagellates, shift the voltage dependency of sodium channel activation and prolong inactivation in a manner similar to Site 2 neurotoxins, but these compounds also exhibit allosteric interactions with compounds acting at Site 2 (Catterall and Risk, 1981; Huang et al., 1984; Atchison et al., 1986; Sharkey et al., 1987). Ciguatoxin, which is also produced by a marine dinoflagellate, similarly prolongs sodium channel activation and synergizes the action of Site 2 neurotoxins (Bidard et al., 1984; Benoit et al., 1986). Binding studies using a tritiated brevetoxin (Poli et al., 1986; Lombet et al., 1987) demonstrate that the brevetoxins and ciguatoxin bind to a single recognition site that is distinct from Sites 1-4 and may be considered to be Site 5.

Other neurotoxins and neuroactive drugs also act at the sodium channel, but the recognition sites for these compounds are less well characterized than those described above. Gonioporatoxin, a polypeptide toxin isolated from coral, prolongs sodium channel inactivation and interacts allosterically with Site 2 neurotoxins in a manner similar to compounds acting at Site 3, but this toxin does not displace labeled scorpion toxin from Site 3 in binding studies (Gonoi et al., 1986). The alkaloid pumiliotoxin B activates sodium channels and is potentiated by scorpion α-toxin in a manner similar to Site 2 neurotoxins, but this compound does not displace [^{3}H]BTX-B from Site 2 and therefore must exert its effects by binding to another site (Gusovsky et al., 1988). Local anesthetics and the anticonvulsants phenytoin and carbamazepine inhibit the activation of sodium channels by Site 2 neurotoxins (Creveling et al., 1983; Willow et al., 1984). For both groups of compounds, this inhibitory action is correlated with an allosteric inhibition of the binding of [^{3}H]BTX-B to Site 2 (Willow and Catterall, 1982; Creveling et al., 1983; Postma and Catterall, 1984) that does not appear to result from an action at any of the other well-characterized binding domains of the sodium channel. Thus, the available data for these neurotoxins and drugs suggest the possible existence of three additional neurotoxin-binding sites on the voltage-sensitive sodium channel.

Insecticides Acting at the Sodium Channel

The actions of DDT and pyrethroids on the voltage-sensitive sodium channels are well-established (for recent reviews, see: Narahashi, 1984, 1986; Ruigt, 1985; Soderlund and Bloomquist, 1989). The principal effect of these compounds, characterized in detail using voltage clamp and patch clamp techniques, is a slowing of sodium channel activation and inactivation (Narahashi 1984, 1986). This effect is observed as the induction of slowly-decaying tail currents that persist after repolarization in voltage clamp experiments and the prolongation of the open times of individual sodium channels in patch clamp studies. Recently, a new group of insecticides has been synthesized (Elliott, 1985) based on naturally-occurring insecticidal lipophilic *N*-alkylamides (Su, 1985). Preliminary physiological studies show that example compounds from this class have pyrethroid-like actions, inducing repetitive activity followed by conduction block in house fly nerves (Blade et al., 1985) and suppressing peak sodium current and inducing slowly-decaying tail currents in voltage-clamped locust nerve cell bodies in culture (Lees and Burt, 1988). These results implicate the sodium channel as the principal target site for this class of insecticides as well. However, these compounds retain high insecticidal activity against *super-kdr* strains of the house fly that exhibit broad cross-resistance to DDT and pyrethroids by virtue of reduced neuronal sensitivity (Elliott et al., 1986). These findings suggest that more than one binding domain may be involved in the action of insecticides on the voltage-sensitive sodium channel.

PHARMACOLOGICAL CHARACTERIZATION OF A DDT/PYRETHROID BINDING DOMAIN

Radiosodium Uptake Studies

Jacques et al. (1980) first applied the technique of radiosodium uptake into cultured neuroblastoma cells to examine the actions of pyrethroids on sodium channels. In these studies, deltamethrin and kadethrin failed to stimulate sodium uptake over control levels when applied alone, but these compounds produced a TTX-sensitive potentiation of the sodium uptake caused by three Site 2 neurotoxins (BTX, VTD, and dihydrograyanotoxin II). Moreover, these pyrethroids also enhanced the sodium uptake caused by α-toxins acting at Site 3. Jacques et al. (1980) interpreted these findings as evidence for a pyrethroid-binding domain on the sodium channel that is distinct from Sites 1-3. Other studies showed that deltamethrin potentiated ciguatoxin-dependent sodium uptake into neuroblastoma (Bidard et al., 1984), thereby demonstrating that the pyrethroid recognition site is separate from that for ciguatoxin and brevetoxins. Although the results with deltamethrin and kadethrin are consistent with the actions of these compounds in biophysical studies, several other neurotoxic pyrethroids failed to enhance sodium uptake stimulated by Site 2 neurotoxins in this system and were only effective as inhibitors of deltamethrin- or kadethrin-stimulated enhancement (Jacques et al., 1980). Subsequent studies (Holan et al., 1985) of the action of DDT analogs and DDT-pyrethroid hybrid structures on α-toxin-stimulated uptake into neuroblastoma cells also failed to detect an effect of these compounds on sodium channel function in this system.

Although these findings with neuroblastoma cells identified methods to explore allosteric interactions between pyrethroids and other sodium channel-directed neurotoxins, they suggested that the insecticide recognition properties of sodium channels in these cells might differ substantially from those of sodium channels in native tissue. We therefore initiated studies of the action of pyrethroids on VTD-stimulated sodium uptake into mouse brain synaptosomes, a system in which results from biochemical assays could be correlated with a substantial body of data on the acute intracerebral toxicity of resolved pyrethroid isomers to mice (Lawrence and Casida, 1982). The methods employed in these studies were based directly on those used to characterize the actions of a variety of neurotoxins on sodium channel function in assays of sodium uptake into rat brain synaptosomes (Tamkun and Catterall, 1981). Our initial studies (Ghiasuddin and Soderlund, 1985) showed that the actions of pyrethroids on synaptosomal sodium channels, as in assays with neuroblastoma, were only evident in combination with an activator such as VTD. This absolute dependence on an activating neurotoxin was particularly evident in studies of sodium uptake at 4 °C (J. R. Bloomquist and D. M. Soderlund, unpublished results). In these experiments, the stimulation of sodium uptake by either VTD or BTX was barely detectable; nevertheless, significant insecticide-dependent sodium uptake was evident even at 4 °C when these activators were included in the assay mixture.

Both deltamethrin and its insecticidal non-cyano analog, NRDC 157, were effective enhancers of VTD-dependent sodium uptake into mouse brain synaptosomes (Ghiasuddin and Soderlund, 1985). Deltamethrin was approximately 10-fold more potent than NRDC 157 as an enhancer of VTD-dependent sodium uptake, a finding that is consistent with its 10-fold greater toxicity to mice by intracerebral injection. In contrast, the noninsecticidal and nontoxic 1*S* enantiomers of deltamethrin and NRDC 157 were ineffective in sodium uptake assays. Thus, the use of this method to document interactions of pyrethroid isomers with the insecticide binding domain of the sodium channel demonstrated the high degree of stereospecificity expected from acute toxicity determinations. In view of the failure of some neurotoxic pyrethroids, DDT analogs, and DDT-pyrethroid hybrid compounds to interact with sodium channels in neuroblastoma, we expanded the scope of these studies to include several of these compounds (Soderlund et al., 1987). Cismethrin, which was ineffective in combination with VTD in assays using neuroblastoma, enhanced VTD-dependent sodium uptake into mouse brain synaptosomes. Similarly, EDO (2,2-*bis*-(4-ethoxyphenyl)-3,3-dimethyloxetane), which was ineffective in combination with sea anemone toxin in neuroblastoma, also enhanced VTD-dependent activation in the mouse brain system. These findings illustrate the value of using native tissue rather than cultured cells in studies that attempt to correlate *in vitro* effects with intoxication.

Subsequent experiments in this system employed DDT, cismethrin, and deltamethrin to define the allosteric interactions of insecticides with several sodium channel activators and with an α-polypeptide toxin acting at Site 3 (Bloomquist and Soderlund, 1988). These insecticides uniformly enhanced sodium channel activation by VTD and BTX, but they inhibited ACN-dependent activation (Fig. 1). The enhancement of BTX-dependent activation by DDT (Fig. 2) and cismethrin resulted from an increase in the potency of this activator with no effect on maximal uptake. This result is consistent with previous studies that have identified BTX as a full agonist at Site 2 of the sodium channel (Tamkun and Catterall, 1981). In contrast, the effects of DDT (Fig. 2) and deltamethrin on activation by VTD, a partial agonist at Site 2, exhibited two distinct components: a slight increase in the potency of VTD and a substantial potentiation of maximal VTD-dependent uptake (Fig. 2). However, cismethrin produced only an increase in potency of VTD with no corresponding potentiation of maximal uptake. The inhibitory effects of insecticides on ACN-dependent activation proved to be more difficult to characterize because of the low levels of sodium uptake produced by ACN, a weak partial agonist in this system. Nevertheless, studies with cismethrin provided evidence for two components of the insecticide-ACN interaction: a slight increase in the potency of ACN and an apparent noncompetitive inhibition of maximal ACN-dependent uptake. These findings demonstrate that modification of sodium channel activation by insecticides in this assay encompasses two separate processes, one which increases the affinity of sodium channels for Site 2 neurotoxins and a second which produces either positive or negative effects on the maximal sodium flux produced by partial agonists at Site 2.

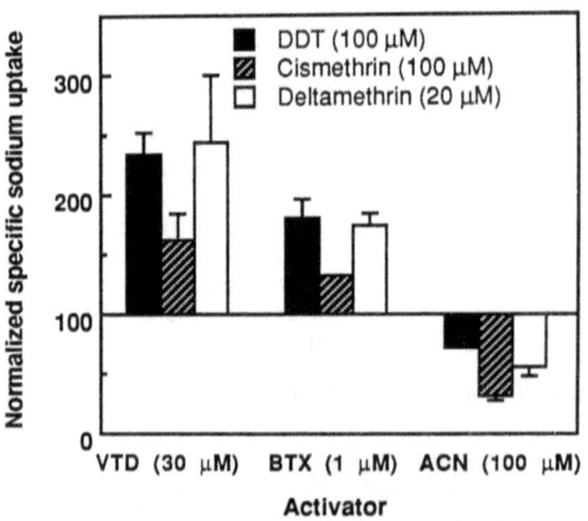

Fig. 1. Insecticide-dependent enhancement or inhibition of sodium uptake into mouse brain synaptosomes. The extent of sodium uptake stimulated by each activator in the absence of insecticides is set equal to 100. Redrawn from Bloomquist and Soderlund (1988).

Since pyrethroids potentiate the activation of sodium channels in neuroblastoma by α-polypeptide toxins (Jacques et al., 1980), we undertook experiments to determine whether these two classes of compounds could act independently to modify sodium channel activation by Site 2 neurotoxins in synaptosomes (Bloomquist and Soderlund, 1988). As illustrated in Fig. 3, DDT was able to further enhance both the potency and efficacy of VTD in the presence of a near-saturating concentration of *Anemonia sulcata* toxin II (ATX), a representative α-polypeptide toxin. In contrast, cismethrin failed to produce a similar effect on VTD-dependent activation in combination with ATX but was nevertheless able to inhibit ACN-dependent activation in the presence of the α-polypeptide toxin (Fig. 4). These results show that the nature of the allosteric modification of sodium channel activation by insecticides and Site 3 neurotoxins depends on the specific activator and insecticide combination used.

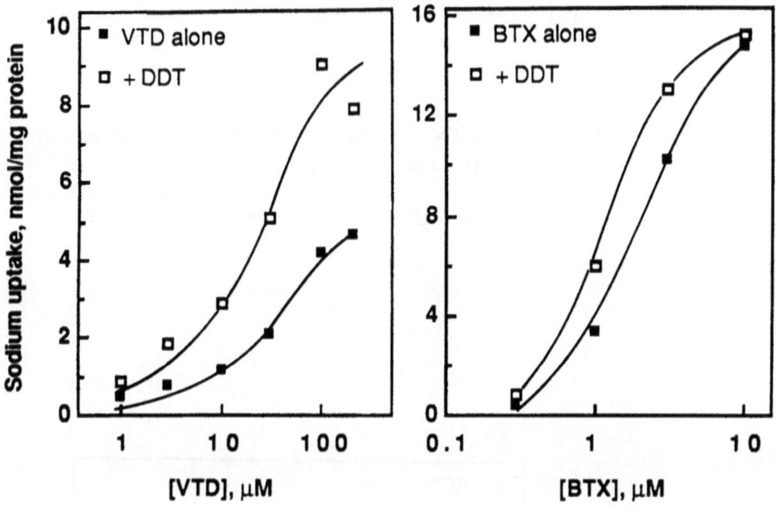

Fig. 2. Effect of DDT (100 μM) on the concentration dependence of sodium uptake into mouse brain synaptosomes stimulated by VTD (left) or BTX (right). Redrawn from Bloomquist and Soderlund (1988).

Radioligand Binding Studies

We have examined further the allosteric effects of DDT and pyrethroids on activator-sodium channel binding interactions using [^{3}H]BTX-B as a radioligand probe to label Site 2 (Catterall et al., 1981). The precedent for this approach was established by Brown et al. (1988), who demonstrated that deltamethrin and the neurotoxic isomers of cypermethrin enhanced the binding of [^{3}H]BTX-B to rat brain sodium channels and that, in the case of deltamethrin, this enhancement was due to an increase in the affinity of sodium channels for [^{3}H]BTX-B. In similar studies with mouse brain synaptoneurosomes, we found that deltamethrin enhanced the specific binding of [^{3}H]BTX-B and that DDT produced a qualitatively similar effect (Fig. 5). In mouse brain preparations, deltamethrin was both more potent (half-maximal stimulation at 4 μM) and more efficacious than DDT (half-maximal stimulation at 14 μM) in enhancing the binding of [^{3}H]BTX-B.

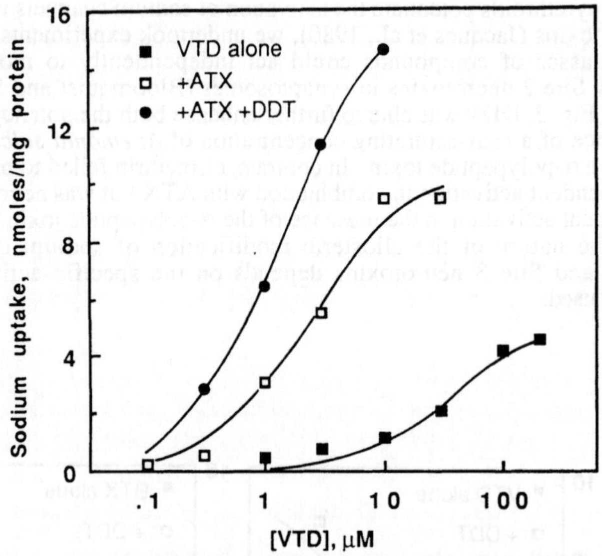

Fig. 3. Effects of ATX (1 µM) alone and in combination with DDT (100 µM) on the concentration dependence of VTD-stimulated sodium uptake into mouse brain synaptosomes. Redrawn from Bloomquist and Soderlund (1988).

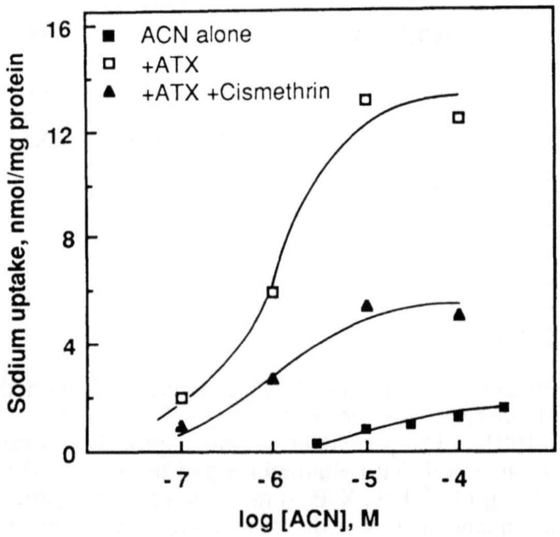

Fig. 4. Effects of ATX (1 µM) alone and in combination with cismethrin (100 µM) on the concentration dependence of ACN-stimulated sodium uptake into mouse brain synaptosomes. Redrawn from Bloomquist and Soderlund (1988).

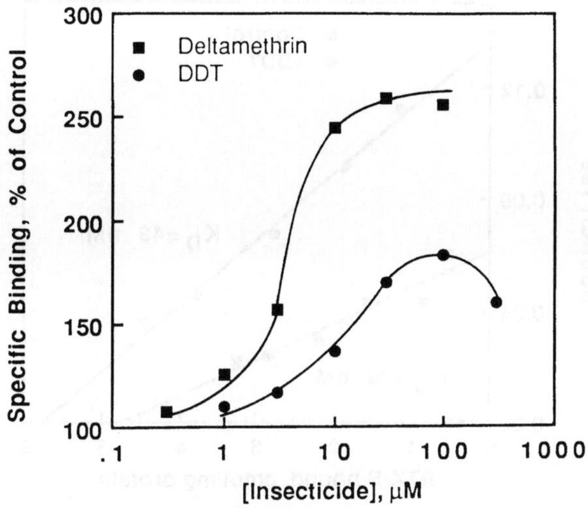

Fig. 5. Enhancement of the binding of [^{3}H]BTX-B (20 nM) to sodium channels in mouse brain vesicles by deltamethrin (G. T. Payne and D. M. Soderlund, unpublished results) and DDT (Payne and Soderlund, 1989).

Further studies with DDT (Payne and Soderlund, 1989) showed that this compound, like deltamethrin, enhanced the binding of [^{3}H]BTX-B by increasing the affinity of sodium channels for this ligand (Fig. 6). Kinetic analyses of binding in the absence and presence of DDT revealed no effects of this compound on the rate of association of [^{3}H]BTX-B with the sodium channel but documented a DDT-dependent decrease in the rate of dissociation of the [^{3}H]BTX-B - sodium channel complex. The magnitude of this effect on dissociation (1.9-fold) was virtually identical to the DDT-dependent increase in affinity for [^{3}H]BTX-B (1.8-fold; Fig. 6) found in equilibrium saturation studies. These studies show that the insecticide-dependent increase in the potency of sodium channel activators is likely to result from allosteric stabilization of the activator-sodium channel complex rather than from an effect on the rate of formation of this complex. It is important to recognize that these experiments describe only the kinetics of the interactions of activators with sodium channels and the allosteric modification of these kinetics by insecticides, thereby reporting indirectly on the occupancy of the insecticide binding site. However, the radioligand binding approach provides no insight into the ion conductance properties of chemically modified sodium channels. Thus, functional assays are still required to define the mechanism that underlies the modification of maximal activator-stimulated sodium uptake, which is the second principal effect of DDT and pyrethroids identified in radiosodium flux studies. This latter effect is likely to be more directly related to the insecticide-dependent prolongation of sodium channel open times observed in biophysical studies (Narahashi 1984, 1986).

ACTION OF *N*-ALKYLAMIDE INSECTICIDES AT SODIUM CHANNEL SITE 2

Although preliminary physiological studies of the action of *N*-alkylamide insecticides implicate an action on the voltage-sensitive sodium channel (Blade et al., 1985; Lees and Burt, 1988), they do not provide insight into the molecular domain that is involved. The observation by Nicholson et al. (1985) that one of these compounds inhibited the TTX-sensitive, VTD-dependent release of [^{3}H]acetylcholine from cockroach synaptosomes led us to investigate the interactions of BTG 502, a representative synthetic *N*-alkylamide (Fig. 7), with sodium channels in mouse brain synaptoneurosomes (Ottea et al., 1989).

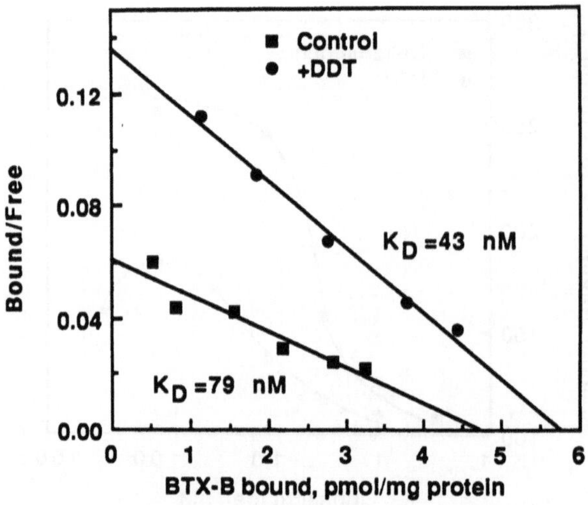

Fig. 6. Scatchard analyses of the specific binding of [^{3}H]BTX-B to sodium channels in mouse brain vesicles in the absence or presence of DDT (100 μM). Redrawn from Payne and Soderlund (1989).

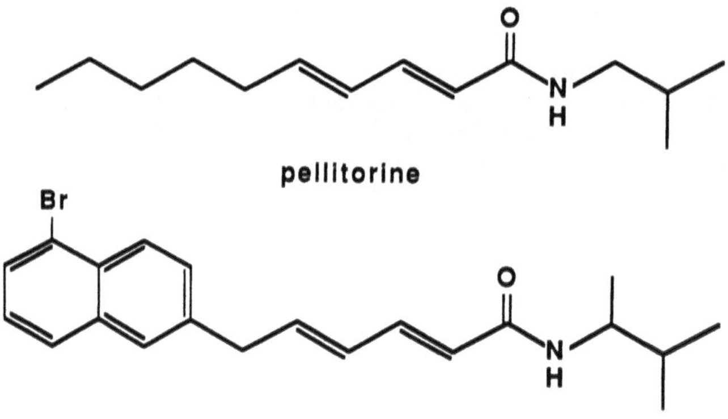

Fig. 7. Structures of pellitorine, a naturally-occurring insecticidal *N*-alkylamide, and BTG 502, a synthetic analog.

Initial sodium flux experiments detected only a very small stimulation of sodium uptake by BTG 502 over background levels. However, when these assays were performed in the presence of a saturating concentration of scorpion (*Leiurus quinquestriatus*) venom, which contains an α-polypeptide toxin acting at Site 3, BTG 502 produced a substantial concentration-dependent stimulation of sodium uptake (Fig. 8). Subsequent studies of the effects of BTG 502 on sodium uptake stimulated by established activators showed that this compound completely inhibited sodium uptake stimulated by BTX in the absence of *Leiurus* venom but only partially inhibited BTX-dependent uptake in the presence of venom (Ottea et al., 1989). These data are consistent with an action of BTG 502 as antagonist or partial agonist at Site 2 of the sodium channel.

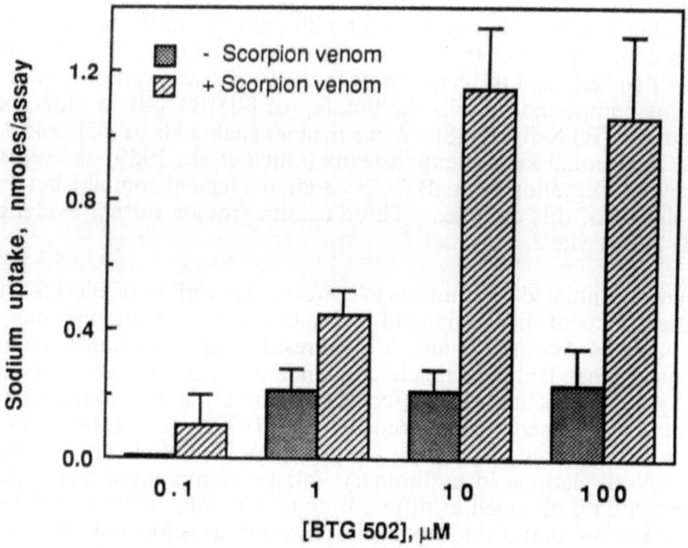

Fig. 8. Stimulation of sodium uptake into mouse brain vesicles by BTG 502 in the absence (J. A. Ottea and D. M. Soderlund, unpublished results) or presence (Ottea et al., 1989) of scorpion venom (25 µg/assay).

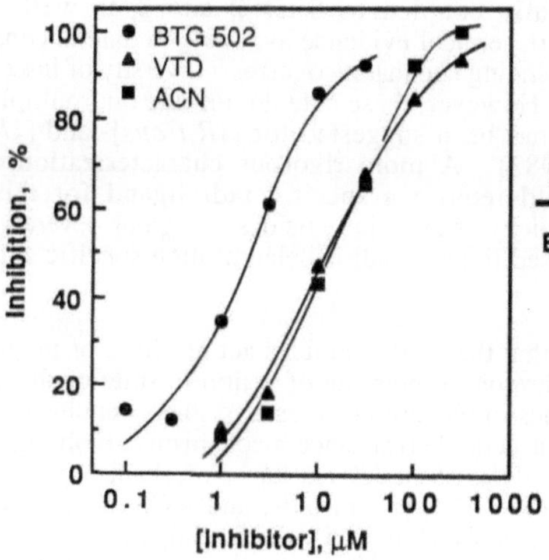

Inhibitor	I_{50}, µM	Hill slope
BTG 502	2	0.83
VTD	17	1.02
ACN	14	1.07

Fig. 9. Inhibition of the specific binding of [^{3}H]BTX-B (20 nM) to sodium channels in mouse brain vesicles by BTG 502 (Ottea et al., 1989), VTD, and ACN (G. T. Payne and D. M. Soderlund, unpublished results).

The results of our sodium uptake experiments led us to explore the effects of BTG 502 on the binding of [^{3}H]BTX-B to Site 2 (Ottea et al., 1989). This compound is a potent and complete inhibitor of the specific binding of [^{3}H]BTX-B (Fig. 9). The I_{50} value obtained in binding experiments is consistent with estimates of the potency of BTG 502 as a stimulator of sodium uptake in the presence of scorpion venom (Fig. 8). The slope of the Hill plot of the displacement of [^{3}H]BTX-B by BTG 502 was sufficiently close to unity to suggest that this compound inhibits the binding of [^{3}H]BTX-B by direct stoichiometric displacement of [^{3}H]BTX-B from Site 2 in a manner analogous to VTD and ACN (Catterall et al., 1981). Additional kinetic experiments (Ottea et al., 1989) showed that BTG 502 affected the rate of formation of the BTX-B - sodium channel complex but did not alter the rate of dissociation of this complex. These results provide further evidence for a direct action of BTG 502 at Site 2.

We have also initiated experiments to correlate that actions of BTG 502 in biochemical assays with the effects of this compound on intact nerves. Preliminary experiments (J. R. Bloomquist and D. M. Soderlund, unpublished results) have documented a time-dependent and concentration-dependent reduction in amplitude of compound action potentials in mouse sciatic nerve by this compound. This effect is qualitatively similar to that of deltamethrin on frog sciatic nerve (Vijverberg and van den Bercken, 1979). The action of BTG 502 in these preparations is also generally consistent with the induction of slowly decaying sodium tail currents by an *N*-alkylamide in preliminary voltage clamp experiments (Lees and Burt, 1988). Although data obtained at different concentrations of BTG 502 are limited, the available results show that a threshold concentration to achieve a 50% reduction in the amplitude of evoked compound action potentials within one hour after bath application lies close to 1 μM, a level of potency consistent with estimates obtained from biochemical assays (Figs. 8 and 9).

CONCLUSIONS

The action of DDT and pyrethroids on the voltage-sensitive sodium channels appears to involve a unique binding domain. The insecticide recognition site is functionally separable from Sites 1 and 4 and is allosterically coupled to Sites 2 and 3 as well as the brevetoxin/ciguatoxin site. All of the biochemical evidence available to date is consistent with the existence of a single insecticide binding site that recognizes a diversity of insecticidal structures in the DDT/pyrethroid series. However, these data do not rule out multiple sites for neuroactive pyrethroid isomers, as has been suggested for [1*R*,*trans*]- and [1*R*,*cis*]-tetramethrin (Lund and Narahashi, 1982). A more rigorous characterization of the DDT/pyrethroid binding domain would require a specific radioligand for this site. Unfortunately, the extremely high lipophilicity of these insecticides as a group severely limits this approach even using insecticide-based ligands radiolabeled at high specific activities (Soderlund et al., 1983).

Our studies with BTG 502 show that the *N*-alkylamides act at Site 2 of the sodium channel and therefore differ from pyrethroids in their site of action in spite of the similar actions of these two classes of insecticides in physiological assays. This conclusion raises several interesting issues with regard to insecticide resistance mechanisms involving altered target site sensitivity. Pharmacological analysis of the *kdr* resistance mechanism in the house fly has shown that it confers resistance to DDT, pyrethroids, and ACN both in toxicity bioassays and in neurophysiological assays (Sawicki, 1985; Salgado et al., 1983; Bloomquist and Miller, 1986). However, the oberved cross-resistance to ACN appears to be peculiar to this compound rather than to all Site 2 neurotoxins, because *kdr* house flies retain full susceptibility to VTD (A. E. Lund, personal communication). It is of interest that the Site 2 neurotoxin to which the *kdr* mechanism confers resistance is also the compound involved in the anomalous inhibitory actions of DDT and pyrethroids in sodium uptake assays (Bloomquist and Soderlund, 1988). These findings identify a qualitative difference between ACN and other Site 2 neurotoxins and suggest that the activator recognition site may consist of at least two partially overlapping subdomains. The excellent activity of *N*-alkylamides against *super-kdr* insects (Elliott et al., 1986), coupled with the apparent action of these compounds at Site 2, broadens the earlier observation that cross-resistance does not extend to VTD and suggests that in the case of the voltage-dependent sodium channel the concept of "target site resistance" must be applied to individual binding domains or even partially overlapping subdomains rather than to the entire target macromolecule. The *N*-

alkylamides provide a provocative reminder that, despite the widespread use of and significant resistance to DDT and pyrethroids, the rich pharmacological diversity of the many binding domains of the voltage-sensitive sodium channel represent unexploited targets for new insecticide chemistry.

ACKNOWLEDGMENTS

Studies from this laboratory were supported by grant number ES 02160 from the National Institutes of Health. We thank the following persons who generously provided neurotoxins and insecticides that were essential to the completion of this research: J. Daly, National Institutes of Arthritis, Metabolism, and Digestive Disease, Bethesda, MD (batrachotoxin); M. Elliott and N. Janes, Rothamsted Experimental Station, Harpenden, U.K. (pyrethroids and *N*-alkylamides); and J. Martel, Roussel-Uclaf Procida, Romainville, France (pyrethroids).

REFERENCES

Atchison, W. E., Luke, V. S., Narahashi, T., and Vogel, S. M., 1986, Nerve membrane sodium channels as the target site of brevetoxins at neuromuscular junctions, Br. J. Pharmacol., 89:731.

Bablito, J., Jover, E., and Couraud, F., 1986, Activation of the voltage-sensitive sodium channel by a β-scorpion toxin in rat brain nerve-ending particles, J. Neurochem., 46:1763.

Benoit, E., Legrand, A. M., and Dubois, J. M., 1986, Effects of ciguatoxin on current and voltage clamped frog myelinated nerve fibre, Toxicon, 24:357.

Bidard, J.-N., Vijverberg, H. P. M., Frelin, C., Chungue, E., Legrand, A.-M., Bagnis, R., and Lazdunski, M., 1984, Ciguatoxin is a novel type of Na^+ channel toxin, J. Biol. Chem., 259:8353.

Blade, R. J., Burt, P. E., Hart, R. J., and Moss, M. D. V., 1985, The action of insecticidal isobutylamide compounds on the insect nervous system, Pestic. Sci., 16:554.

Bloomquist, J. R., and Miller, T. A., 1986, Sodium channel neurotoxins as probes of the knockdown resistance mechanism, NeuroToxicology, 7:217.

Bloomquist, J. R., and Soderlund, D. M., 1988, Pyrethroid insecticides and DDT modify alkaloid-dependent sodium channel activation and its enhancement by sea anemone toxin, Mol. Pharmacol., 33:543.

Brown, G. B., Gaupp, J. E., and Olsen, R. W., 1988, Pyrethroid insecticides: stereospecific, allosteric interaction with the batrachotoxinin-A benzoate binding site of mammalian voltage-sensitive sodium channels, Mol. Pharmacol., 34:54.

Catterall, W. A., 1984, The molecular basis of neuronal excitability, Science, 223:653.

Catterall, W. A., Morrow, C. S., Daly, J. W., and Brown, G. B., 1981, Binding of batrachotoxinin A 20-α-benzoate to a receptor site associated with sodium channels in synaptic nerve ending particles, J. Biol. Chem., 256:8922.

Catterall, W. A., and Risk, M., 1981, Toxin $T4_6$ from *Ptychodiscus brevis* (formerly *Gymnodinium breve*) enhances activation of voltage-sensitive sodium channels by veratridine, Mol. Pharmacol., 19:345.

Creveling, C. R., McNeal, E. T., Daly, J. W, and Brown, G. B., 1983, Batrachotoxin-induced depolarization and [^{3}H]batrachotoxinin-A 20-α-benzoate binding in a vesicular preparation from guinea pig cortex. Inhibition by local anesthetics, Mol. Pharmacol., 23:350.

Elliott, M., 1985, Lipophilic insect control agents, in: "Recent Advances in the Chemistry of Insect Control," N. F. Janes, ed., pp. 73-102, Royal Society of Chemistry, London.

Elliott, M., Farnham, A. W., Janes, N. F., Johnson, D. M., Pulman, D. A., and Sawicki, R. M., 1986, Insecticidal amides with selective potency against a resistant (*super-kdr*) strain of house flies (*Musca domestica* L.), Agric. Biol. Chem., 50:1347.

Ghiasuddin, S. M., and Soderlund, D. M., 1985, Pyrethroid insecticides: potent, stereospecific enhancers of mouse brain sodium channel activation, Pestic. Biochem. Physiol., 24:200.

Gonoi, T., Ashida, K., Feller, D., Schmidt, J., Fujiwara, M., and Catterall, W. A., 1986, Mechanism of action of a polypeptide neurotoxin from the coral *Goniopora* on sodium channels in mouse neuroblastoma cells, Mol. Pharmacol., 29:347.

Gusovsky, F., Rossignol, D. P., McNeal, E. T., and Daly, J. W., 1988, Pumiliotoxin B binds to a site on the voltage-dependent sodium channel that is allosterically coupled to other binding sites, Proc. Natl. Acad. Sci. USA, 85:1272.

Holan, G., Frelin, C., and Lazdunski, M., 1985, Selectivity of action between pyrethroids and combined DDT-pyrethroid insecticides on Na^+ influx into mammalian neuroblastoma, Experientia, 41:520.

Huang, J. M. C., Wu, C. H., and Baden, D. G., 1984, Depolarizing action of a red-tide dinoflagellate brevetoxin on axonal membranes, J. Pharmacol. Exp. Ther., 229:615.

Jacques, Y., Romey, G., Cavey, M. T., Kartalovski, B., and Lazdunski, M., 1980, Interaction of pyrethroids with the Na^+ channel in mammalian neuronal cells in culture, Biochem. Biophys. Acta, 600:882.

Lawrence, L. J., and Casida, J. E., 1982, Pyrethroid toxicology: mouse intracerebral structure-toxicity relationships, Pestic. Biochem. Physiol., 18:9.

Lees, G., and Burt, P. E., 1988, Neurotoxic actions of a lipid amide on the cockroach nerve cord and on locust somata maintained in short-term culture: a novel preparation for the study of Na^+ channel pharmacology, Pestic. Sci., 24:189.

Lombet, A., Bidard, J.-N., and Lazdunski, M., 1987, Ciguatoxins and brevetoxins share a common receptor site on the neuronal voltage-dependent Na^+ channel, FEBS Lett., 219:355.

Lund, A. E., and Narahashi, T., 1982, Dose-dependent interaction of the pyrethroid isomers with sodium channels of squid axonal membranes, NeuroToxicology, 3:11.

Narahashi, T., 1984, Nerve membrane sodium channels as the target of pyrethroids, in: "Cellular and Molecular Neurotoxicology," T. Narahashi, ed., pp. 85-108, Raven Press, New York.

Narahashi, T., 1986, Mechanisms of action of pyrethroids on sodium and calcium channel gating, in: "Neuropharmacology and Pesticide Action," M. G. Ford, G. G. Lunt, R. C. Reay, and P. N. R. Usherwood, eds., pp. 36-60, Ellis Horwood, Chichester.

Nicholson, R. A., Botham. R. P., and Blade, R. J., 1985, The interaction of sodium channel directed neurotoxicants and a novel insecticidal isobutylamide with central nerve terminals prepared from the cockroach (*Periplaneta americana*), Pestic. Sci., 16:554.

Ottea, J. A., Payne, G. T., Bloomquist, J. R., and Soderlund, D. M., 1989, Activation of sodium channels and inhibition of [^{3}H]batrachotoxinin A-20-α-benzoate binding by an *N*-alkylamide neurotoxin, Mol. Pharmacol., in press.

Payne, G. T., and Soderlund, D. M., 1989, Allosteric enhancement by DDT of the binding of [^{3}H]batrachotoxinin A 20-α-benzoate to sodium channels, Pestic. Biochem. Physiol., 33:276.

Poli, M. A., Mende, T. J., and Baden, D. G., 1986, Brevetoxins, unique activators of voltage-sensitive sodium channels, bind to specific sites in rat brain synaptosomes, Mol. Pharmacol., 30:129.

Postma. S. W., and Catterall, W. A., 1984, Inhibition of binding of [^{3}H]batrachotoxinin A 20-α-benzoate to sodium channels by local anesthetics, Mol. Pharmacol., 25:219.

Ruigt, G. S. F., 1985, Pyrethroids, in: "Comprehensive Insect Physiology, Biochemistry and Pharmacology," G. A. Kerkut and L. I. Gilbert, eds., vol. 12, pp. 183-262, Pergamon, Oxford.

Salgado, V. L., Irving, S. N., and Miller, T. A., 1983, Depolarization of motor nerve terminals by pyrethroids in susceptible and *kdr*-resistant house flies, Pestic. Biochem. Physiol., 20:100.

Sawicki, R. M., 1985, Resistance to pyrethroid insecticides in arthropods, in: "Insecticides," Progress in Pesticide Biochemistry and Toxicology, Vol. 5, D. H. Hutson and T. R. Roberts, eds., pp.143-192, Wiley, New York.

Sharkey, R. G., Jover, E., Couraud, F., Baden, D. G., and Catterall, W. A., 1987, Allosteric modulation of neurotoxin binding to voltage-sensitive sodium channels by *Ptychodiscus brevis* toxin 2, Mol. Pharmacol., 31:273.

Soderlund, D. M., and Bloomquist, J. R., 1989, Neurotoxic mechanisms of pyrethroid action, Ann. Rev. Entomol., 34:77.

Soderlund, D. M., Bloomquist, J. R., Ghiasuddin, S. M., and Stuart, A. M., 1987, Enhancement of veratridine-dependent sodium channel activation by pyrethroids and DDT analogs, in: "Sites of Action for Neurotoxic Pesticides," R. M. Hollingworth and M. B. Green, eds., pp. 251-261, American Chemical Society, Washington.

Soderlund, D. M., Ghiasuddin, S. M., and Helmuth, D. W., 1983, Receptor-like stereospecific binding of a pyrethroid insecticide to mouse brain membranes, Life Sci., 33:261.

Strichartz, G., Rando, T., and Wang, G. K., 1987, An integrated view of the molecular toxinology of sodium channel gating in excitable cells, Ann. Rev. Neurosci., 10:237.

Su, H. C. F., 1985, *N*-Isobutylamides, in: "Comprehensive Insect Physiology, Biochemistry and Pharmacology," G. A. Kerkut and L. I. Gilbert, eds., vol. 12, pp. 273-290, Pergamon, Oxford.

Tamkun, M. M., and Catterall, W. A., 1981, Ion flux studies of voltage-sensitive sodium channels in synaptic nerve-ending particles, Mol. Pharmacol., 19:78.

Vijverberg, H. P. M., and van den Bercken, J., 1979, Frequency-dependent effects of the pyrethroid insecticide decamethrin in frog myelinated nerve fibres, Eur. J. Pharmacol., 58:501.

Willow, M., and Catterall, W. A., 1982, Inhibition of binding of [^{3}H]batrachotoxinin A 20-α-benzoate to sodium channels by the anticonvulsant drugs diphenylhydantoin and carbamazepine, Mol. Pharmacol., 22:627.

Willow, M., Kuenzel, E. A., and Catterall, W. A., 1984, Inhibition of voltage-sensitive sodium channels in neuroblastoma cells and synaptosomes by the anticonvulsant drugs diphenylhydantoin and carbamazepine, Mol. Pharmacol., 25:228.

DROSOPHILA SODIUM CHANNEL MUTATIONS AFFECT PYRETHROID SENSITIVITY

Linda M. Hall and Durgadas P. Kasbekar

Department of Biochemical Pharmacology
State University of New York at Buffalo
317 Hochstetter Hall, Buffalo, NY

ABSTRACT

The primary target of pyrethroids is the sodium channel in neuronal membranes. We have used mutations in *Drosophila melanogaster* to define two different types of target site alterations which affect pyrethroid sensitivity. The first alteration is illustrated by the $para^{ts}$ locus on the X chromosome. Gene cloning experiments by the laboratory of B. Ganetzky suggest that this locus is the structural gene for a protein homologous to the α-subunit of vertebrate sodium channels. We demonstrate that mutations at the *para* locus show changes in pyrethroid sensitivity when assayed by either a knockdown assay or a lethality assay. Some alleles show increased resistance and some show decreased resistance to fenvalerate. These changes are probably due to changes in the primary structure of one class of sodium channels in *Drosophila*. Such mutational changes can be used to define specific amino acid residues in the sodium channel which affect pyrethroid binding.

A second type of target site alteration is exemplified by the nap^{ts} mutation which maps to the second chromosome and which has recently been shown by Kernan, Kreber and Ganetzky to be an allele of the maleless (*mle*) locus. This locus is involved in the regulation of genes on the X chromosome. The *nap* mutation shows resistance to all pyrethroids tested (fenvalerate, permethrin, and Pay-Off). Although there is no evidence for structural changes in *nap* sodium channels, the density of saxitoxin-binding sites is reduced. In this mutant strain the *nap* mutation appears to down regulate the class of sodium channels coded for by the X-linked *para* locus. This reduction in target site density causes increased resistance to pyrethroids.

INTRODUCTION

Hereditary resistance to pyrethroid insecticides has been reported for a number of different insect species (Gammon, 1980; Miller *et al*., 1980; Omer *et al*., 1980; DeVries and Georghiou, 1981; Scott and Matsumura, 1983; Malcolm, 1988; Miller, 1988). In order to develop strategies for preventing or circumventing pyrethroid resistance, it is important to define the different types of genes which are involved in producing this resistance. There are two general strategies for developing laboratory models for use in elucidating the mechanisms of pyrethroid resistance which may arise in natural insect populations following widespread use of pyrethroid insecticides. One approach is to mutagenize a genetically homogeneous laboratory population and then select for single gene mutations which alter pyrethroid sensitivity. In essence, this approach is similar to what happens in nature when insect populations are exposed to pyrethroids. The advantage of studying model laboratory populations is that the genetic background can be controlled. Thus, single gene changes can be studied in the absence of complicating multifactorial interactions that accumulate in response to repeated exposure to the pyrethroid. This approach should allow the isolation of mutations in genes affecting all of the common types of insecticide resistance including: permeability mechanisms involving the exoskeleton, gut, and other tissue barriers; metabolism and excretion mechanisms; behavioral modifications including increased fecundity and avoidance of the toxic substance; and target site changes.

An alternative approach is to study mutations in genes which are known to affect sodium channels in a variety of ways (Ganetzky and Wu, 1985; Hall, 1986). Since the toxic effects of pyrethroids on insects are the result of their interaction with voltage-sensitive sodium channels, this approach focuses directly on target site changes which lead to resistance. In general, there are at least three types of target site changes which could lead to increased resistance to pyrethroids. Thus, **changes in the primary amino acid sequence of structural components** of the voltage-sensitive sodium channel could lead to resistance due to reduced affinity of the pyrethroids for their binding site. We present evidence that mutations in the para^{ts} locus are likely candidates for this type of target site resistance. A second type of target site resistance could arise from **changes in sodium channel density**. We present evidence that the nap^{ts} mutation is a likely candidate for this type of resistance. Finally, **changes in posttranslational modifications** could also lead to target site resistance if these changes affected interaction of pyrethroids with sodium channels. To date no mutants have been shown to cause resistance by this mechanism. However, it should be considered as a possible mechanism since sodium channels are heavily glycosylated and are good substrates for phosphorylation.

MUTANTS IN DROSOPHILA WHICH AFFECT SODIUM CHANNELS

Pyrethroid resistance due to mutations at the para^{ts} locus

The first mutant alleles at the X-chromosome locus paralytic-temperature sensitive (para^{ts}) were isolated on the basis of their phenotype of temperature-induced paralysis (Suzuki *et al*., 1971; Grigliatti *et al*., 1973). Temperature-induced blockade of nerve conduction by this mutant suggested an effect on sodium channels (Wu and Ganetzky, 1980). Recent gene cloning and sequencing studies have shown that the para gene product has a high degree of sequence homology with the vertebrate sodium channel alpha subunit (Loughney and Ganetzky, personal communication). Thus, para is an excellent structural gene candidate for one class of sodium channels in *Drosophila*.

As shown in Figure 1 in a lethality assay, para^{ts1} mutants are substantially more resistant to the pyrethroid fenvalerate than are the wild-type control flies. Because we see sex specific differences in susceptibility to pyrethroids, data are presented for males and females separately. In addition to testing for lethality differences, we also tested for differences in rates of knockdown. As shown in Figure 2, para^{ts}1 flies are also more resistant than wild-type flies to fenvalerate in the knockdown assay. Because para appears to be a structural gene for one class of sodium channels, we propose that the resistance seen in the para^{ts1} flies is due to an alteration in the primary amino acid sequence of the sodium channel component. This mutational change may decrease the affinity of the sodium channel for fenvalerate.

To provide further evidence that the pyrethroid resistance demonstrated in the para^{ts1} stock is due to a mutation in the para gene and not due to some other factor in the genetic background, it is necessary to demonstrate that the resistance phenotype cosegregates with the para behavioral and physiological phenotypes which previously had been shown to be due to a single gene mutation. Since para is an X chromosome linked gene, the first step in such an analysis is to do the reciprocal crosses illustrated in Figure 3. All males and females from both crosses will be identical with respect to their autosomes so if the resistance is due to an autosomal factor, the resistance patterns for progeny from both crosses would be the same. Alternatively, if the resistance is due to an X-linked factor, the progeny males would phenotypically resemble their mothers since they receive their single X chromosome from their mothers. (Note that the Y chromosome does not generally carry alleles of X

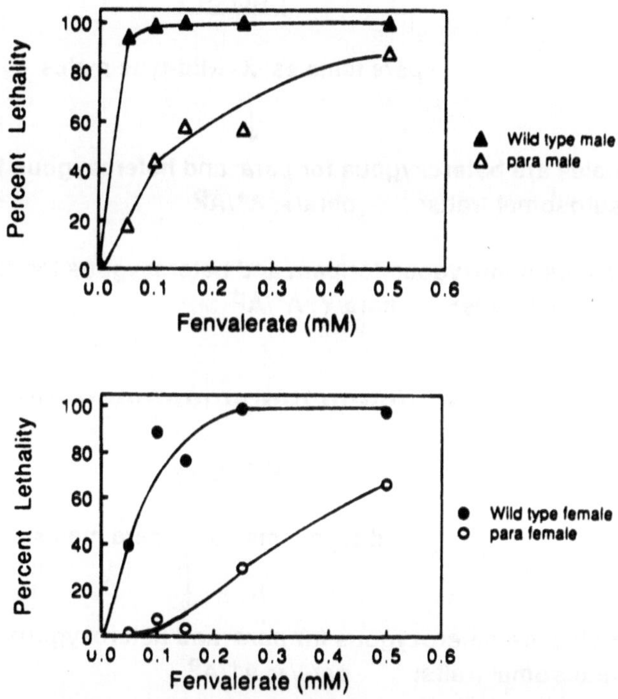

Fig. 1. Fenvalerate-induced lethality for para^{ts1} and wild-type (Canton-S) flies. Adult flies aged 2-6 days were exposed to filters impregnated with 150 μl of the indicated concentrations of fenvalerate for 18 hours at 21°C. They were then transferred to standard medium and lethality was scored 24 hours after transfer. Each point represents 47-122 flies.

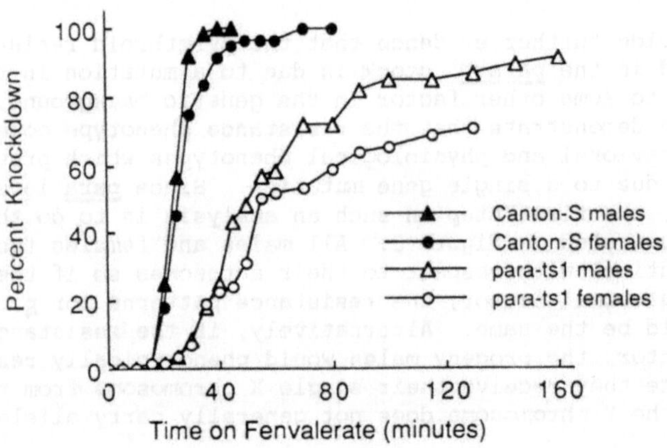

Fig.2. Kinetics of knockdown of $\underline{para}^{ts1}$ and wild-type flies by fenvalerate. Adult flies aged 3-5 days were exposed to filters impregnated with 150 µl of 10 mM fenvalerate. Each curve represents 55-60 flies.

CROSS 1

para females X wild-type males

↓

Females are heterozygous for *para* and heterozygous for all autosomal traits: para/+; A^+/A^p

Males are hemizygous for para and heterozygous for all autosomal traits: para/Y; A^+/A^p

****

CROSS 2

wild-type females X para males

↓

Females are heterozygous for *para* and heterozygous for all autosomal traits: para/+; A^+/A^p

Males are hemizygous for the wild-type X chromosome and heterozygous for all autosomal traits: +/Y; A^+/A^p

Fig. 3. Reciprocal crosses to test for X-linkage.

chromosome genes and can be considered genetically "silent" for the purpose of this discussion.) Thus, for an X-linked trait male progeny from para mothers and wild-type fathers should show the resistance phenotype while male progeny from wild-type mothers and para fathers should show wild-type sensitivity to fenvalerate. The male progeny results shown in Figure 4 demonstrate that the resistance is, in fact, X linked. The heterozygous females from both crosses show the para resistance phenotype. This shows that the para^{ts1} mutation is dominant with respect to the wild type for pyrethroid resistance.

The sodium channel is a large molecule with several pharmacologically distinct toxin binding domains (see reviews by Catterall, 1980, 1986, 1988; Strichartz et al., 1987). Mutational alterations in different domains of the molecule should have distinct effects on pyrethroid resistance. Mutations at or near the pyrethroid binding domain could either decrease or increase the affinity of the channel for pyrethroids and thus result in an increase or decrease, respectively, in resistance to pyrethroids. Mutations in parts of the molecule which are not involved in pyrethroid interactions would have little or no effect on pyrethroid resistance. By correlating the effects of different para alleles on pyrethroid resistance with the location of these mutations within the sodium channel, it should be possible to identify the pyrethroid binding domain within the channel molecule. With this goal in mind, we have begun to compare the pyrethroid sensitivity of different para alleles.

As summarized in Figure 5 in the lethality assay, para^{ts1} and para^{ts4} are both resistant to the pyrethroid fenvalerate while para^{ts2} is more sensitive than wild-type flies to this pyrethroid. These same sensitivity differences are also seen in the knockdown assay shown in Figure 6. We are continuing this analysis with additional alleles. Future work will

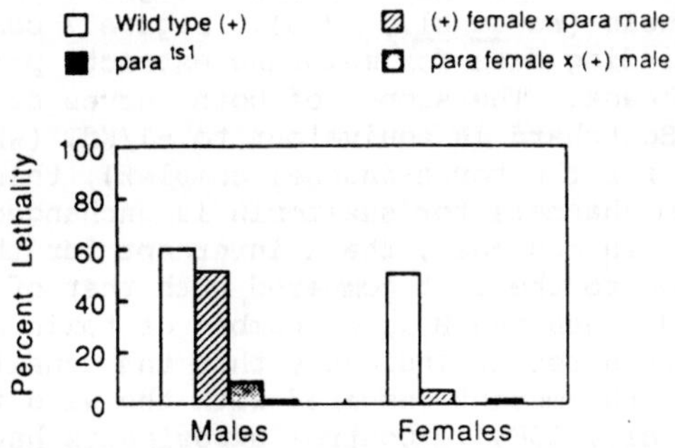

Fig. 4. Fenvalerate resistance of para^{ts1} segregates with the X chromosome and is dominant. Sons of para^{ts1} mothers mated with wild-type males differ genetically from sons from the reciprocal cross with respect to their X chromosome but have the same set of autosomes. The daughters from the two crosses are genetically identical. F_1 progeny from the two sets of crosses were exposed to filters impregnated with 0.1 mM fenvalerate for 18 hours at 21°C. They were then transferred to standard medium and lethality was scored 24 hours after transfer. Wild-type flies and para^{ts1} were tested in parallel.

also include an analysis of the effects of these alleles on sensitivity to different pyrethroids. Ultimately, it will be possible to define differences in binding domains within the sodium channel for various pyrethroids.

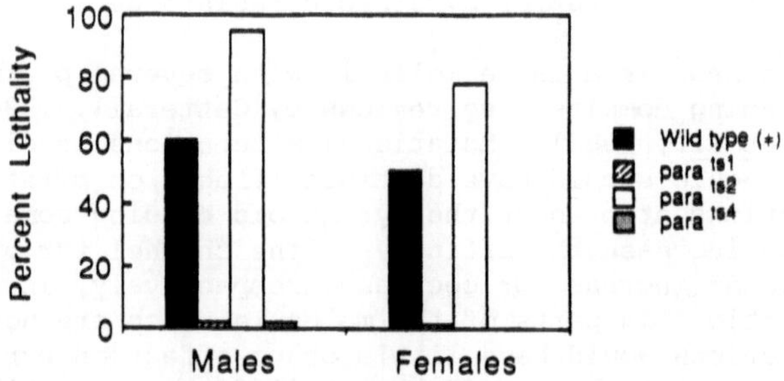

Fig. 5. Pyrethroid sensitivities of different para alleles. Adult flies aged 2-5 days were exposed to filters impregnated with 150 µl of 0.1 mM fenvalerate for 18 hours at 21°C. They were then transferred to standard medium and lethality was scored 24 hours after transfer. Each bar represents 50-117 flies.

Pyrethroid resistance due to the nap^{ts} mutation

Like the para mutations, the no action potential temperature-sensitive (nap^{ts}) mutation was originally isolated on the basis of its temperature-sensitive paralytic phenotype (Wu *et al.*, 1978). This mutation, which maps to the second chromosome, also shows a temperature-induced blockade in nerve conduction and was thought to have a direct effect on sodium channels (Wu *et al.*, 1978). Figure 7 compares Scatchard plots of saturation binding data for membrane extracts prepared from mutant and wild-type heads. The slopes of both curves are parallel. Since the slope of a Scatchard is equivalent to -1/Kd, (where Kd = dissociation constant for the toxin-channel complex), this indicates that the affinity of sodium channels for saxitoxin is unchanged in the mutant versus the wild type. In contrast, the X intercept for the mutant extract is reproducibly shifted to the left compared with that of the wild type. Since this intercept defines the B_{max} or number of toxin binding sites per mg membrane protein, this result indicates that the density of sodium channels is reduced in the mutant compared with the wild type (Hall *et al.*, 1982; Jackson *et al.*, 1984). Control experiments have shown that this reduction in sodium channels does not extend to other neuronal membrane protein markers since the levels of α-bungarotoxin binding acetylcholine receptors (Hall *et al.*, 1982) and (Na^+-K^+)ATPase (Hall, unpublished observations) are the same in neuronal membranes from the nap^{ts} mutant and the wild-type. In addition, there is no change in the pH sensitivity of saxitoxin binding in mutant versus wild-type extracts (Jackson *et al.*, 1984).

These results taken together provide no evidence for mutant-induced structural changes in sodium channels and have led us to suggest that

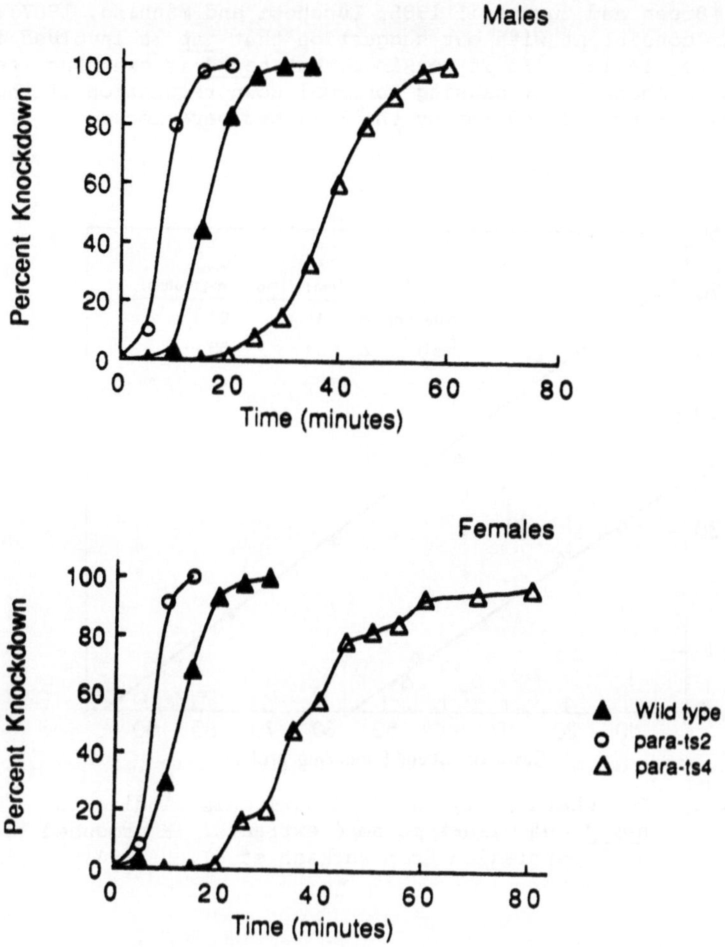

Fig. 6. Kinetics of knockdown of wild-type $\underline{para}^{ts2}$ and $\underline{para}^{ts4}$ flies by 10 mM fenvalerate. The number of flies knocked down was scored as a function of time. Each curve represents 60 flies.

nap^{ts} is involved in sodium channel regulation rather than having direct effects on channel structure. Recently, the laboratory of Barry Ganetzky has shown that nap is an allele of the maleless locus (mle) (Fukunaga et al., 1975) which is involved in dosage compensation, i.e., regulation of X-linked genes. Maleless appears to be one of at least four loci whose wild-type state is necessary for the two-fold hyperactivation of X-linked gene expression in D. melanogaster. This hyperactivation equalizes the output of the single X chromosome in the male with the two X chromosomes in the female (Breen and Lucchesi, 1986; Lucchesi and Manning, 1987). This finding is consistent with our suggestion that nap is involved in sodium channel regulation. It is possible that nap^{ts} is exerting its effects on sodium channels by causing abnormal down regulation of the class of sodium channels coded for by the X-linked para locus.

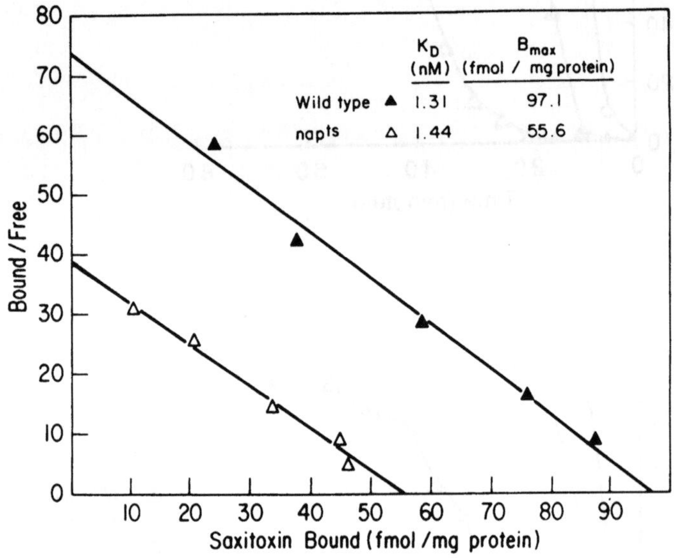

Fig. 7. Scatchard analysis of ^{3}H-saxitoxin binding to nap^{ts} and wild-type head extracts. Reproduced with permission from Jackson et al. (1984).

We were interested in determining how reduction in the number of target sites would affect pyrethroid sensitivity. As shown in Figure 8, nap^{ts} shows resistance to the pyrethroid fenvalerate in the lethality assay. A similar resistance is found in the knockdown assay shown in Figure 9.

We used the lethality assay to compare the effects of the nap^{ts} mutation on sensitivity to other types of pyrethroids. Fenvalerate is a type II pyrethroid. We also tested sensitivity to another type II pyrethroid (Pay-Off) and to a type I pyrethroid (permethrin).

As shown in Figure 10 the nap^{ts} mutation shows increased resistance for all the different pyrethroids tested. In each case, the females are always much more resistant than the males. We do not know what the molecular basis is for the sex difference in resistance. It may be related to different effects of the nap gene product in regulation of X-linked genes in males and females.

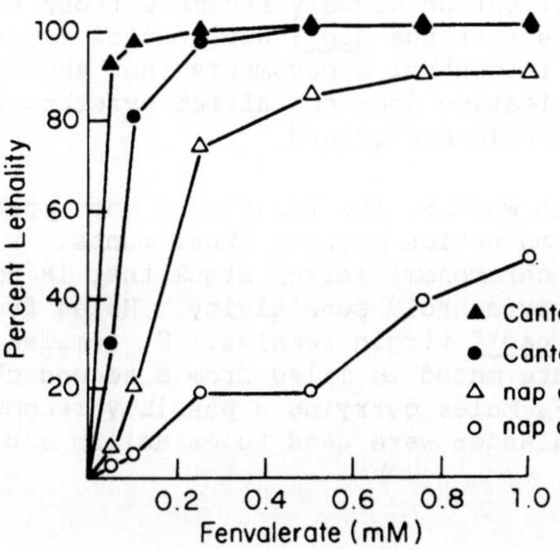

Fig. 8. Fenvalerate-induced lethality for nap^{ts} cn and wild-type Canton-S flies. The protocol was the same as that used in Fig. 1 legend. Each point represents 193-341 flies. Reproduced with permission from Kasbekar and Hall (1988).

It is important to establish rigorously that the effects on pyrethroid sensitivity seen in the nap^{ts} stock are due to the mutation at the nap locus and not to other mutations in the genetic background. The first step in such a genetic analysis is to establish by segregation analysis that the resistance phenotype segregates with the second chromosome on which the nap^{ts} mutation resides. A genetic scheme for this analysis is shown in Figure 11. In the F_2 generation we found that the nap type resistance appears only when the second chromosome is homozygous for the nap chromosome. Thus, like the biochemical and paralytic phenotypes of the nap^{ts} mutation (Jackson *et al*., 1984), the resistance factor is completely recessive and segregates with the second chromosome.

To further localize the resistance mutation, we used a translocation (designated as $nap^{+}Y$), which includes a wild-type allele of the nap gene (Wu *et al*., 1978). Flies that are homozygous for the nap^{ts} chromosome but carry one copy of the nap^{+} allele on the Y chromosome are wild-type with respect to behavior (Wu *et al*., 1978) and to saxitoxin-binding parameters (Jackson *et al*., 1984). Figure 12 shows that such flies also display wild-type sensitivity to pyrethroids. The dose-response curves for males and females carrying this duplication closely resemble those for wild-type males and females. This shows that the $nap^{+}Y$ duplication suppresses the nap^{ts} resistance phenotype. In control experiments (not shown) we have confirmed that the $nap^{+}Y$ duplication does not affect pyrethroid sensitivity in a wild-type genetic background.

Furthermore, to establish whether the resistance maps precisely to the nap locus, we set up recombination mapping experiments. For these experiments we used a second chromosome marker stock that is homozygous for nap^{+} and shows wild-type pyrethroid sensitivity. Males from the marker stock were crossed to nap^{ts} virgin females. F_1 females that were nap^{ts} /(marker chromosome) were mated to males from a second chromosome balancer stock. Individual F_2 males carrying a possibly recombinant second chromosome over the balancer were used to establish a balanced line

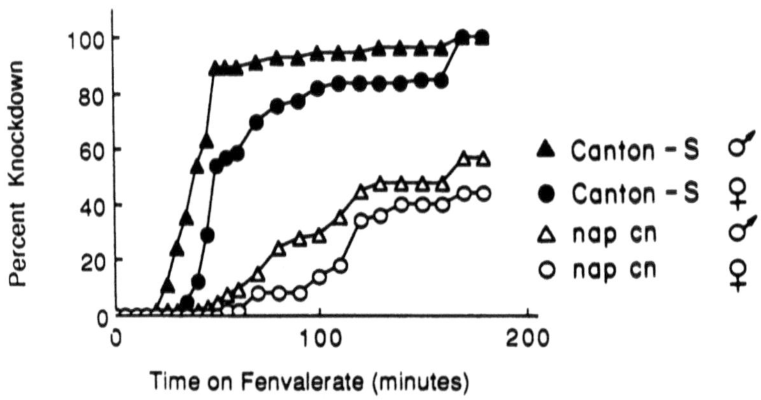

Fig. 9. Kinetics of knockdown of nap^{ts} cn and wild-type Canton-S flies by fenvalerate. The protocol was the same as that used in Figure 2. Each curve represents 40-70 flies. Reproduced with permission from Kasbekar and Hall (1988).

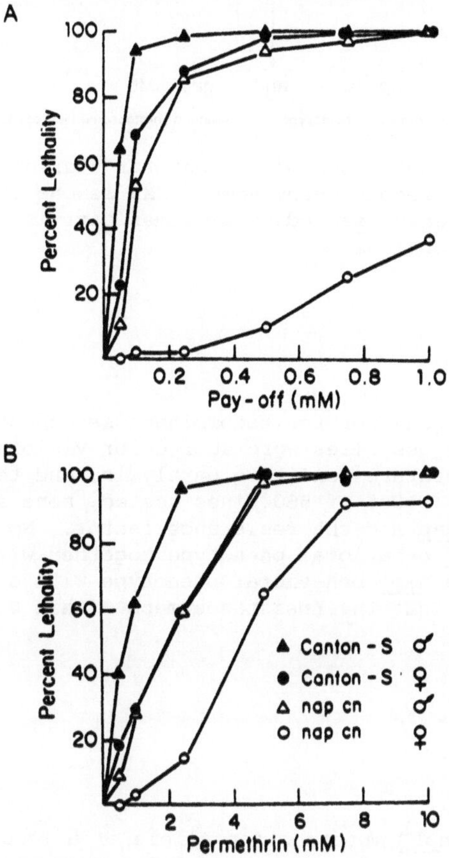

Fig. 10. Pay-Off (A) and permethrin (B) induced lethality for <u>nap^{ts} cn</u> and wild-type Canton-S flies. The protocol was the same as that used for fenvalerate described in Figure 1 legend. Each point represents 33-77 flies in A and 181-289 flies in B. Note that the permethrin concentrations used were ten-fold higher than those for Pay-Off. Reproduced with permission from Kasbekar and Hall (1988).

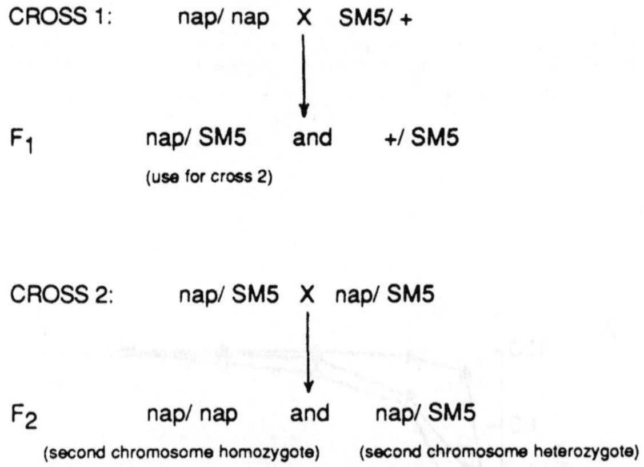

Fig. 11. A genetic scheme to test for segregation with the second chromosome. SM5 is one of several balancer second chromosomes carrying multiple inversions.

from which flies homozygous for the recombinant second chromosome were obtained. These homozygous flies were scored for visible markers, tested for nap^{ts}-induced temperature-sensitive paralysis, and tested for fenvalerate sensitivity. Out of 860 lines tested, none showed recombination between *nap* and the resistance factor. No lines were obtained with the nap^{ts} behavioral phenotype together with drug sensitivity or with the nap^{+} behavioral phenotype with drug resistance. Therefore, we conclude that the resistance maps within 0.12 map units of the *nap* locus.

SUMMARY OF MUTANT WORK

In summary, the nap^{ts} mutation (an allele of a locus known to be involved in the regulation of expression of X-linked genes in males versus females) appears to cause pyrethroid resistance based on possible down regulation of the class of sodium channels encoded by the X-linked *para* locus. This type of target site resistance could be particularly problematical in the field since it would affect all classes of insecticides which act on the same target. Since this mutational effect does not change the channel structure, altering the structure of the killing agent so that it "sees" a different region of the target will not effectively combat this type of resistance.

In contrast, mutations of the *para* type involve changes in the primary structure of the sodium channel. Since the sodium channel is a very large protein molecule with many pharmacologically distinct and independent binding domains, changes in the structure of the killing agent in some cases would be expected to overcome this type of resistance. Different mutant alleles at this locus should be useful for defining the sites of interaction of structurally different pyrethroids with the channel.

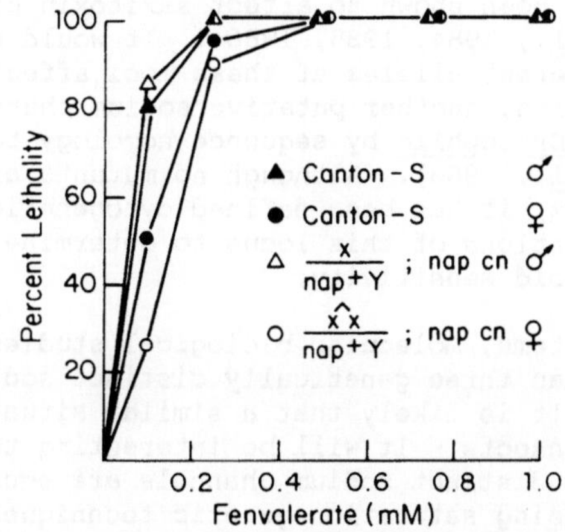

Fig. 12. The fenvalerate resistance of the nap^{ts} cn strain is suppressed by a duplication that includes the nap^{+} allele. Wild-type (Canton-S) males and females were compared with males and females homozygous for the nap^{ts} cn chromosome and also carrying a nap^{+} allele on the Y chromosome ($nap^{+}Y$). The genotype XXY in *Drosophila* is a phenotypically normal female. Each point represents 31-81 flies. Compare these results with those in Figure 8. Reproduced with permission from Kasbekar and Hall (1988).

DIRECTION OF FUTURE RESEARCH

With respect to future research directions, the well-defined, single gene mutations which we have described should help to answer a number of questions concerning pyrethroid action. These questions include:

(a) Which amino acids in the sodium channel primary structure are involved in pyrethroid interactions?
(b) How different are the binding sites for structurally different pyrethroids?
(c) How do the two different resistance mechanisms (i.e., direct structural effects versus changes in channel density) affect cross resistance to other sodium channel poisons such as DDT?

We have focused on two different genetic loci which affect pyrethroid resistance by altering the structure or density of sodium channels. Are these the only loci in the genome to have such effects? The answer to this question is almost certainly **no.** Mutations in two other genetic loci (*tip-E* and *sei*^{ts}) have been shown to affect saxitoxin binding to sodium channels (Jackson *et al*., 1984, 1985, 1986). It would be interesting to determine whether different alleles at these loci affect sensitivity to pyrethroids. In addition, another putative sodium channel structural gene has been described in *Drosophila* by sequence homology to vertebrate sodium channels (Salkoff *et al*., 1986). Although no mutants are currently available at this locus, it has been defined cytogenetically and it will be possible to use deletions of this locus to determine the effects of gene dosage on pyrethroid sensitivity.

In vertebrate systems, molecular biological studies have established that there are more than three genetically distinct sodium channel classes (Noda *et al*., 1986). It is likely that a similar situation will exist in *Drosophila* and other insects. It will be interesting to determine whether all of the genetically distinct sodium channels are equally susceptible to pyrethroid actions. Using saturation genetic techniques, it would also be interesting to determine how many different genetic loci can be mutated to give the phenotype of pyrethroid resistance. These questions can be addressed in a genetically tractable system such as *Drosophila*.

ACKNOWLEDGEMENTS

We thank Michael Charles for excellent technical assistance. This work was supported by NIH Grant NS16204. L.M.H. is a Jacob Javits Neuroscience Investigator.

REFERENCES

Breen, T. R., and Lucchesi, J. C., 1986, Analysis of the dosage compensation of a specific transcript in *Drosophila melanogaster*, *Genetics* 112:483-491.

Catterall, W. A. M., 1980, Neurotoxins that act on voltage-sensitive sodium channels in excitable membranes, *Annu. Rev. Pharmacol. Toxicol.* 20:15-43.

Catterall, W. A. M., 1986, Molecular properties of voltage-sensitive sodium channels, *Annu. Rev. Biochem.* 55:953-985.

Catterall, W. A. M., 1988, Structure and function of voltage-sensitive sodium channels, *Science* 242:50-61.

DeVries, D. H., and Georghiou, G. P., 1981, Decreased nerve sensitivity and decreased cuticular penetration as mechanisms of resistance to pyrethroids in a (1R)-*trans*-permethrin-selected strain of the housefly, *Pestic. Biochem. Physiol.* 15:234-241.

Fukunaga, A., Tanaka, A., and Oishi, K., 1975, Maleless, a recessive autosomal mutant of Drosophila melanogaster that specifically kills male zygotes, Genetics 81:135-141.

Gammon, D. W., 1980, Pyrethroid resistance in a strain of Spodoptera littoralis is correlated with decreased sensitivity of the CNS in vitro, Pestic. Biochem. Physiol. 13:53-62.

Ganetzky, B., and Wu, C.-F., 1985, Genes and membrane excitabiilty in Drosophila, Trends in Neurosci. 8:322-326.

Grigliatti, T. A., Hall, L., Rosenbluth, R., and Suzuki, D. T., 1973, Temperature-sensitive mutations in Drosophila melanogaster. XIV. A selection of immobile adults, Mol. Gen. Genet. 120:107-114.

Hall, L. M., 1986, Genetic variants of voltage-sensitive sodium channels, in: "Tetrodotoxin, Saxitoxin and the Molecular Biology of the Sodium Channel", Ann. N.Y. Acad. Sci. 479:313-324.

Hall, L. M., Wilson, S. D., Gitschier, J., Martinez, N., and Strichartz, G. R., 1982, Identification of a Drosophila melanogaster mutant that affects the saxitoxin receptor of the voltage-sensitive sodium channel, in: "Neuropharmacology of Insects", Ciba Foundation Symposium 88, Evered, D., O'Connor, M. and Whelen, J., eds., pp. 207-220, Pitman, London.

Jackson, F. R., Wilson, S. D., Strichartz, G. R. and Hall, L. M., 1984, Two types of mutants affecting voltage-sensitive sodium channels in Drosophila melanogaster, Nature 308:189-191.

Jackson, F. R., Gitschier, J., Strichartz, G. R., and Hall, L. M. 1985, Genetic modifications of voltage-sensitive sodium channels in Drosophila: gene dosage studies of the seizure locus, J. Neurosci. 5:1144-1151.

Jackson, F. R., Wilson, S. D., and Hall, L. M., 1986, The tip-E mutation of Drosophila decreases saxitoxin binding and interacts with other mutations affecting nerve membrane excitability, J. Neurogenet. 3:1-17.

Kasbekar, D. P. and Hall, L. M., 1988, A Drosophila mutation that reduces sodium channel number confers resistance to pyrethroid insecticides, Pestic. Biochem. Physiol. 32:135-145.

Lucchesi, J. C., and Manning, J. E., 1987, Gene dosage compensation in Drosophila melanogaster, Adv. Genet. 24:371-429.

Malcolm, C. A., 1988, Current status of pyrethroid resistance in Anophelines, Parasitol. Today 4:s13-s15.

Miller, T. A., 1988, Mechanisms of resistance to pyrethroid insecticides, Parasitol. Today 4:s8-s12.

Miller, T. M., Kennedy, J. M. and Collins, C., 1980, CNS insensitivity to pyrethroids in the resistant kdr strain of house flies, Pestic. Biochem. Physiol. 12:224-230.

Noda, M., Ikeda, T., Kayano, T., Suzuki, H., Takeshima, H., Kurasaki, M., Takahashi, H., and Numa, S., 1986, Existence of distinct sodium channel messenger RNAs in rat brain, Nature 320:188-192.

Omer, S. M., Georghiou, G. P. and Irving, S. N., 1980, DDT/pyrethroid resistance inter-relationships in Anopheles stephensi, Mosquito News 40:200-209.

Salkoff, L., Butler, A., Wei, A., Scarvada, N., Giffen, K., Ifune, C., Goodman, R., and Mandel, G., 1987, Genomic organization and deduced amino acid sequence of putative sodium channel gene in Drosophila, Science 237:744-749.

Scott, J. G., and Matsumura, F., 1983, Evidence for two types of toxic actions of pyrethroids on susceptible and DDT-resistant German cockroaches, Pestic. Biochem. Physiol. 19:141-150.

Strichartz, G., Rando, T., and Wang, G. K., 1987, An integrated view of the molecular toxicology of sodium channel gating in excitable cells, Annu. Rev. Neurosci. 10:237-267.

Suzuki, D. T., Grigliatti, T., and Williamson, R. M. J., 1971, Temperature-sensitive mutations in Drosophila melanogaster. VII. A mutation (para-ts) causing reversible adult paralysis, Proc. Natl. Acad. Sci. USA 68:890-893.

Wu, C.-F., and Ganetzky, B., 1980, Genetic alteration of nerve membrane excitability in temperature-sensitive paralytic mutants of Drosophila melanogaster, Nature 286:814-816.

Wu, C.-F., Ganetzky, B., Jan, L. Y., Jan, Y.-N., and Benzer, S., 1978, A Drosophila mutant with a temperature-sensitive block in nerve conduction, Proc. Natl. Acad. Sci. USA 75:4047-4051.

ACTIONS OF INSECTICIDES IN MAMMALIAN CENTRAL NERVOUS SYSTEM - USE OF IN VIVO HIPPOCAMPAL EVOKED POTENTIALS

T.E. Albertson[1] and R.M. Joy[2]

Department of Internal Medicine[1]
School of Medicine and
Department of Veterinary Pharmacology and Toxicology[2]
School of Veterinary Medicine
University of California, Davis
Davis, CA 95616

ABSTRACT

The dentate gyrus of intact, urethane-anesthetized rats was employed to evaluate the effects of different pesticide agents on granule excitability. Compounds such as the convulsants pentylenetetrazol, picrotoxin, bicuculline, and the anticonvulsant agent diazepam were also examined for comparison purposes. The data demonstrates that picrotoxin, pentylenetetrazol, bicuculline, the cyclodiene insecticide dieldrin, and the hexachlorocyclohexane (HCH) insecticide lindane (gamma-HCH) all increased granule cell excitability and decreased $GABA_A$-mediated recurrent early inhibition. A small decrease in recurrent inhibition was seen with the beta-HCH isomer without altering granule cell excitability. By comparison, the delta-HCH isomer was associated with an increase in recurrent inhibition and a small decrease in granule cell excitability. Diazepam greatly increased recurrent inhibition and resulted in a small decrease in granule cell excitability. Both allethrin (a prototype type I pyrethroid) and deltamethrin (a prototype type II pyrethroid) greatly increased recurrent inhibition. Analysis indicated that the pyrethroid-induced effects were primarily associated with an increase in interneuronally mediated inhibition. Deltamethrin moderately reduced granule cell excitability. These effects are unlikely to be related to direct $GABA_A$ agonism but may be secondary to an increase in tonic inhibition evoked by the same mechanisms responsible for the increase in phasic inhibition. Together these studies offer an in vivo mammalian model to study pesticide mechanisms of action and allow a comparison to proposed mechanisms established from in vitro studies.

INTRODUCTION

Several general approaches have been taken to evaluate the mechanisms of insecticide-induced neurotoxicity. Historically, the first approach has been to observe the effects of insecticides in lower mammals. In an attempt to dissect these effects, specific studies looking at alterations in sensory, motor, cognitive, physiological, and consummatory responses have been performed. Additional electrophysiological studies have examined global brain activity, complex motor sensory responses, brain slice, spinal cord reflexes and responses, and nerve conduction velocities in various mammalian or insect species. This was done in an attempt to better understand insecticide-induced neurotoxicity. These in vivo whole animal or whole animal isolate approaches have been augmented by the use of the second general approach which is best described as a cellular or in vitro approach. The cellular approach to understanding mechanisms ranges from neuroelectrophysiologic intracellular studies of the neurotoxic consequences of insecticide on single cells of invertebrate organisms or mammalian cultured cell lines to biochemical and biophysical studies examining alterations in receptor characteristics, cell ion channels, cell membrane fluidity, or cell enzymes. Studies examining the neurotoxic consequences of a specific agent can then be classified as utilizing an "integrative approach" (the effect of the agent on neuron communications) or a "cellular approach" (the effect of the agent on neuron morphology, metabolism, receptors, enzymes, etc.). Although the overlapping of approaches occurs, most paradigms that are designed to evaluate the neurotoxic consequence of a specific insecticide are cellular or integrative in nature. The results of integrative paradigms can often be too complex to generate a uniform insecticide mechanism of action while the cellular approach may result in a theory that is too simplistic and fails to predict what happens when a compound is introduced into a complex neural environment. In attempting to understand the site and mechanism of action of a specific insecticide, the information obtained from both cellular and integrative approaches must be consistent and complementary if a uniform theory is to emerge.

This laboratory has examined in detail the effects of several insecticides on the dentate gyrus of the hippocampus of the rat. This electrophysiological paradigm is best described as an integrative in vivo approach for understanding toxic mechanisms. The dentate gyrus is a highly laminated structure in the rat with well defined afferents, efferents and neurotransmitters (Cowan et al., 1980; Fonnum, 1984; Storm-Mathisen, 1977). This laminated structure yields evoked field responses elicited from activation of the perforant path that are monosynaptic, primarily glutamate mediated, large in amplitude, stable and easy to locate and measure. After the previous potentials were analyzed extensively the the origin of the components of the evoked response was accepted (Andersen et al., 1964, 1971; Bliss and Lomo, 1973; McNaughton and Barnes, 1977). The amplitude of the field population spike (PS) reflects the number and synchrony of the monosynaptically activated granule cell discharges (Andersen et al., 1971; Lomo, 1971). Changes in field excitatory postsynaptic potential (EPSP) amplitudes or slopes reflect changes in the number of perforant path fibers activated by the stimulus or changes in the efficacy of the synaptic process through altered transmitter release and/or receptor properties (Bliss and Lomo, 1973; Lomo, 1971).

Stimulation of the perforant path using a classical paired-pulse paradigm leads to a series of changes in granule cell excitability to the second stimulation, the sources of which are largely dependent upon the paired pulse interval. At short interpulse intervals (10-40 msec) granule cell responsiveness to the second stimulus is decreased due primarily to the activation by recurrent collaterals of interneurons subserving $GABA_A$-mediated inhibition (Albertson and Joy, 1987; Matthews et al., 1981; Oliver and Miller, 1985; Thalman and Ayala, 1982). At intermediate intervals (40-200 msec) granule cell responsiveness to the second stimulus is enhanced. In this model, the second elicited EPSP amplitudes at these intervals are not facilitated, suggesting that presynaptic actions are not involved in the increase in excitability (Joy and Albertson, 1987b) (See Figure 10, for an example). The primary basis for the increase in excitability at these time intervals appears to be either activation of excitatory interneurons or, less likely, disinhibition. At long intervals (400-1000 msec) granule cell responsiveness to the second stimulus is again depressed. In our model the reduction in responsiveness appears to be largely presynaptically mediated and may involve the activation of $GABA_B$ receptors (Peet and McLennan, 1986). An increase in calcium-dependent potassium conductance evoked by spike activation may, however, also be involved (Thalman and Ayala, 1982). This system, then, provides an opportunity to examine the mammalian central effects in vivo of a number of insecticides and to compare these findings with known mechanisms obtained from other in vitro and cellular studies.

METHODS

Subjects and surgical preparation

Male Sprague-Dawley rats, 250-350 g, were anesthetized with urethane (1.2 g/kg) and placed into a stereotaxic device (incisor bar 5 mm above the intra-aural line). The skull was exposed, and 2 mm diameter holes were drilled to place electrodes. The concentric stimulating electrode consisted of a tip made from sharpened 34-gauge nichrome wire and a barrel made from 21 gauge stainless steel tubing with a tip-to-hub spacing of 1 mm. The stimulating electrode was placed into the perforant path in the right medial entorhinal cortex (near the angular bundle) using coordinates of 7.0 mm posterior to bregma, 4.1 mm lateral to midline and 3.2 mm below the cortical surface. The recording electrode, a tungsten microelectrode with tip resistance of 1.2 Mohm was lowered into the dentate gyrus of the right hippocampus using surface coordinates of 2.0 mm posterior to bregma and 1.6 mm lateral to the midline. The final depth placement of the recording electrode was made while stimulating the perforant path input to achieve the largest and most identifiable dentate response near 3.5 mm below the surface of the exposed cortex. In some experiments a second stimulating electrode, of similar construction, was lowered into the left hippocampus using coordinates of 2.0 mm posterior to bregma, 1.6 mm lateral to the midline and 3.5 mm below the cortical surface. A stainless steel screw electrode was placed over the frontal sinus and served as a ground. Body temperature was kept at 38 ± 1 °C with constant temperature heating pads and wraps.

Recording and stimulation

Stimuli were square-wave pulses formed by a WPI Model 830 stimulator delivered through a WPI Model 850 isolation unit. Pulses were 0.05 msec in duration and varied in constant current depending on the experiment. Pulses or pairs of pulses were used. Repetition frequency was always 0.2 Hz.

Dentate responses were amplified using Grass Model 7P511 amplifiers with low and high frequency cutoffs at 1Hz and 10 kHz. Responses were displayed on a Nicolet 3091 oscilloscope. Responses for analysis were digitized using a Keithly DAS Series 500 A/D converter coupled to an IBM PC XT computer. The conversion rate used was 10 kHz.

Analysis of Responses

For purposes of data analysis, six sample responses were collected, averaged and stored as a single record. Typical coefficients of variation for such records were 2-5%. The computer-aided analysis of wave forms (Figure 1) followed closely those previously described (Joy and Albertson, 1987a, b, 1988). EPSP amplitudes were determined as the maximum rate of rise or slope of the EPSP between its onset and the onset of the population spike (PS). PS amplitudes were measured from the EPSP baseline to the peak of the PS. Thresholds were defined as the lowest stimulus intensities producing a consistently identifiable response on the oscilloscope.

Dose-Response studies

The procedures used to obtain dose-response data were similar to those described previously (Albertson and Joy, 1987; Joy and Albertson, 1987a, b, 1988). Because the intensity chosen for perforant path stimulation has been demonstrated to be a significant variable in previous exposure studies (Joy and Albertson, 1987a, b) (see Figure 9, for example), the analysis chosen was designed to include both variable and constant stimulus amplitude approaches. After electrode placement and time for stabilization of responses, subjects were exposed to five different exposure conditions (one control and four sequential doses of test substance). Doses were administered at approximately 40 minute intervals, and data were collected between minutes 20-40 after each dosing. For each exposure condition the following measures were made: 1) EPSP and PS thresholds; 2) recording an input/output (I/O) series of responses to pairs of increasing amplitude perforant path stimuli using an interpulse interval of 15 msec with stimuli ranging from 1-6 times the PS threshold; 3) recording a similar I/O series using an interpulse interval of 60 msec (note: a similar I/O series using an interpulse interval of 400 msec was also recorded in some studies); and 4) recording pairs of responses at a fixed stimulus intensity (80% of maximum) at variable interpulse intervals of 10, 15, 20, 25, 30, 40, 60, 100, 200, 400 and 1000 msec. Responses were computer subtraction corrected when portions of the responses overlapped in time (Joy and Albertson, 1987a). The stability of these responses have been previously demonstrated in control studies under urethane anesthesia over the time periods used in these studies (Albertson and Joy, 1987; Joy and Albertson, 1987a, b). All drugs and insecticides used in this study were dissolved in dimethylsulfoxide (DMSO) to give a total of 2 ml/kg DMSO exposure for the entire study. DMSO was also used as a the initial control exposure for each compound. The following test compounds and cumulative doses were used: 1) DMSO (0.25, 0.5, 1.0 and 2.0 ml/kg); 2) diazepam (0.5, 1.0, 2.0 and 4.0 mg/kg); 3) pentylenetetrazol (20, 40, 80 and 160 mg/kg); 4) picrotoxin (1.0, 2.0, 4.0, and 8.0 mg/kg); 5) bicuculline (1.0, 2.0, 4.0 and 8.0 mg/kg); 6) lindane or gamma-hexachlorocyclohexane (5, 10, 20 and 40 mg/kg); 7) delta-hexachlorocyclohexane (5, 10, 20 and 40 mg/kg); 8) beta-hexachlorocyclohexane (5, 10, 20 and 40 mg/kg); 9) dieldrin (5, 10, 20 and 40 mg/kg); 10) allethrin (10, 20, 40 and 80 mg/kg); and 11) deltamethrin (2.5, 5, 10, and 20 mg/kg). Unless stated, each animal received only one test compound at the four cumulative doses noted after the initial DMSO control exposure.

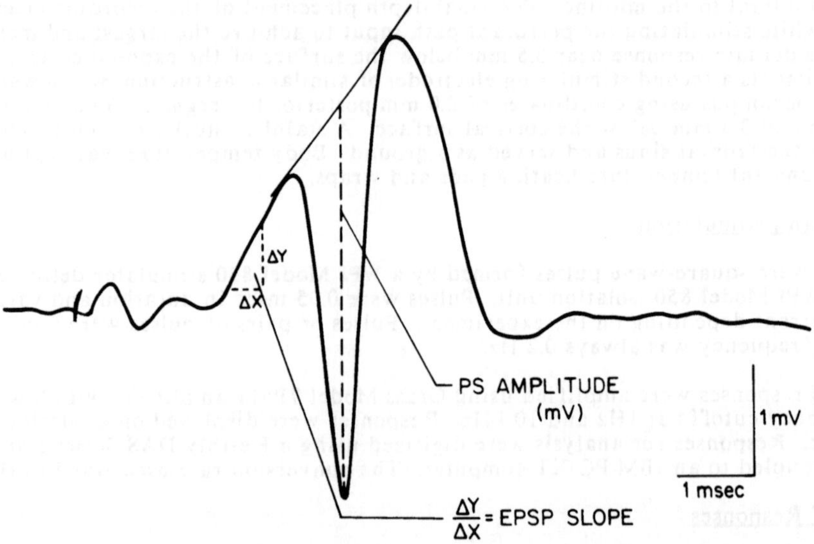

Fig. 1. The field response evoked in dentate gyrus by perforant path stimulation. Low intensity stimulation of the perforant path initially leads to a pure excitatory postsynaptic potential (EPSP). When the stimulus intensity is increased a field population spike (PS) is elicited. The amplitude of the EPSP can be measured directly when no PS is elicited. Otherwise it can be measured as the slope of the waveform before the onset of the PS. The PS is measured from the deepest portion of the PS to the intercept of a line drawn between the two wave forms made up from the PS onset to the PS peak. The initial line and wave represents stimulus artifact. This is followed by a short latency perforant path fiber response.

Other studies

Other studies were done to determine the site of action of the pyrethroids within the dentate gyrus and the time of maximum effect. In these studies a fixed stimulus amplitude and variable interval series was employed with stimulus intensity set at 80% of maximum using interpulse intervals of 10, 15, 20, 25, 30, 40, 60, 100, 200, 400 and 1000 msec. After stabilization, a control series was generated and then a single dose of allethrin (40 mg/kg) or deltamethrin (10 mg/kg) was administered. Subsequent series were generated at different times to determine the time of onset of insecticide effect. In similar studies fixed stimulus intensities were selected to produce either a pure EPSP response or a near maximal PS, and the influence of allethrin (40 mg/kg) or deltamethrin (10 mg/kg) on the two responses was compared. Finally, in some additional studies, stimuli were applied via a contralateral hippocampal electrode before those via the perforant path electrode using the same interpulse intervals as used in the typical paired-perforant pulse series. The inhibitory effects of prior contralateral hippocampal stimulation on granule cell excitability at various intervals were then compared to paired-pulse series with the prior inhibitory stimulation from the perforant path at the same time intervals. This allowed comparison of drug effects (allethrin 40 mg/kg) from recurrent inhibition to be compared to direct or feed-forward inhibition.

Data analysis

Threshold data were obtained as current intensity (in μA), and all values within an experiment were normalized to the value obtained during the control period. Data for

EPSP and PS amplitudes and latencies were obtained from the first responses of the 60 msec variable amplitude, fixed interval paired pulse series (I/O). Mean values over the I/O series were calculated during control and after each dose of a test compound. Values at each stimulus intensity were also used to assess potential interactions between stimulus intensity and drug effects. Paired pulse (variable amplitude, fixed interval) I/O data were compared as previously described (Joy and Albertson, 1987b, 1988). Amplitudes of the second responses (PS2) were plotted as functions of the amplitudes of the first responses (PS1) for each of the stimuli at a given interpulse interval before and after a test compound administration. Where these curves overlapped along the PS1 amplitude axis, the areas under the curves were determined using an IBM PC XT computer (see Figure 4 for an example). Changes in these areas reflect changes in the excitability of the granule cells at the interpulse interval employed corrected for drug-induced changes in the initial PS amplitude. Paired pulse (variable interval, fixed amplitude stimulus) data were quantified by determining the ratio of the amplitude of the second PS (PS2) to that to the first PS (PS1) at each time interval.

For comparative purposes, raw data from the first control (DMSO) period and after the various doses of test compounds were analyzed using a repeated measures ANOVA. Two- and three-way ANOVA designs were used to determine main and interactive effects in the various situations. Where a significant effect was detected with the ANOVA, Dunnet's test was used to compare individual dose means to the control. For presentation in the tables and figures, data have been normalized with respect to the control period and presented as percent change from this control. All doses reported in the results section are cumulative (total) doses.

Drugs

Dimethylsulfoxide, pentylenetetrazol (PTZ) picrotoxin, bicuculline, and urethane were purchased from Sigma Chemical Company. Allethrin (S-1R-trans-allethrin) and deltamethrin were gifts from Drs. J. Casida and J. Hawkinson. Dieldrin (95% pure) was a gift of Shell Chemical Company. Purified lindane was provided by Dr. J.N. Seiber. Delta- and beta-hexachlorocyclohexane isomers were provided by Dr. W. Ost, Merck Co., FRG. Diazepam was provided by Hoffmann La Rouche Co.

RESULTS

An example of a series of evoked potentials produced with increasing stimulus intensity (I/O curve) that started with a "pure" EPSP and ended with a large PS are demonstrated in Figure 2. When a second stimulus was added (paired-pulse) and an evoked potential I/O series performed, a resulting relationship was developed that related the first evoked potential to the second across several stimulus intensities. Figure 3 demonstrates the effects of three stimulus intensities on 15 msec paired-pulse evoked potentials during control and after 40 mg/kg pentylenetetrazol treatments. When the PS amplitude of the first evoked potential is graphed as a function of the second evoked PS, a curve can be constructed (Figure 4).

After drug treatment, the overlap of the generated curves allowed comparison of drug dose effects over an entire I/O curve. The areas under the curves that overlap allowed quantification of drug effects for paired-pulse-inhibition (15 msec) or facilitation (60 msec). Figure 4 demonstrates that the ten stimulus intensities at a paired-pulse interval of 15 msec resulted in a small increase in amplitude in the first PS after 40 mg/kg pentylenetetrazol, but more importantly the second elicited PS tended to be larger for a given first PS amplitude. The areas under the overlapping curves from various individual animals at each tested paired-pulse interval can be averaged and compared across doses (Figures 5-7).

Modification of granule cell excitability determined from analysis of single evoked potentials over an I/O series is demonstrated in Table 1.

Picrotoxin, pentylenetetrazol, gamma-hexachlorocyclohexane and dieldrin increased PS amplitude and tended to decrease PS thresholds to a greater degree than they increased EPSP slopes and decreased EPSP thresholds. Bicuculline increased PS amplitudes and decreased PS thresholds without significant effects on the elicitied EPSP. Both diazepam and delta-hexachlorocyclohexane reduced granule cell excitability by decreasing PS amplitude and increasing PS threshold without altering the EPSP. Beta-hexachloro-

Table 1. Effects of Various Compounds on Granule Cell Excitability and Responsiveness Single Pulse Analysis[1]

Compound	EPSP	PS
DMSO	-	-
Diazepam	-	↓
Picrotoxin	↑	↑↑
Bicuculline	-	↑
Pentylenetetrazol	↑	↑↑
gamma-HCH	↑	↑↑
delta-HCH	-	↓
beta-HCH	↓	-
dieldrin	↑	↑↑
allethrin	-	-
deltamethrin	↓	↓↓

DMSO = dimethylsulfoxide; HCH = hexachlorocyclohexane; EPSP = excitatory postsynaptic potential; PS = field population spike; - = no change; ↑, ↑↑ = dose dependent increase, large increase; ↓, ↓↓ = dose dependent decrease, large decrease. [1] Includes summary of effects on thresholds and amplitudes or slopes.

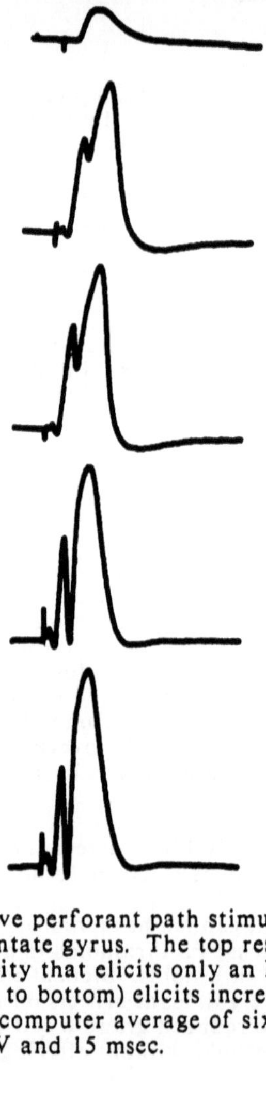

Fig. 2. Examples of five perforant path stimulated evoked responses (I/O) in the dentate gyrus. The top response represents a stimulus intensity that elicits only an EPSP. Increasing stimulus intensities (top to bottom) elicits increasing PS amplitudes. Each waveform is a computer average of six responses. Scale represents 1 mV and 15 msec.

Table 2. Comparison of Various Compounds on Granule Cell Inhibition and Facilitation Paired Pulse Analysis

Compound	Inhibition (15 msec)	Facilitation (60 msec)
DMSO	-	-
Diazepam	↑↑	-
Picrotoxin	↓	-
Bicuculline	↓	-
Pentylenetetrazol	↓↓	-
gamma-HCH	↓↓	-
delta-HCH	↑	-
beta-HCH	↓	-
dieldrin	↓↓	-
allethrin	↑↑	-
deltamethrin	↑↑	-

Abbreviations are the same as Table 1.

cyclohexane decreased EPSP slope and raised EPSP threshold without altering the PS. The reduction of granule cell excitability by deltamethrin resulted in both reductions in EPSP slope and PS amplitude and increases in EPSP and PS thresholds. Table 2 summarizes the effects from paired-pulse studies of the various compounds on early granule cell inhibition and facilitation.

Diazepam, delta-hexachlorocyclohexane, allethrin, and deltamethrin increased inhibition at 15 msec in a dose-dependent manner (see Figures 6 and 7) without consistently altering facilitation (60 msec). Reductions were seen in inhibition with picrotoxin, bicuculline, pentylenetetrazol, gamma-hexachlorocyclohexane, beta-hexachlorocyclohexane and dieldrin without consistently modifying facilitation (see Figures 5 and 6). Figure 7 also dmonstrates a lack of modification by both deltamethrin and allethrin of the late (400 msec) inhibitory phase of paired-pulse stimulation.

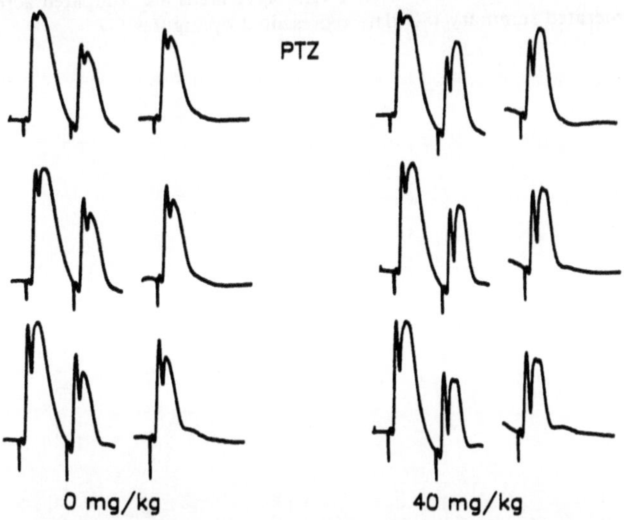

Fig. 3. Effects of 40 mg/kg pentylenetetrazol (PTZ) on paired PS elicited by three stimulus (I/O) intensities at 15 msec intervals. The left column represents control responses. The right column represents responses elicited by the same stimulus intensities after 40 mg/kg PTZ. Each group of three responses show a PS1 followed by a PS2. The third response is the computer "subtracted" PS2 wave form. Scale represents 5 mV and 15 msec.

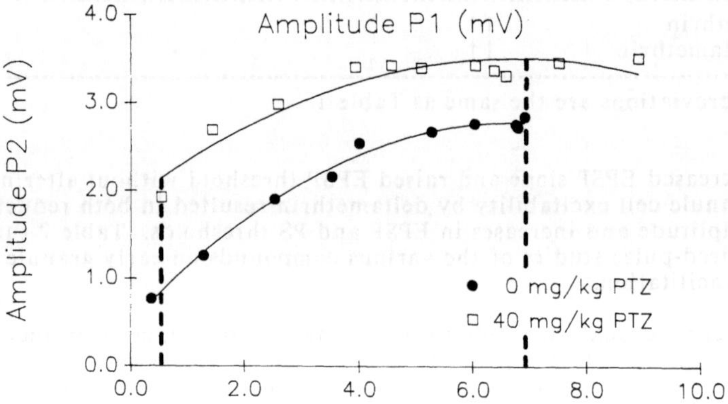

Fig. 4. The I/O before and after 40 mg/kg PTZ. When the relationship of PS1 (P1) amplitude is graphed as a function of PS2 (P2) amplitude over ten stimulus intensities an I/O curve is generated before and after PTZ. The change in area under the overlap curves are then determined. The dotted lines represent the portions of the two curves that overlap in this animal. PS2 is increased for any given PS1 after PTZ. These areas are compared across doses and animals to generated summary I/O data represented by Figures 5-7.

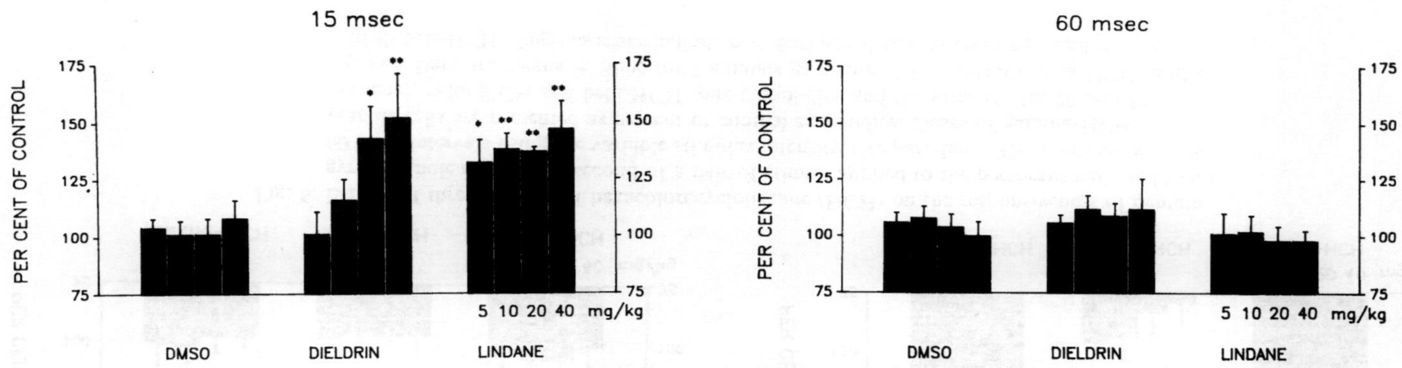

Fig. 5. Effects of DMSO, dieldrin and lindane on the responsiveness of dentate gyrus granule cells to the second of a pair of stimuli applied to the perforant path at 15 and 60 msec intervals using the variable stimulus intensity I/O paradigm. The excitability of the granule cells are presented as percent of control excitability. Doses of DMSO (0.25, 0.5, 1.0 and 2.0 ml/kg), dieldrin and lindane are cumulative doses and indicate the total administered to the subject at the time the data were obtained. Doses of dieldrin and lindane were the same (5, 10, 20 and 40 mg/kg). Data are means ± SEM for 14 animals DMSO, 7 animals lindane and 6 animals dieldrin. Single asterisks indicate $p < 0.05$ and double asterisks represent $p < 0.01$.

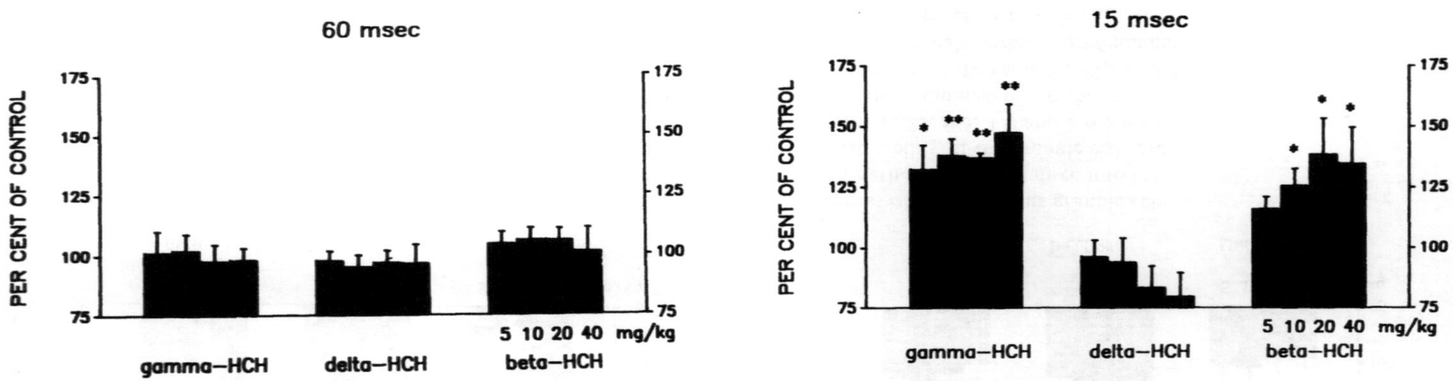

Fig. 6. Effects of three isomers of hexachlorocyclohexane (HCH) on the responsiveness of dentate gyrus granule cells to the second of a pair of stimuli applied to the perforant path at 15 and 60 msec intervals using the variable stimulus intensity I/O paradigm. The excitability of the granule cells are presented as percent of control excitability. Doses of gamma-HCH (lindane), delta-HCH and beta-HCH were cumulative and the same (5, 10, 20 and 40 mg/kg). Data are means ± SEM for 7 animals gamma-HCH, 6 animals delta-HCH, and 6 animals beta-HCH. Single asterisks indicate $p < 0.05$ and double asterisks represent $p < 0.01$.

Figure 8 also offers a typical control or DMSO variable interpulse series with the constant stimulus intensity selected to generate a first PS amplitude that is 80% maximum. The degree of early inhibition and facilitation is a function of stimulus intensity in the variable interpulse series (Figure 9). As stimulus intensities are increased, the amount of early inhibition is increased and facilitation is reduced. The changes seen in synaptic or perforant path terminal excitability when stimulus intensities that result in only EPSP are used, do not correlate with the changes in granule cell excitability seen when stimulus intensities result in PS generation (Figure 10). When paired low intensity stimuli were used, the amplitudes of generated EPSP were facilitated over the short interpulse intervals which inhibited a second PS and returned to baseline during the intervals that facilitated a second PS.

As expected from the I/O data, DMSO had no significant effect on the ratio of PS2/PS1 using the fixed stimulus intensity variable interpulse interval paradigm. As can be seen from Figures 11 and 12, diazepam increased early paired PS inhibition and either reduced the paired PS facilitation or extended paired PS early inhibition phase into the facilitation phase.

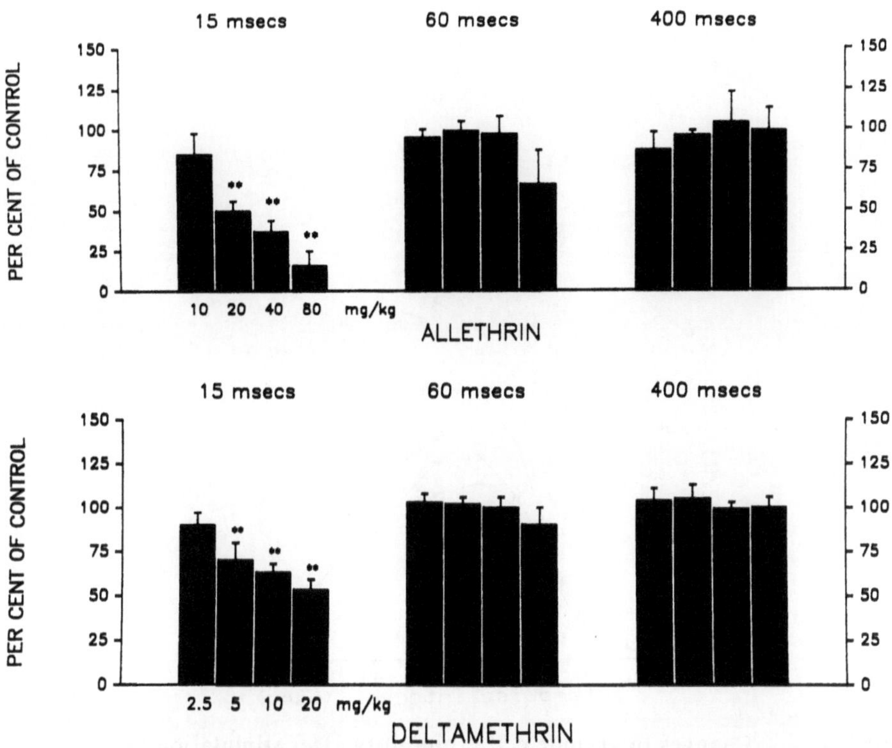

Fig. 7. Effects of allethrin and deltamethrin on the responsiveness of dentate gyrus granule cells to the second of a pair of stimuli applied to the perforant path at 15, 60 and 400 msec intervals using the variable stimulus intensity I/O paradigm. The excitability of the granule cells are presented as percent of control excitability. Doses of pyrethroids were cumulatively increased and indicated the total administered to the subject at the time the data were obtained. Data area means ± SEM for 5 subjects for each pyrethroid. Asterisks indicate significant effects: (** = $p < 0.01$).

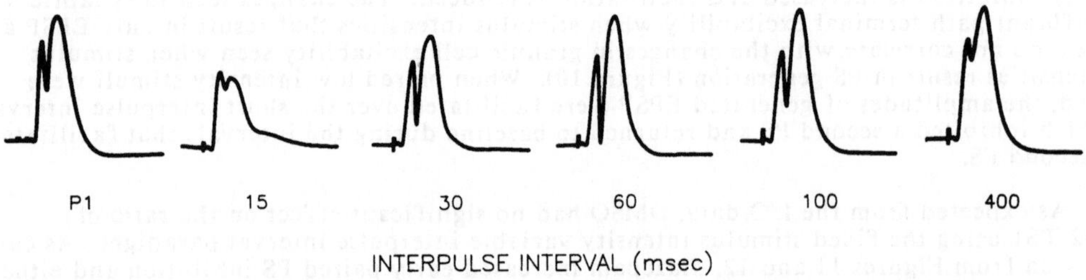

Fig. 8. An example of evoked paired pulse responses with variable interpulse intervals and constant stimulus intensity. The first pulse P1 represents the first elicited PS (PS1). The next 5 evoked responses occurred at increasing interpulse intervals that demonstrate early recurrent inhibition (15 msec), followed by facilitation (30 and 60 msec) and late inhibition (400 msec). Each represented response is the average of six evoked responses in one animal.

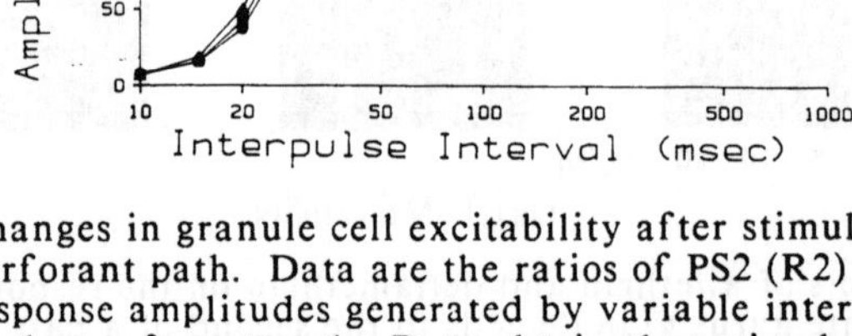

Fig. 9. Changes in granule cell excitability after stimulation of perforant path. Data are the ratios of PS2 (R2) to PS1 (R1) response amplitudes generated by variable interpulse intervals to the perforant path. Data obtained at stimulus intensities producing a 60% (filled triangles), 80% (filled squares) or 100% (filled circles) maximal PS. Each point of each curve is the mean from six different subjects. Higher stimulus intensities increased early inhibition and decreased facilitation.

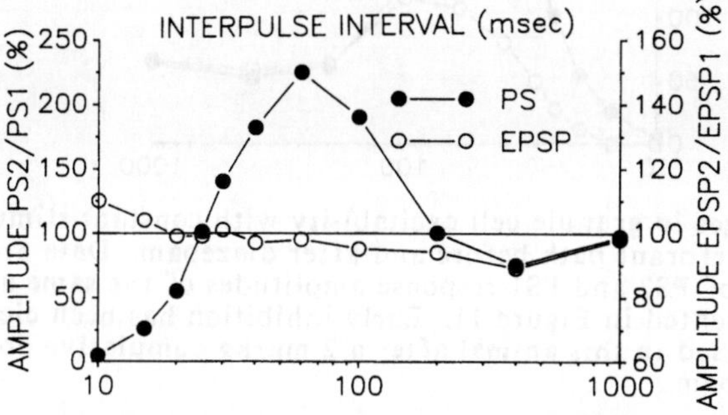

Fig. 10. Changes in the ratio of paired-pulse granule cell excitability generated at two stimulus intensities using the variable interpulse interval paradigm. The first intensity was low enough not to produce a PS. Early facilitation of the amplitude of the second EPSP was seen followed by late inhibition. The ratio of PS amplitudes demonstrated early inhibition of PS2. This was followed by facilitation of PS2 then a late inhibition. Each point represents the average of six responses during the control phase of one animal.

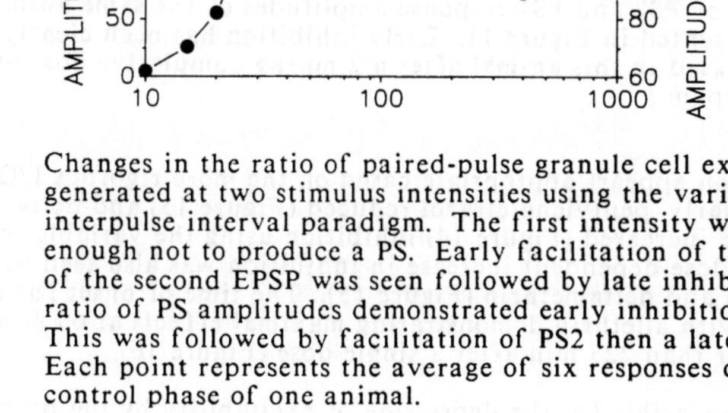

Fig. 11. An example of evoked paired pulse responses with variable interpulse intervals and constant stimulus intensity. The first pulse (PS 1) is followed by 5 second evoked responses at increasing interpulse intervals. The top set of responses was obtained after DMSO control treatment. The bottom set of responses was obtained after a cumulative dose of 2 mg/kg diazepam. The prolongation of the early inhibition phase in this animal into the facilitatory phase was observed. Each represented response is the average of six evoked responses in one animal.

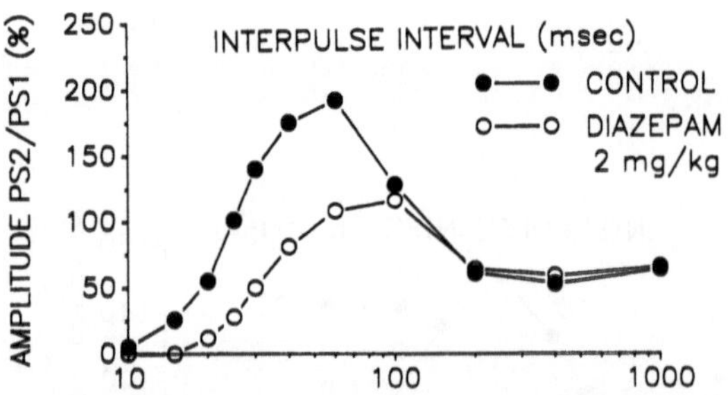

Fig. 12. Changes in granule cell excitability with constant stimulation of the perforant path before and after diazepam. Data are the ratio of PS2 and PS1 response amplitudes of the same animal represented in Figure 11. Early inhibition has been clearly increased in this animal after a 2 mg/kg cumulative dose of diazepam.

The latter conclusion appears appropriate based on the more rigorous I/O paradigm discussed earlier. Simiarly, pentylenetetrazol reduced (Figure 13) and delta-hexachlorocyclohexane increased (Figure 14) inhibition using the variable interpulse interval paradigm. A dose-dependent increase in inhibition was also seen using this paradigm for allethrin and deltamethrin (Figure 15). The time of onset for these two pyrethroids differed with allethrin demonstrating maximal effects at 60 min and deltamethrin at greater than 225 min after a single dose (Figure 16).

The mechanism responsible for the depression of excitability by the pyrethroids over the intervals of 10-100 msec after an initial stimulus to the perforant path was tested in a number of ways. Conduction block of axons within the perforant path, failure of transmitter release from perforant path terminals, or reduced sensitivity of the receptors for that transmitter would all affect the amplitude of the EPSP component of the response as well as the amplitude of the PS component. This was tested by delivering pairs of pulses to the perforant path at an intensity well below threshold for the generation of a PS and measuring the amplitude of the EPSP responses produced.

Alternatively the same procedure was employed using stimulus intensities that evoked a PS of 80% maximal amplitude, and PS responses were measured. A comparison of these two responses (EPSP and PS) (Figure 17) indicated that the pyrethroid-induced depression in granule cell discharge (amplitude of PS) over the first seven interpulse intervals ($F_{6,18} = 26.12$, $p < 0.01$) was not accompanied by a similar depression in the EPSP response ($F_{6,18} = 1.75$, N.S.). Additionally, the insert in Figure 17 shows the first three EPSP rsponses produced by 50 Hz stimulation of the perforant path before and 60 minutes after allethrin. Exposures significantly depressing granule cell discharge did not reduce the ability of the perforant path terminals to respond to high frequency stimulation. In one additional experiment, latencies to reach half maximal EPSP amplitudes and EPSP durations were measured as the width between half amplitude points. These were compared before and 60 minutes after exposure to 40 mg/kg allethrin i.p. No significant differences were observed (control latency = 7.71 $\pm$ 0.05 msec, control duration = 4.89 $\pm$ 0.06 msec; allethrin latency = 7.69 $\pm$ 0.02 msec, allethrin duration = 4.75 $\pm$ 0.11 msec). Similar findings were obtained in four other experiments using deltamethrin rather than allethrin. These data indicate that the locus of action of the pyrethroids within this system is not the perforant path axon, the perforant path terminal or the granule cell receptors activated by the perforant path.

Two other possibilities could be involved. When pairs of pulses are used to stimulate the perforant path, the first pulse activates the granule cells (producing PS1) and then through recurrent collaterals activates inhibitory interneurons. The effects on the second response (PS2) could result from pyrethroid-induced granule cell inactivation following PS1 or pyrethroid-induced intensification of the recurrent collateral inhibition produced

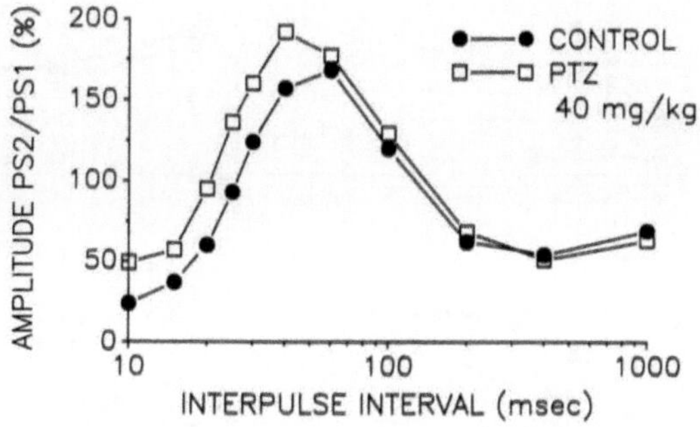

Fig. 13. Changes in granule cell excitability with constant intensity stimulation of the perforant path before and after pentylenetetrazol. Data are the ratios of PS2 to PS1 response amplitudes of the same animal represented in Figure 3 and 4. Early inhibition has been decreased after treatment, with a cumulative dose of 40 mg/kg pentylenetetrazol in this animal.

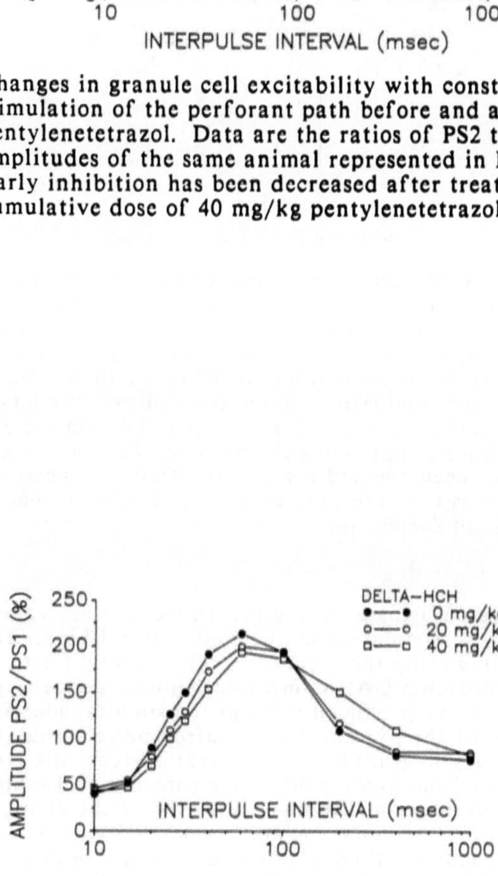

Fig. 14. Average changes in granule cell excitability with constant intensity stimulation of the perforant path before and after delta-hexachlorocyclohexane (delta-HCH). Data are the ratios of PS2 to PS1 response amplitudes. Increased early inhibition can be seen after 20 and 40 mg/kg cumulative doses of delta-HCH.

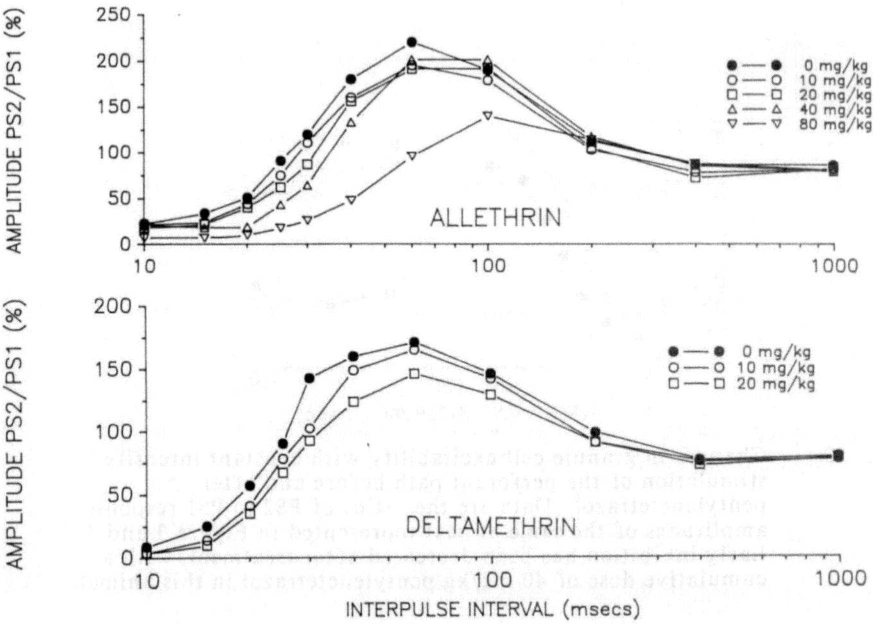

Fig. 15. Dose-dependence and time course of pyrethroid-induced depression of granule cell excitability. Top: Control curve indicates the relative excitability of the granule cells (PS2/PS1) at interpulse intervals of 10-1000 msec. Other curves indicate excitability at the same intervals after allethrin. Doses indicated are cumulative values and indicate the total amount of pyrethroid administered at the time the data were obtained. Bottom: Similar data for deltamethrin. Data at 2.5 and 5 mg/kg doses have been omitted for clarity. Data are mean values from four subjects for each pyrethrin. Standard error bars have also been omitted for clarity.

by PS1. These can be separated using a testing procedure that activates the interneuronal pool subserving inhibition via stimulation of the contralateral hippocampus (Douglas et al., 1983) rather than by stimulating the recurrent collaterals by first stimulating the perforant path. By this approach a $GABA_A$-mediated, interneuronally generated inhibition of granule cells can be produced without the simultaneous activation of the granule cells. If the effects of the pyrethroids on paired pulse stimulation of the perforant path were secondary to granule cell inactivation resulting from PS1, then there should be little or no effect of the pyrethroids when contralateral hippocampal stimulation precedes a perforant path stimulus. However, if the changes produced by the pyrethroids after perforant path stimulation were associated with an increase in the response of inhibitory interneurons, then a similar depression would be expected regardless of whether the first pulse of the pair was delivered to the perforant path or to the contralateral hippocampus. Figure 18 compares the effects of allethrin under these two conditions. As shown the increased depression is observed in both situations. A three way ANOVA using the first seven interpulse intervals demonstrated a significant pyrethroid effect on depression ($F_{1,4} = 36.50$, $p < 0.01$). The ANOVA was unable to distinguish between the two paths based on their ability to exhibit pyrethroid-induced depression ($F_{1,4} = 0.17$, N.S.). These data suggest that the primary action of the pyrethroids at this locus is to enhance the intensity and duration of interneuronally mediated inhibition.

DISCUSSION

The perforant path to hippocampal dentate evoked potential offers a model for examining the in vivo effects of pesticides on granule cell excitability. The current study has utilized this paradigm to examine the neurotoxic mechanisms of several classes of pesticides and isomers of a pesticide. The various compounds and pesticides used in this study were examined over an exposure range that is of clinical relevance. Baseline granule cell excitability is increased by exposure to the variously GABA-mediated convulsants including pentylenetetrazol, picrotoxin and bicuculline and depressed by the benzodiazepine agonist diazepam. Since all four of these compounds are thought to in part act through mechanisms associated with $GABA_A$ receptors, it would appear that tonic or basal excitability at this site is in part modulated by direct or indirect GABA inputs. Of interest, the chlorinated hydrocarbon pesticides lindane (gamma-HCH) and dieldrin both increased while delta-HCH and the type II pyrethroid deltamethrin decreased baseline granule cell excitability.

Several limitations arise if over extrapolation of the results of in vivo testing occur. One cannot assume that the etiology of an altered evoked response is by the same mechanisms as a known compound simply because they modify the response in the same way. For example, increasing or decreasing baseline granule cell excitability has been demonstrated in this study to be modulated by $GABA_A$-mediated compounds but other mechanisms that involve the primary transmitter (thought to be glutamate) for these synapses or other non-receptor mediated processes could give similar overall modulation of the evoked response. Therefore, the findings of increased or decreased granule cell excitability by a test compound, does not by itself invoke a $GABA_A$-mediated mechanism of action. Only when these data are integrated with other in vitro and in vivo data does a proposed mechanism emerge.

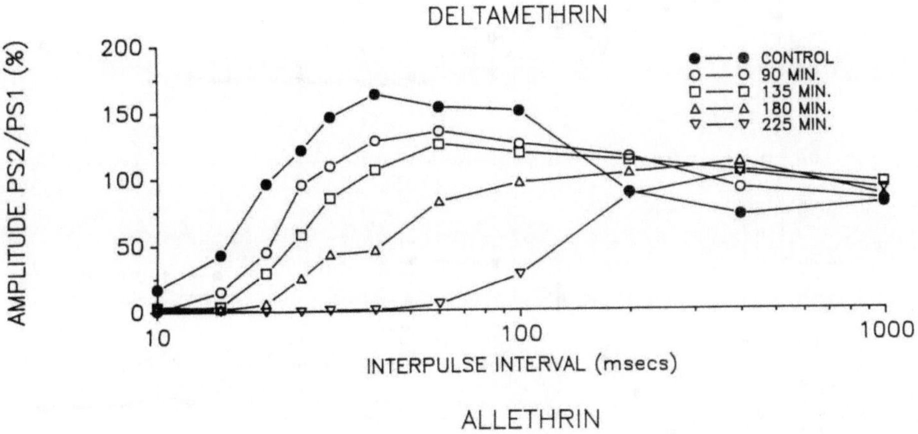

Fig. 16. Time required for maximal development of the effects of pyrethroids on granule cell paired pulse excitability. Top: Control curve indicates the relative excitability of the granule cells (PS2/PS1) at interpulse intervals of 10-1000 msec. Allethrin, 40 mg/kg ip was administered immediately after control data were obtained. Bottom: Similar graph for a different subject given deltamethrin, 10 mg/kg ip.

Another problem that was initially seen with this model was that the test compound can alter the underlying granule cell excitability and this can have indirect effects on a paired-pulse paradigm that specifically examines recurrent $GABA_A$-mediated inhibition. When initial studies by this laboratory examined the effects of lindane on this evoked potential, no decrease in recurrent inhibition was demonstrated (Joy and Albertson, 1985). This was because the baseline granule cell excitability was increased by lindane exposure resulting in an in increased first PS. As noted in Figure 4, when the stimulus intensity is increased, a corresponding increase in recurrent inhibition occurs. Lindane exposure acts as an increased stimulus with the resulting increased PS amplitude recruiting additional recurrent inhibitory circuits that thus prevents the demonstration of true decreased inhibition. When a complete I/O curve is constructed after lindane exposure, the PS amplitudes can be matched and, a decrease in inhibition is demonstrated. By more rigorous testing, this in vivo paradigm is able to demonstrate significant lindane-induced decreased recurrent inhibition.

The relationship of GABA receptor modulation to synaptic activity is complex, but can be demonstrated by in vitro techniques such as examining GABA receptor binding affinity, [^{35}S] t-butylbicyclophosphorothionate (TBPS) channel site inhibition, [^{3}H] flunitrazepam (Flu) inhibition and GABA stimulated [^{36}Cl] uptake inhibition. Table 3 offers a correlation between the current study findings with these compounds and early granule cell recurrent inhibition compared to GABA stimulated [^{36}Cl] uptake results.

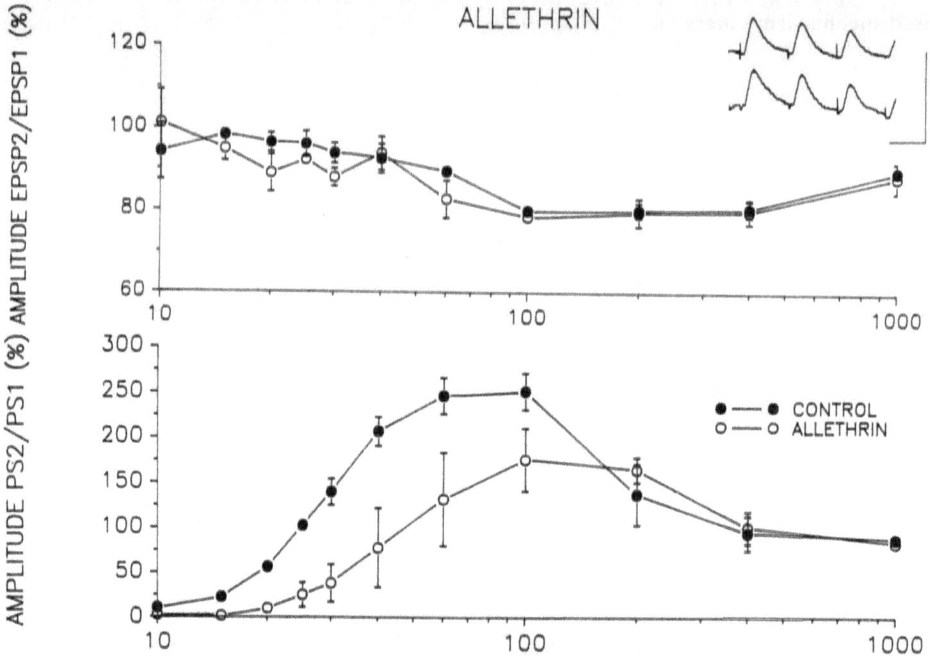

Fig. 17. Comparison of the effects of allethrin on EPSP and PS paired pulse responses. Top: Filled circles indicate the relative amplitude of the EPSP response (EPSP2/EPSP1) generated in granule cells by perforant path stimulation subthreshold for PS before administration of allethrin. Open circles indicate relative amplitudes 60 minutes after administration of 40 mg/kg allethrin ip. Bottom: Filled circles indicate the relative amplitude of the PS response (PS2/PS1) generated by granule cells to a perforant path stimulation of 80% maximal intensity in the same subjects as top. Open circles indicate relative amplitudes 60 minutes after administration of 40 mg/kg allethrin ip. Data are means ± SE for four subjects. Insert at top: EPSPs generated by 50 Hz stimulation of the perforant path before (top) and 60 minutes after 40 mg/kg allethrin (bottom).

Table 3.

Compound	GABA stimulated {^{36} uptake}	Early granule cell recurrent Inhibition
Diazepam	↑ a	↑↑
Pentylenetetrazol	↓ b	↓↓
Picrotoxin	↓↓ a,b,c	↓
Bicuculline	↓↓ a,b,c	↓↓
Gamma-HCH	↓ c or -- b,d	↓↓
delta-HCH	-- (↓) b	↑
beta-HCH	-- (↑) c	↓
dieldrin	↓↓ a,b,d	↓↓
allethrin	-- (↓) c	↑↑
deltamethrin	-- (↓) d	↑↑

a (Gant et al., 1987) rat brain microsacs
b (Fishman and Gianutsos, 1988) mouse brain microsacs
c (Abalis, et al., 1986) rat brain microsacs
d (Bloomquist, et al., 1986) mouse brain vesicles
↑, ↑↑ = increase, large increase; ↓, ↓↓ = decrease, large decrease; -- = no change or small increase (↑) or decrease (↓) only at high concentrations ($\gtrsim$ 1000 μM).

A strong correlation appears between inhibition or stimulation of GABA induced chloride uptake and in vivo effect on recurrent inhibition. Two major divergences are noted. The first is that the HCH isomers generally appear to have weak depressant effects on chloride uptake in brain microsacs and they modify recurrent granule cell inhibition. In the case of delta-HCH the recurrent granule cell inhibition is actually increased despite having little or even depressant effects on chloride uptake. The second divergency noted is with the pyrethroids. Both the type I pyrethroid allethrin and the type II pyrethroid deltamethrin caused a marked increase in recurrent granule cell inhibition while having small or depressant effects on chloride influx.

The effects of the various isomers of HCH on other aspects of GABA receptor binding in vitro brain microsacs are of interest. Using [^{3}H] tert-butyl-bicycloorthobenzoate (TBOB) binding Fishman and Gianutsos (1988) have found that picrotoxin and lindane are nearly equal in their ability to inhibit binding while alpha- and delta-HCH are less potent and pentylenetetrazol the least potent. Lindane is at least five times more potent than the alpha-, beta- or delta-HCH isomers in inhibiting [^{35}S] TBPS channel site binding in brain synaptic membranes (Lawrence and Casida, 1984). In the same study, dieldrin was found to be several times more potent in inhibitory [^{35}S] TBPS binding than lindane. Thus it would appear that although a general correlation exists between GABA-stimulated chloride uptake inhibition and GABA receptor bindings studies, several divergence persist. Some local brain differences in drug distribution, receptor types and receptor numbers may contribute to this divergence. Other in vivo studies have shown whole animal depressant effects from alpha- and delta-HCH treatments (Fishman and Gianutsos, 1988). In addition, pentylenetetrazol-induced seizures in mice have been made worse by lindane treatment and improved with alpha- or delta-HCH treatments (Fishman and Gianutsos, 1988). Similarly, lindane and dieldrin have been shown to increase and beta-HCH decrease the rate of amygdaloid kindling in rats (Stark et al., 1986).

Together the data from the current study is consistent with most existing in vivo and in vitro data suggesting a GABA-mediated mechanism explaining the findings that lindane, dieldrin, pentylenetetrazol, bicuculline and picrotoxin increase baseline granule cell excitability and decrease early recurrent inhibition. The finding that the delta-HCH isomer decreases granule cell excitability and increases early recurrent inhibition is consistent with in vivo data from other seizure models but this finding is not easily explained by in vitro data on $GABA_A$ related mechanisms.

Both representative type I (allethrin) and type II (deltamethrin) pyrethroids produced similar qualitative effects on recurrent granule cell inhibition despite reducing GABA stimulated chloride uptake (Abalis et al., 1986; Bloomquist and Soderlund, 1985; and Bloomquist et al., 1986) and inhibition of [^{35}S] TBPS binding in brain microsacs (Lawrence and Casida, 1983 and 1984; and Crofton et al., 1987). The current studies demonstrated that the pyrethroids differed from each other in potency and in time required to produce their effects but not in the direction of the effects. These findings are the opposite to what would have been predicted from the known effects of pyrethroids on GABA in vitro mechanisms. A comparison of the consequences of activating the inhibitory interneurons by a preceding stimulus to the perforant path (recurrent inhibition) or the contralateral hippocampus (feed forward inhibition) indicated that a pyrethroid-induced inactivation of granule cells was not responsible for the increased inhibition observed. Because granule excitability was not greatly altered by the pyrethroids, the site of action would appear to be the inhibitory interneurons subserving the granule cells. Gilbert et al. (1987) have reported a similar effect on granule cell excitability in chronically implanted, unanesthetized rats following exposure to deltamethrin. They reported a depression of granule cell responsiveness that persisted for up to 500 msec after the first stimulus to the perforant path, using a paired pulse paradigm similar to the one described in this study. When the same subjects were given cismethrin (a type I pyrethroid) no comparable changes in granule cell excitability occurred, even though symptoms of toxicity (tremoring) were observed.

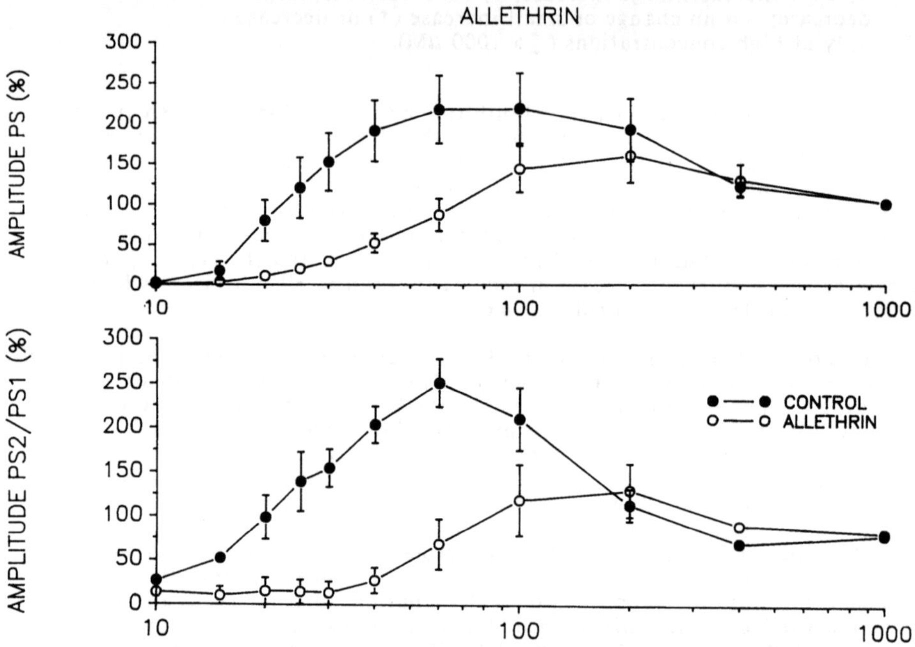

Fig. 18. Effects of prior stimulation of contralateral hippocampus or perforant path on responsiveness of granule cells to a second perforant path stimulation. Top: Filled circles indicate excitability of the granule cells to a perforant path stimulation occurring after contralateral hippocampal stimulation at intervals of 10-1000 msec. Open circles indicate excitability to the same stimulation paradigm 60 minutes after administering 40 mg/kg allethrin ip. Bottom: Data from the same subjects using paired pulse stimulation of the perforant path. Filled circles indicate excitability of the granule cells to a perforant path stimulation occurring at intervals of 10-1000 msec after a prior perforant path stimulus. Open circles indicate excitability to the same stimulation paradigm 60 minutes after administering 40 mg/kg allethrin ip. Data are means ± SE for 3 subjects.

While our studies lack intracellular data and cannot define the specific mechanisms responsible for the increase in interneuronal activity produced by the pyrethroids, several possibilities exist. One possibility is that the interneurons are more susceptible than other neuronal elements in the dentate gyrus to a prolongation of sodium current by pyrethroids (Narahashi, 1971, 1985; Vijverberg and deWeille, 1985). One unique difference between the interneurons in the dentate gyrus and the other neuronal components (the perforant path terminals or dentate granule cells) is that the normal mode of function of the interneurons is to produce repetitive discharges when synaptically activated. The duration and frequency of the burst of action potentials is correlated with the intensity of synaptic activation or the intensity of depolarization of their soma (Andersen et al., 1964). In contrast, the perforant path axons and the granule cells tend to fire individually when activated (Fricke and Prince, 1984).

Allethrin induces repetitive activity in vertebrate sensory axons (van den Bercken et al., 1973; Wouters et al., 1977) and presynaptic nerve endings (Wouters et al., 1977; van den Bercken et al., 1979). It also induces or intensifies repetitive discharge in free sensory nerve endings and sense organs (Akkermans et al., 1975). Allethrin may directly intensify repetitive discharge of interneurons in the dentate gyrus which could directly lead to an increase in effective inhibition. This would occur at exposure levels not affecting perforant path axons or granule cells. Deltamethrin produces sustained depolarization and repetitive discharge in sensory nerve endings and sense organs (Vijverberg and van den Bercken, 1982; Vijverberg et al., 1986). These are neuronal systems which normally respond to depolarization with repetitive discharges as do hippocampal interneurons. If deltamethrin prolongs the depolarization in these interneurons due to its effects on somatic sodium channels, the result would both prolong and intensify repetitive discharge. The time constant for closure of deltamethrin-affected sodium channels in mammalian neuroblastoma cells in culture has been reported to be on the order of 400-800 msec at 18° C (Ruigt et al., 1987) and 2-7.7 seconds at 11° C (Chinn and Narahashi, 1986) both much shorter than those reported in invertebrate preparations. If hippocampal interneurons display similar time constants, then at 37° C deltamethrin-induced depolaraization may be relatively transient, being sufficiently long to prolong repetitive discharge but short enough to allow recovery over the next 1-2 seconds before the next pair of stimuli. This would then explain the findings of the current study.

Other alternative mechanisms of action may explain the prolongation of recurrent inhibition seen in this study and in the study by Gilbert et al. (1987). Both allethrin and deltamethrin compete for binding to the peripheral benzodiazepine receptor in the mammalian CNS (Devaud and Murray, 1988). For deltamethrin, this interaction has been correlated with its convulsant activity in rats. Pyrethroids also affect other types of ion channels, including the voltage-dependent potassium channel (Narahashi and Anderson, 1967), the rapidly inactivating calcium channel in neuroblastoma cells (Narahashi, 1986), and the calcium-dependent potassium channel in the leech Retzius cell (Leake et al., 1986). Comparable channels in hippocampal interneurons may also be involved in the action of pyrethroids at this loci. The results of the current and other (Gilbert et al., 1987; Smith, 1980) in vivo studies using both type I and II pyrethroids have failed to demonstrate an antagonism of GABA-mediated inhibition that has been suggested to be the mechanism of action of these compounds by in vitro studies.

These studies have demonstrated the usefulness and importance of in vivo testing of pesticides using the perforant path to dentate evoked potential model. An attempt has been made to integrate the complex neurotoxicological results obtained by these studies with current available data from more cellular and biochemically oriented studies to establish consistent theories for the mechanism of action of these pesticides.

REFERENCES

Abalis, I. M., Eldefrawi, M. D., and Eldefrawi, A. T., 1986, Effects of insecticides on GABA-induced chloride influx into rat brain microsacs, J. Toxicol. Env. Hlth., 18:13-23.

Akkermans, L. M. A., van den Bercken, J., and Versluijs-Helder, M., 1975, Comparative effects of DDT, allethrin, dieldrin and aldrin-transdiol on sense organs of *Xenopus laevis*, Pestic. Biochem. Physiol., 5:451-457.

Albertson, T. E., and Joy, R. M., 1987, Increased inhibition in dentate gyrus granule cells following exposure to GABA-uptake blockers, Brain Res., 13:208-221.

Andersen, P., Bliss, T. V. P., and Skrede, K. K., 1971, Unit analysis of hippocampal population spikes. Exp. Brain Res., 13:208-221.
Andersen, P., Eccles, J. C., and Loyning, Y., 1964, Pathway of postsynaptic inhibition in the hippocampus, J. Neurophysiol., 27:608-619.
Anderson, P., Homquist, B, and Voorhoeve, P. E., 1966, Entorhinal activation of dentate granule cells, Acta Physiol. Scand., 66:448-460.
Bliss, T. V. P., and Lomo, T., 1973, Long lasting potentiation of synaptic transmission in the dentate area of the anesthetized rabbit following stimulation of the perforant path, J. Physiol. (Lond.), 232:357-374.
Bloomquist, J. R., Adams, P. M., and Soderlund, D. M., 1986, Inhibition of gamma-aminobutyric acid-stimulated chloride flux in mouse brain vesicles by polychlorocycloalkane and pyrethroid insecticides, Neurotoxicology, 7:11-20.
Bloomquist, J. R., and Soderlund, D. M., 1985, Neurotoxic insecticides inhibit GABA-dependent chloride uptake by mouse brain vesicles, Biochem. Biophys. Res. Commun., 133:37-43.
Chinn, K., and Narahashi, T., 1986, Stabilization of sodium channel states by deltamethrin in mouse neurobastoma cells, J. Physiol. (Lond.), 380:191-207.
Cowan, W. M., Stanfield, B. B., and Kishi, K, 1980, The development of the dentate gyrus, in: "Current Topics in Developmental Biology, Vol. 2" Hunt, R. D., ed., Academic Press, New York, 103-157.
Crofton, K. M., Reiter, L. W., and Mailman, R. B., 1987, Pyrethroid insecticides and radioligand displacement from GABA receptor chloride ionophore complex, Toxicol. Letts., 35:183-190.
Devaud, L. L., and Murray, T. F., 1988, Involvement of peripheral-type benzodiazepine receptors in the proconvulsant actions of pyrethroid insecticides. J. Pharmacol. Exptl. Therap., 247:14-22.
Douglas, R. M., McNaughton, B. L., and Goddard, G. V., 1983, Commissural inhibition and facilitation of granule cell discharge in Fascia Dentata. J. Comp. Neur., 219:285-294.
Fishman, B. E., and Gianutsos, 1988, CNS biochemical and pharmacological effects of the isomers of hexachlorocyclohexane (lindane) in the mouse, Toxicol. Appl. Pharmacol., 93:146-153.
Fonnum, F., 1984, Glutamate: A neurotransmitter in mammalian brain, J. Neurochem., 42:1-11.
Fricke, R. A., and Prince, D. A., 1984, Electrophysiology of dentate gyrus granule cells, J. Neurophysiol., 51:195-209.
Gant, D. B., Eldefrawi, M. E., and Eldefrawi, A. T., 1987, Cyclodiene insecticides inhibit $GABA_A$ receptor-regulated chloride transport, Toxicol. Appl. Pharmacol., 88:313-321.
Gilbert, M. E., Mack, C., and Crofton, K. M., 1987, The effects of pyrethroids on paired pulse inhibition in the hippocampus of the rat, Soc. Neurosci. Abstr., 13:931.
Joy, R. M., and Albertson, T. E., 1985, Effects of lindane on excitation and inhibition evoked in dentate gyrus by perforant path stimulation, Neurobehav. Toxicol. Teratol., 7:1-8.
Joy, R. M., and Albertson, T. E., 1987a, Factors responsible for increased excitability of dentate gyrus granule cells during exposure to lindane, Neurotoxicology, 8:517-528.
Joy, R. M., and Albertson, T. E., 1987b, Interactions of lindane with synaptically mediated inhibition and facilitation in the dentate gyrus, Neurotoxicology, 8:529-542.
Joy, R. M., and Albertson, T. E., 1988, Convulsant-induced changes in perforant pathway dentate gyrus excitability in urethane-anesthetized rats, J. Pharmacol. Exptl. Therap., 246:887-895.
Lawrence, L. J., and Casida, J. E., 1983, Stereospecific action of pyrethroid insecticides on the gamma-aminobutyric acid receptor-ionophore complex, Science, 221:1399-1401.
Lawrence, L. J., and Casida, J. E., 1984, Interactions of lindane, toxaphene and cyclodienes wth brain-specific t-butylbicyclophosphorothionate receptor, Life Sci., 35:171-178.
Lawrence, L. J., Gee, K. W., and Yamamura, H. I., 1985, Interactions of pyrethroid insecticides with chloride ionophore-associated binding sites, Neurotoxicology, 6(2):87-98.
Leake, L. D., Dean, J. A., and Ford, M. G., 1986, Pyrethroid action and cellular activity in invertebrate neurons, in: "Neuropharmacology and Pesticide Action," Ford, M. G., Lunt, G. G., Reay, R. C. and Usherwood, P. N. R., eds., Ellis Horwood, Ltd, Chichester (England), 244-266.
Lomo, T., 1971, Patterns of activation in a monosynaptic cortical pathway: The perforant path input to the dentate area of the hippocampal formation, Exp. Brain Res., 12:18-45.
Matthews, W. O., McCafferty, G. P., and Setler, P. E., 1981, An electrophysiological model of GABA-mediated neurotransmission, Neuropharmacology, 20:561-565.
McNaughton, B. L., and Barnes, C. A., 1977, Physiological identification and analysis of dentate granule cell responses to stimulation of the medial and lateral perforant

pathways in the rat, J. Comp. Neur., 175:439-454.
Narahashi, T., 1971, Mode of action of pyrethroids, Bull. Wld. Hlth. Org., 44:337-345.
Narahashi, T., 1985, Nerve membrane ionic channels as the primary target of pyrethroids, Neurotoxicology, 6(2):3-22.
Narahashi, T, 1986, Mechanisms of action of pyrethroids on sodium and calcium channel gating, in: "Neuropharmacology and Pesticide Action," Ford, M. G., Lunt, G. G., Reay, R. C., and Usherwood, P. N. R., eds., Ellis Horwood, Ltd., Chichester (England), 36-60.
Narahashi, T, and Anderson, N. C., 1967, Mechanism of excitation block by the insecticide allethrin applied externally and internally to squid giant axons. Toxicol. Appl. Pharmacol., 10:529-547.
Oliver, M. W., and Miller, J. J., 1985, Alterations of inhibitory processes in dentate gyrus following kindling-induced epilepsy. Exp. Brain Res., 57:443-447.
Peet, M. J., and McLennan, H., 1986, Pre- and postsynaptic actions of baclofen: blockade of the late synaptically-evoked hyperpolarization of CA1 hippocampal neurones. Exp. Brain Res., 61:567-574.
Ruigt, G. S., Neyt, H. C., Van der Zalm, J. M., and Van den Bercken, J., 1987, Increase of sodium current after pyrethroid insecticides in mouse neuroblastoma cells, Brain Res., 437:309-322.
Smith, P. R., 1980, The effect of cismethrin on the rat dorsal root potentials, Europ. J. Pharmacol., 66:125-128.
Stark, L. G., Albertson, T. E., and Joy, R. M., 1986, Effects of hexachlorocyclohexane isomers on the acquisiton of kindled seizures, Neurobehav. Toxicol. Teratol., 8:487-491.
Storm-Mathisen, J., 1977, Localization of transmitter candidates in the brain: The hippocampal formation as a model, Prog. Neurobiol, 8:119-181.
Thalman, R. H., and Ayala, G. F., 1982, A late increase in potassium conductance follows synaptic stimulation of granule neurons in the dentate gyrus, Neurosci. Letts., 29:243-248.
van den Bercken, J., Akkermans, L. M. A., and van der Zalm, J. M., 1973, DDT-like action of allethrin in the seonsory nervous system of Xenopus laevis, Europ. J. Pharmacol., 21:95-106.
van den Bercken, J., Kroese, A. B. A., and Akkermans, L. M. A., 1979, Effects of insecticides on the sensory nervous system, in: "Neurotoxicology of Insecticides and Pheromones," T. Narahashi, ed., Plenum Press, New York, 183-210.
Vijverberg, H. P. M. and de Weille, J. R., 1985, The interaction of pyrethroids with voltage-dependent Na channels, Neurotoxicology, 6(2):23-34.
Vijverberg, H. P. M., de Weille, J. R., Ruigt, G. S. F., and van den Bercken, J., 1986, The effect of pyrethroid structure on the interaction with the sodium channel in the nerve membrane, in: "Neuropharmacology and Pesticide Action," Ford, M. G., Lunt, G. G., Reay, R. C., and Usherwood, P. N. R., eds., Ellis Horwood, Ltd., Chichester (England), 268-285.
Vijverberg, H. P. M. and van den Bercken, J., 1982, Action of pyrethroid insecticides on the vertebrate nervous system, Neuropath. Appl. Neurobiol., 8:421-440.
Wouters, W., van den Bercken, J., Ginneken, A. V., 1977, Presynaptic action of the pyrethroid insecticide allethrin in the frog motor end plate, Europ. J. Pharmacol., 43:163-171.

ENHANCED NEUROTRANSMITTER RELEASE BY PYRETHROID INSECTICIDES

J. Marshall Clark and Jacques R. Marion

Department of Entomology
University of Massachusetts
Amherst, MA

ABSTRACT

Using a perfusion system, synaptosomes prepared from rat brain released ^{3}H-norepinephrine in a Ca^{2+}-dependent manner when pulse depolarized by briefly elevating external potassium concentrations. Tetrodotoxin (10^{-7}M) inhibited 48% of the pulsed release and D595 (10^{-5}M), a phenethylamine-type calcium channel blocker, inhibited 27%. In combination, these two specific ion channel antagonists appear to function independently of each other in an additive fashion. Pretreatment of this preparation with deltamethrin resulted in an enhanced release of norepinephrine which occurred in a biphasic fashion with a greater enhancement evident in the second or tailing peak when compared to release in both the presence and absence of tetrodotoxin. Addition of deltamethrin to D595-pretreated synaptosomes resulted in no significant enhancement of release from either peak. This indicates that deltamethrin may be enhancing norepinephrine release by increasing the uptake of Ca^{2+} via other voltage-gated channels (e.g., calcium) or ion-selective transporters in addition to its action at voltage-gated sodium channels. To determine whether deltamethrin may also have an effect on intraterminal Ca^{2+} homeostasis, external Ca^{2+} was replaced with Ba^{2+} and synaptosomes depolarized with pentylenetetrazol. Deltamethrin enhanced neurotransmitter release over that induced by pentylenetetrazol alone with an estimated ED_{50} value calculated to be 2.4 x 10^{-10}M. It is concluded that Type II pyrethroids may be acting as voltage-gated channel agonists in a manner consistent with their structure and activity relationship to the phenethylamine-type calcium channel blockers.

INTRODUCTION

It is well established that DDT and pyrethroid insecticides are neurotoxins that are highly convulsive in the physiological expression of their mechanisms of action. In the development of photostable pyrethrin analogs, pyrethroids elicited enhanced mammalian toxicity, increased environmental persistence and diverging signs of poisoning quite distinct from their natural pyrethrum counterparts (Gray, 1985). Consequently, there has been intense efforts to understand their mechanism of action.

In mammals, it is now generally agreed that toxic pyrethroids that possess both a halogenated acid component and alpha cyano 3-phenoxybenzyl alcohol (i.e., Type II) will induce choreoathetotic writhing and salivation or CS syndrome. Toxic pyrethroids which lack either of these two entities (i.e., Type I) induce tremor or the T syndrome which in most cases is identical to that caused by DDT (Staatz et al., 1982; Cremer et al., 1980; Gray and Richard, 1982; Ray and Cremer, 1979; Verschoyle and Aldridge, 1980; Verschoyle and Barnes, 1972; Gray, 1985). The T syndrome produced by Type I pyrethroids and DDT is remarkably unlike the CS syndrome and has been highly correlated to the repetitive action of Type I pyrethroids at the nerve axon, an action not produced by Type II pyrethroids. The CS syndrome of Type II pyrethroid poisoning, however, is remarkably similar to an epileptic and pentylenetetrazol-induced seizure.

In the 25 year interval since the classic electrophysiological studies first described pyrethroid mode of action on voltage-gated sodium channels (Narahashi and Anderson, 1967), the sodium channel theory has been enlarged upon using a multitude of nerve preparations with essentially the same result (Narahashi, 1986). Thus, while there is little disagreement that pyrethroids are likely to have a principal action at or near the sodium channel in the nerve, there is no reason to believe that this is the only site of action.

A number of additional mechanisms have been suggested recently including the perturbation of other voltage-gated ion channels such as the calcium channel (Cremer et al., 1983; Berlin et al., 1984; Gammon and Sander, 1985; Leake et al., 1985; Narahashi, 1986; Clark and Brooks, 1988; Orchard and Osborne, 1979). Moreover, from an organismal viewpoint, the variety of toxic symptoms which are produced by pyrethroids makes it difficult to ascribe all of them to a single mechanism (Leake et al., 1986; Gray, 1985; Clark and Matsumura, 1986; Gammon and Sander, 1985;

Lawrence and Casida, 1982). Indeed, the lack of repetitive activity in nerve axons poisoned by the more potent neurotoxic Type II pyrethroids seems in contradiction to their obvious convulsive nature (Narahashi, 1986). The production of such distinctly different symptomology due to alterations in pyrethroid structure raises the questions of whether a single mechanism or site of action is completely justifiable at this time.

In response to this uncertainty, we have developed a functional bioassay which consists of isolated presynaptic nerve terminals (i.e., synaptosomes) in continuous perfusion. This technique allows us to investigate the action of pyrethroids and other ion channel directed neurotoxins at multiple biochemical mechanisms which remain functional in the isolated synaptosome by monitoring their effects on neurotransmitter release. Synaptosomes are uniquely suited for this purpose because so much of their endogenous physiology remains functional after isolation. Synaptosomes maintain energy metabolism which supports electrochemical gradients. They have various ion selective channels (sodium, potassium, calcium and chloride) which allows synaptosomes to depolarize and repolarize in a controlled fashion when either electrically- or chemically-stimulated. Because of these functional aspects, synaptosomes actively load radiolabeled neurotransmitters, package them in presynaptic vesicles and can be induced to release them upon depolarization in a Ca^{2+}-dependent fashion. This provides a most useful and easily measureable biochemical endpoint (e.g., release of radiolabeled neurotransmitters).

Finally, we have taken the approach that pyrethroids may interact with other aspects within the nerve in addition to the voltage-gated sodium channel. To test this hypothesis, we have developed a pulse-depolarized method of neurotransmitter release using elevated external K^+ concentration rather than the more often used veratridine-induced release. In this way, we can evaluate the effects of pyrethroids on other neural aspects in addition to the voltage-gated sodium channel. Also, we are able to evaluate the effects of these insecticides during both depolarization and repolarization processes.

MATERIALS AND METHODS

Animals

Male Sprague-Dawley rats, 56 to 60 d old, were obtained from Charles River Breeding Laboratories Inc., Wilmington, MA.

Chemicals

All insecticides were gifts or were purchased from Chem Services, West Chester, PA (Brooks and Clark, 1987). L-[7,8-^{3}H]-Norepinephrine (^{3}H-NE) (30-40 Ci/mmol) and $^{45}CaCl_2$ (17 mCi/mg) were obtained from Amersham Radiochemicals, Arlington Hts., IL, and New England Nuclear, Boston, MA, respectively. D600 (5-[(3,4-dimethoxyphenylethyl) methylaminol]-2-isopropyl-2-(3,4,5-trimethoxyphenyl)-valeronitrile hydrochloride) and D595 (4-[3,4-dimethoxyphenylethyl) methylamino]-2-(3,4-dichlorophenyl)-2-isopropyl valeronitrile hydrochloride) were gifts of Prof. Kretzchmar and Prof. Oberdorf, Knoll Ag, West Germany, respectively. All other organic and inorganic chemicals were obtained from Sigma Chemical Co., St. Louis, MO.

Preparation of synaptosomes

Synaptosomes were prepared from whole rat brains (2-4 rats per preparation) by the method of Whittaker et al. (1964) as modified by Hajos (1975). The P_2 fraction (crude mitochondrial fraction) was layered onto a discontinuous sucrose density gradient and centrifuged at 50,000xg for 1 hr. The material at the 0.8-1.2 M interface was returned to a more physiological normotonic environment by addition of small amounts of ice cold normal superfusion media (NSM, Table 1) over a 30-min period to a final volume equal to 4 times the collected synaptosomal volume (Blaustein et al., 1978). The equilibrated synaptosomes were then pelleted at 15,000xg for 10 min and resuspended in 2.0 ml of ice cold NSM (Brooks and Clark, 1987).

Uptake and release of ^{3}H-NE

The procedure for uptake and release of ^{3}H-NE was similar to that developed by Levi and Raiteri (1975) and West and Fillenz (1980) as modified previously by Brooks and Clark (1987). Synaptosomal aliquots (0.1 ml) were incubated at 37°C with 0.139 uM ^{3}H-NE (approx. 111, 446 cpm per aliquot) for 15 min with constant shaking. The uptake was terminated by adding 1.0 ml ice-cold NSM. Background radiation was determined from blanks which consisted of identical incubations carried out at 0°C. All solutions were filtered with suction onto 0.65-um cellulose acetate filters (Millipore Co., Bedeford, MA) and washed with 10 ml of ice-cold NSM to remove any unbound ^{3}H-NE. Uptake of ^{3}H-NE was determined by extracting filters overnight in 1 M perchloric acid (1.0 ml) and measuring ^{3}H in the acid-solubilized extract and non-extractable filter residue by liquid scintillation spectrometry. Specific uptake was determined by subtracting the total cpm value of the 0°C incubation from that which occurred at 37°C.

Release of ^3H-NE was measured using a continuous perfusion device that consisted of eight 10-ml polypropylene syringes fixed in a container which was maintained at 37 ± 2°C (Clark and Brooks, 1988). Synaptosomal aliquots on cellulose acetate filters were housed in 25-mm Swinnex propylene filters (Millipore Co.) which were attached to the bottom of each syringe barrel. Tygon tubing carried perfused liquid from filters to scintillation vials via peristaltic pumps. There existed a 3-ml dead volume in the tubing. Hence at a flow rate of 0.5 ml/min, it would take approximately 6 min for fluid to appear in the scintillation vial. Twenty-four 0.5 ml-fractions were collected over 24 min period from each barrel. At fraction 6 (6 min), selected barrels were pulse depolarized by addition of 2.0 ml K^+ depolarizing media (DM, Table 1). Depolarization was terminated after 3 min by diluting the remaining solution 10-fold by adding NSM. The radioactivity (cpm ^3H-NE) in each fraction was determined by liquid scintillation spectrometry.

Table 1. Representative buffer compositions of external solutions[a]

Solution[b]	NaCl	KCl	D-Glucose	$CaCl_2$[c]	$MgCl_2$
			(mmol/liter)		
NSM	128	5	16	1	---
DM	77	56	16	1	---
0 CaNSM	128	5	16	--	1
0 CaDM	77	56	16	--	1
LCNSM	128	5	17	0.01	---
LCDM	77	56	17	0.01	---

[a] In addition, all solutions contained 12 uM nialamide and 20 mM HEPES. All solutions were adjusted to pH 7.35 with Tris base.

[b] NSM, normal superfusion media; DM, depolarization media; LCNSM and LCDM; low calcium buffers.

[c] In experiments where $SrCl_2$, $BaCl_2$ and $MnCl_2$ were substituted for $CaCl_2$, they were done so on an equimolar basis (1 mM) for both NSM and DM.

In some cases, synaptosomes were loaded with ^3H-NE, washed and perfused exactly as above except that Ca^{2+} was replaced with Ba^{2+} and the synaptosomal aliquots were never treated with high K^+ depolarizing media.

Instead at min 6 (fraction 6), selected synaptosomal aliquots received BaNSM containing various concentrations of pentylenetetrazol ($5x10^{-5}$M PTZ usually). After 3 min, PTZ was removed and perfusion continued with BaNSM until 24 fractions (i.e., 0.5ml per fraction) were collected. Synaptosomal aliquots serving as controls received BaNSM without PTZ for the entire assay. To analyze any synergistic effect between PTZ and deltamethrin, synaptosomes were incubated at 0°C with various concentrations of deltamethrin for 10 min prior to loading with ^{3}H-NE. Some deltamethrin-treated synaptosomes received only BaNSM without PTZ for the entire assay and served as nondepolarized controls for those assays.

Calculation of ^{3}H-NE released

Release of ^{3}H-NE was expressed as either a fractional rate constant or as a summation of fractional rate constant differences. The fractional rate constant is the amount of ^{3}H-NE released in each 1 min fraction (i.e., 0.5 ml) as a percentage of the radioactivity remaining in synaptosomes during the preceding min (Orrego et al., 1974; Orrego and Miranda, 1976; Levi et al., 1980; Orrego and Sanchez-Armass, 1981). After determining fractional rate constants for all barrels, an average of all fractional rate constants from similarly treated barrels was calculated. The average determined for nondepolarized or non-PTZ treated barrels was then subtracted from the average determined for K^{+}-pulse depolarized or a PTZ-treated barrels to give an average difference per fraction. Average differences for fractions 12 through 20 were summed and these summed average differences (e.g., summations of ^{3}H-NE released) were used to compare the effects of various insecticides, convulsants and inhibitors against non-treated controls.

Calcium-45 uptake studies

Synaptosomes were preincubated with either deltamethrin or 1.0 ul of 95% ethanol prior to addition of unlabeled norepinephrine (0.139 uM NE). The incubation with unlabeled NE was performed as described for ^{3}H-NE. The uptake of unlabeled NE was terminated after 15 min by adding of 10.0 ml of ice cold low calcium normal superfusion media (LCNSM, Table 1). Aliquots were filtered with suction onto 0.65 um cellulose acetate filters and perfused as described with an additional 3.0 ml of LCNSM, Table 1. Selected barrels received 2.0 ml of low calcium depolarization media (LCDM) containing 0.1 uCi $^{45}CaCl_2$ per ml. The remaining barrels received LCNSM which also contained 0.1 uCi$^{45}CaCl_2$ per ml. After a 3 min ^{45}Ca perfusion period, all filters were washed with 15.0 ml ice cold LCNSM and the ^{45}Ca present on filters determined by liquid scintillation spectrometry. The difference in cpm between the depolarized filters and non-depolarized filters was used to calculate fmoles ^{45}Ca uptake/mg protein/3 min K^+-depolarization.

Addition of inhibitors and insecticides

Water-insoluble inhibitors (i.e., D595) and pyrethroid insecticides were dissolved in 95% ethanol at concentrations such that the final assay concentration was attained by adding 1.0 ul to a 0.1 ml synaptosomal aliquot. Synaptosomes were pre-equilibrated for 10 min on ice prior to the addition of ^3H-NE. Ethanol was included in all control tubes and accounted for not more than 1% of the total volume. Tetrodotoxin was solubilized in 6.0 x 10^{-3}M sodium citrate (pH 4.8) and was added to appropriate control tubes.

When pyrethroid insecticides were used (see Fig. 3), incubation tubes were coated with Carbowax PEG 20000 (Helmuth et al., 1983). Insecticides were added to carbowax treated tubes and incubated under standard conditions. Solutions were transferred to non-coated tubes prior to loading with ^3H-NE. This transfer was necessary in that carbowax (polyethylene glycol) readily absorbs amines such as norepinephrine (Brooks and Clark, 1987).

Statistics

T-tests and least squares regression analyses were performed on the Cyber CDC mainframe computer located at the University of Massachusetts using the Statistical Package for the Social Sciences (SPSS).

RESULTS

Effect of deltamethrin on ^3H-NE release

Functionally intact synaptosomes can be artificially depolarized by increasing external concentration of K^+ ions which in the presence of Ca^{2+} leads to release of vesicle bound neurotransmitters (West and Fillenz, 1980; Blaustein, 1975; Maura et al., 1984; Raiteri et al., 1984). When viable synaptosomes are continuously perfused and pulse depolarized by K^+ in the presence of Ca^{2+}, they release neurotransmitter in a measureable and reproducible pattern (Brooks and Clark, 1987). Fig. 1 shows the fractional rate constants calculated for ^3H-NE release from synaptosomes treated with either 10^{-5}M deltamethrin or ethanol. Addition of K^+

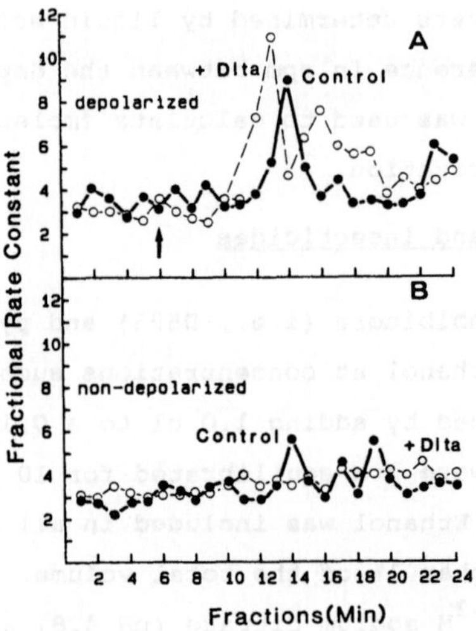

Fig. 1. Effect of deltamethrin on ^{3}H-NE efflux. Synaptosomal aliquots were incubated with either 10^{-5}M deltamethrin (Dlta, open circles) or 95% ethanol (control, solid circles) prior to loading with ^{3}H-NE. In panel A, synaptosomes were pulse depolarized by addition of DM for 3 min (at fraction 6 as indicated by arrow). In panel B, synaptosomes were treated with NSM for the entire assay. The results represent an average of 4 ethanol-treated control experiments and 4 deltamethrin-treated experiments with each type of experiment consisting of 4 non-depolarized replicates and 4 depolarized replicates (Reproduced with permission from Brooks and Clark, 1987. Copyright 1987 Academic Press).

depolarization media to deltamethrin treated synaptosomes produced a large initial spike of ^{3}H-NE release compared to ethanol treated synaptosomes (Fig. 1A). The initial enhanced spike following K^+ depolarization suggests that deltamethrin facilitates Ca^{2+} entry into the cytosol (Orrego and Sanchez-Armass, 1981). Furthermore, pulsed depolarization of deltamethrin-treated synaptosomes produced a prolonged tailing effect of ^{3}H-NE release (second or tail peak) which continued through fraction 20. Enhanced release of ^{3}H-NE was not evident in synaptosomes which did not receive K^+ depolarization regardless of the presence of deltamethrin or ethanol (Fig. 1B). Thus, enhanced ^{3}H-NE release due to the addition of deltamethrin was only evident during depolarization and in the combined presence of Ca^{2+}.

Effect of deltamethrin on ^{45}Ca uptake

In order to directly compare the effect of deltamethrin on ^{3}H-NE release and ^{45}Ca uptake during pulsed K^{+}-depolarization, deltamethrin-enhanced values of ^{3}H-NE release and ^{45}Ca uptake due to increasing the concentration of deltamethrin from 10^{-10}M to 10^{-7}M were subjected to a regression analysis and co-plotted (Fig. 2). The effective dose of deltamethrin which resulted in a 50% maximal response (i.e., ED_{50}) for both these phenomena was estimated from these regressed lines. In the case of enhanced ^{3}H-NE release, the ED_{50} for deltamethrin was calculated

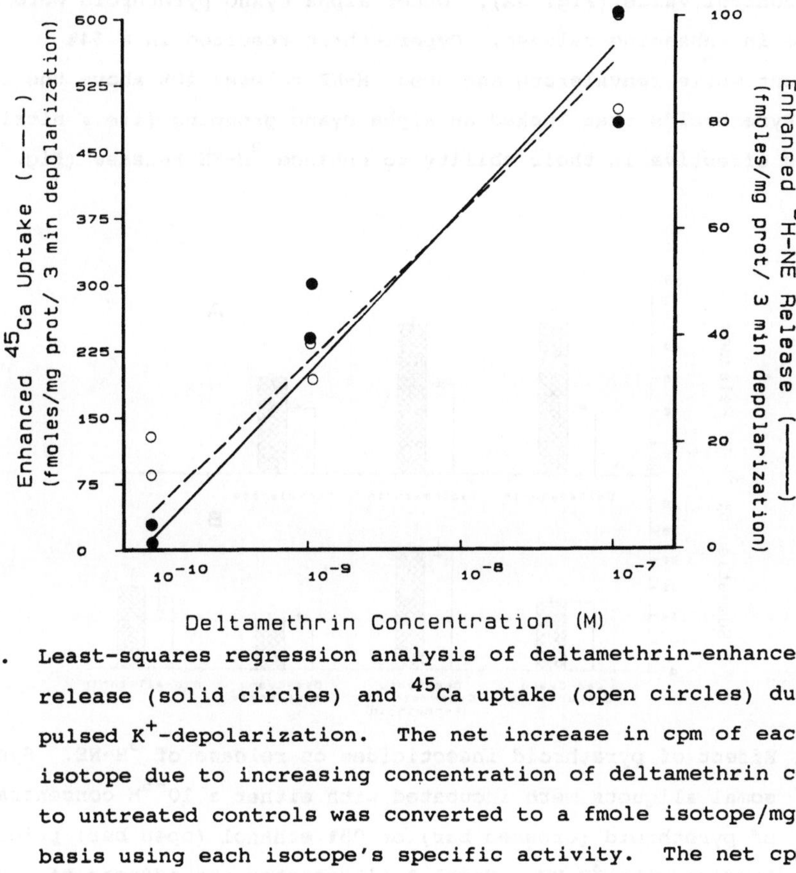

Fig. 2. Least-squares regression analysis of deltamethrin-enhanced ^{3}H-NE release (solid circles) and ^{45}Ca uptake (open circles) during pulsed K^{+}-depolarization. The net increase in cpm of each isotope due to increasing concentration of deltamethrin compared to untreated controls was converted to a fmole isotope/mg protein basis using each isotope's specific activity. The net cpms in fraction 12 to 20 were used to calculate ^{3}H-NE release and the net cpms between depolarized and non-depolarized filters were used to calculate ^{45}Ca uptake (see method section). The regression analysis for enhanced ^{3}H-NE release (—) due to the addition of deltamethrin gave an r^2 value of 0.82, and 0.98 for ^{45}Ca uptake (----). Data points represent the mean of 2 experiments for each isotope with each consisting of 4 non-depolarized and 4 depolarized replicates. (Reproduced with permission from Brooks and Clark, 1987. Copyright 1987 Academic Press).

to be 2.9 x 10^{-9}M compared to 2.4 x 10^{-9}M for enhanced ^{45}Ca uptake. The apparent stoichiometry between ^{45}Ca uptake and ^{3}H-NE release estimated from these lines is approximately 5.8 Ca^{2+} to 1 NE.

Comparison of the effect of pyrethroids on ^{3}H-NE release

As judged by the summation value of ^{3}H-NE release, deltamethrin (10^{-5}M) was the most effective pyrethroid in enhancing ^{3}H-NE release. Deltamethrin enhanced ^{3}H-NE release 114% compared to a concurrent non-treated control value (Fig. 3A). Other alpha cyano pyrethroid were also effective in enhancing release. Cypermethrin resulted in a 64% enhancement while fenvalerate enhanced ^{3}H-NE release 40% above the control value. Pyrethroids that lacked an alpha cyano grouping (i.e., nitrile) were less effective in their ability to enhance ^{3}H-NE release (Fig. 3B).

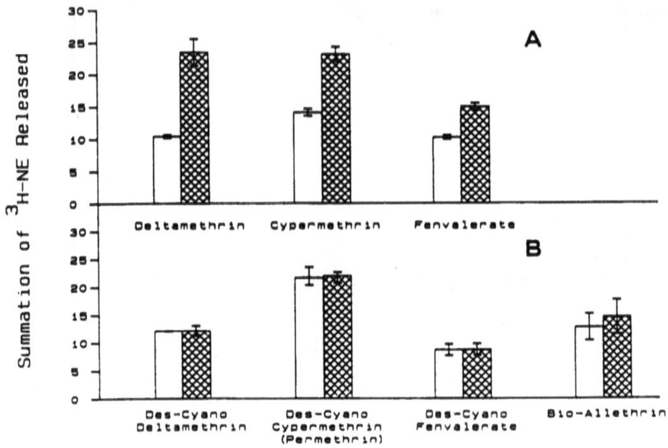

Fig. 3. Effect of pyrethroid insecticides on release of ^{3}H-NE. Synaptosomal aliquots were incubated with either a 10^{-5}M concentration of pyrethroid (crossed bar) or 95% ethanol (open bar) prior to loading with ^{3}H-NE. Panel A illustrates the effects of alpha cyano pyrethroids (Type II) while panel B illustrates the effects of pyrethroids lacking such a moiety (Type I). For each insecticide examined, summations expressed are an average of 2 control experiments and 2 insecticide-treated experiments with each type of experiment consisting of 4 non-depolarized replicates and 4 depolarized replicates. (Reproduced with permission from Brooks and Clark, 1987. Copyright 1987 Academic Press).

Addition of bioallethrin resulted in a 4.7% enhancement and des-cyano cypermethrin in a 1.2% enhancement but neither was significantly different ($p>0.05$) from their concurrent control values. Des-cyano deltamethrin and des-cyano fenvalerate were also without an effect under these experimental conditions. The ability of pyrethroids to enhance ^{3}H-NE release during pulsed K^+ depolarization is also stereospecific. As illustrated in Fig. 4, the toxic enantiomer of deltamethrin (1R, alpha S) at 10^{-7}M resulted in a 106% enhancement of neurotransmitter release above an ethanol control summation value (18.7 ± 5.4 versus 9.1 ± 1.5, respectively). These two

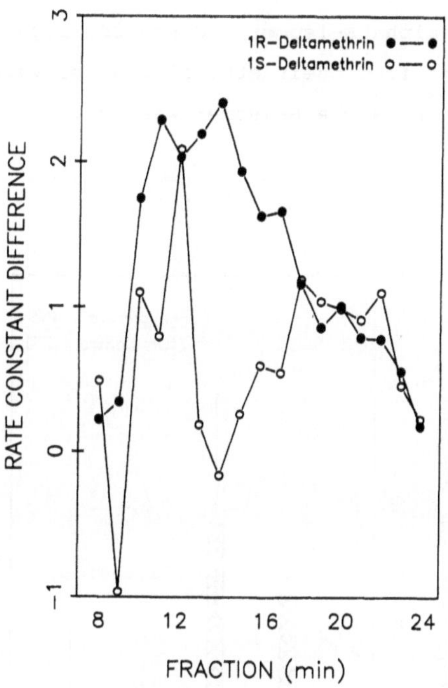

Fig. 4. Enhancement of Ca^{2+}-dependent K^+-stimulated ^{3}H-NE release by the deltamethrin enantiomers, 1R-deltamethrin (1R, alpha S) and 1S-deltamethrin (1S, alpha S). Synaptosomal aliquots were incubated with either 10^{-7}M 1R-deltamethrin (closed circles), 10^{-7}M 1S-deltamethrin (open circles) or 1.0 ul 95% ethanol (control) for 10 min prior to loading with ^{3}H-NE. Results are expressed as the differences of fractional rate constant averages (i.e., difference values calculated between depolarized and non-depolarized fractions treated with deltamethrin substracted from the difference values calculated between depolarized and non-depolarized control fractions). The results represent average of 3 separate experiments.

summation values are statistically different at the 95% level of significance ($p<0.05$, t-test). The non-toxic 1S enantiomer was less effective in enhancing ^{3}H-NE release (28.5% enhanced net release) but this summation value (11.9 ± 3.0) was not statistically different from the ethanol control value ($p>0.05$). Also, it should be noted that in the presence of 1S-deltamethrin, the following tail peak of ^{3}H-NE release is virtually absent. Stereoisomers of fenvalerate produced similar results (Fig. 5). As judged by the treated summation values versus ethanol control values, the most toxic isomer 2S, alpha S was the most stimulatory (28.5 ± 11.2 _vs_. 7.3 ± 3.4), followed by 2S, alpha R-isomer (14.7 ± 2.1 _vs_. 9.6 ± 2.8), and the non-toxic 2R, alpha R-isomer (13.5 ± 6.0 _vs_. 11.8 ± 2.2). Both the 2S, alpha S-isomer and the 2S, alpha R-isomer were statistically different from their ethanol control values ($p<0.05$, t-test) whereas the non-toxic 2R, alpha R-isomer was not.

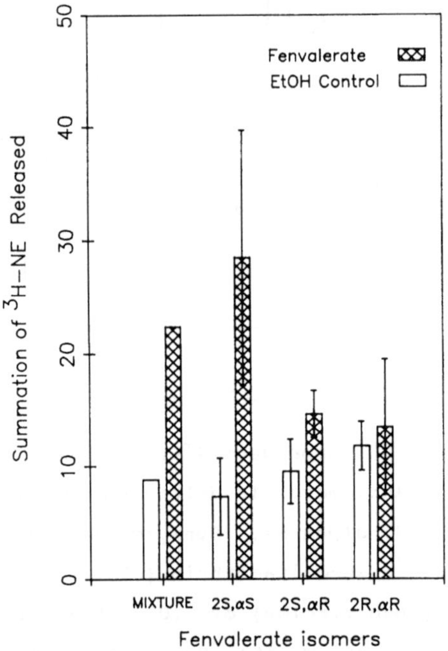

Fig. 5. Effect of stereoisomers of fenvalerate on release of ^{3}H-NE. Synaptosomal fractions were incubated with fenvalerate (racemic mixture; 2S, alpha S; 2S, alpha R or 2R, alpha R as indicated by cross bars) at 10^{-7}M or ethanol (open bars) 10 min prior to loading with ^{3}H-NE. The results represent an average of three experiments ± standard error of mean (SEM) for each stereoisomer.

These results are in accordance with the theory that voltage-gated ionic channels in the nerve are the principal target of pyrethroids (Lawrence and Casida, 1982). The most pronounced effect of alpha cyano pyrethroids on nerve cells under voltage-clamp conditions is the prolongation of the time constants calculated for the decay of sodium tail currents (Vijverberg and de Weille, 1985). All alpha cyano pyrethroids (i.e., Type II) result in greatly enhanced time constants reflecting extremely slow kinetics of the tail currents. Prolonged opening of voltage-gated ionic channels, both sodium and calcium, could enhance calcium influx, result in enhanced neurotransmitter release, and ultimately in the depletion of synaptic vesicles and neurotransmitters in presynaptic nerve terminals. Our data on enhancement of norepinephrine release by alpha cyano pyrethroids when converted to a relative ratio compare well to Vijverberg's data on time constant enhancement by alpha cyano pyrethroids and to the relative mammalian toxicity of these compounds (Table 2). It is not expected that non-alpha cyano pyrethroid (i.e., Type I) would be as effective in creating a prolonged depolarization as are the alpha cyano pyrethroids.

Table 2. Ability of alpha cyano pyrethroids (Type II) to produce toxicity, to enhance neurotransmitter release and to prolong sodium tail currents

Pyrethroid	LD_{50}[a] (mg/kg)	Toxicity Ratio[d]	^{3}H-NE Release[b]	Release Ratio[d]	Time constant(t)[c] (ms)	Time constant ratio
Deltamethrin	135	1.0	114%	1.0	1772	1.0
Cypermethrin	251	0.54	64%	0.56	1114	0.63
Fenvalerate	451	0.30	40%	0.35	603	0.34

[a]Acute oral LD_{50} in rat [53].

[b]Enhanced amount of Ca2+-dependent, K+-stimulated ^{3}H-NE release in the presence of 10^{-5}M pyrethroid compared to the release measured in non-treated controls (Brooks and Clark, 1987).

[c]Time constants reflect the decay rates of sodium tail currents upon repolarization after step depolarization under voltage clamp conditions (Vijverberg and de Weille, 1985).

[d]Toxicity and release ratios are relative to the effect of deltamethrin.

SOURCE: Reproduced with permission from ref. (Clark and Brooks, 1989)

Effects of ion channel antagonists on ^{3}H-NE release

The concentrations utilized for tetrodotoxin (TTX) and D595 in the following experiments produced no interfering action on ^{3}H-NE uptake (Brooks and Clark, 1987). The ability of TTX and D595 to block ^{3}H-NE release is detailed in Table 3. TTX at 10^{-7}M inhibited 48% of ^{3}H-NE release. This level of TTX inhibition is consistent with previous findings (Blaustein, 1975). At 10^{-5}M, D595 inhibited 27% of ^{3}H-NE release versus control values. Addition of both TTX and D595 inhibited 74% of ^{3}H-NE release compared to controls. Thus, these agents apparently block sodium and calcium channels in an additive fashion independent of each other.

Table 3. Effect of deltamethrin on Ca^{2+}-dependent, K^+-stimulated ^{3}H-NE release from rat brain synaptosomes treated with D595 or TTX

	^{3}H-NE Release		
Treatment	Control[a]	Treated	Percent inhibited[b]
	(Summation of 3H-NE Release ± SEM)[c]		
D595[d]	16.7 ± 1.4	13.2 ± 0.6*	21 ± 4
D595 + Deltamethrin	16.7 ± 1.4	12.6 ± 1.4*	24 ± 8
TTX[e]	12.1 ± 2.5	6.3 ± 0.3*	48 ± 3
TTX + Deltamethrin	12.1 ± 2.5	12.7 ± 3.0	0[f]
D595 ± TTX	19.8 ± 1.1	5.1 ± 1.4*	74 ± 7

[a]Controls were treated with 1.0 ul ethanol.

[b]Percent inhibited refers to the difference between the 3H-NE released by treated synaptosomes compared to control synaptosomes.

[c]Values represent means of 2 separate experiments for each treatment. Means followed by an asterisk are significantly different from control means ($p<0.05$, one-tailed t-test).

[d]D595 concentration was 10^{-5}M, deltamethrin concentration was 10^{-7}M.

[e]TTX concentration was 10^{-7}M, deltamethrin concentration was 10^{-7}M.

[f]No significant difference between control and treated ($p<0.05$, one-tailed t-test).

SOURCE: Reproduced with permission from ref. (Clark and Brooks, 1989)

As illustrated in Fig. 6, deltamethrin failed to significantly enhance release of ^{3}H-NE from D595-pretreated synaptosomes (i.e., from 13.2 ± 0.6 to 12.6 ± 1.4, Table 3) but did significantly increase release of ^{3}H-NE from TTX-pretreated synaptosomes (i.e., from 6.3 ± 0.3 to 12.7 ± 3.0, Table 3). Since TTX is a specific sodium channel blocker, deltamethrin is apparently enhancing the uptake of Ca^{2+} via other channels or ion-selective transporters in addition to TTX-sensitive sodium channels. As in the absence of TTX, the addition of deltamethrin in the presence of of TTX (10^{-7}M, Clark and Brooks, 1989) enhanced both the initial spike at fraction 12 as well as producing substantial tailing from fractions 16-20.

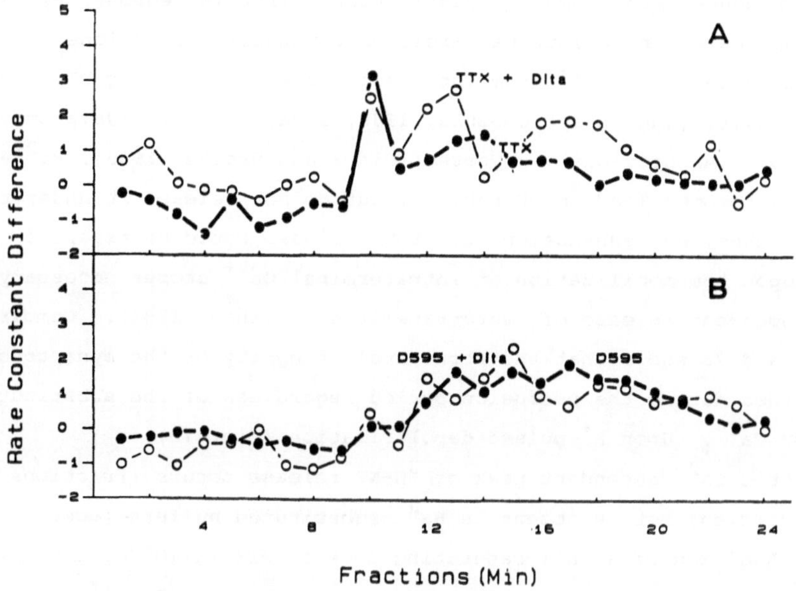

Fig. 6. Effect of deltamethrin on ^{3}H-NE release from synaptosomes pre treated with TTX and D595. In panel A, synaptosomal aliquots were incubated with 10^{-7}M TTX for 5 min. In panel B, synaptosomal aliquots were incubated with 10^{-5}M D595 for 5 min. At the conclusion of this period, synaptosomal aliquots were incubated an additional 5 min with either 10^{-7}M deltamethrin (open circles) or 95% ethanol (closed circles). All aliquots were then incubated with ^{3}H-NE for 15 min. Results for both panels represent the average of 2 control experiments and 2 deltamethrin-treated experiments with each type of experiment consisting of 4 non-depolarized replicates and 4 depolarized replicates.

Effect of deltamethrin on PTZ-induced ^3H-NE

The convulsive drug PTZ has been used extensively in elucidating the mechanism of epileptogenesis and recently has been shown to be involved in the phosphorylation of synapsin I associated with Ca^{2+} influx into synaptosomes (Onozuka, 1987). In experiments which examined the pharmacological effects of deltamethrin on rats, this pyrethroid was found to prolong the convulsive seizures induced by PTZ and potentiate PTZ toxicity (Chanh et al., 1984). In order to determine whether deltamethrin may have an effect on intraterminal Ca^{2+} homeostasis, the convulsive drug PTZ was used as a depolarization agent (Matsumoto and Ajmone, 1964; Matsumoto, 1969; Doerner et al., 1984). Besides its well established ability to produce paraoxysmal depolarization shifts in neurons, PTZ also has been implicated in various Ca^{2+}-related cytoplasmic reactions resulting in release of Ca^{2+} from intracellular stores causing PTZ-induced bursting activity (Sugaya and Onozuka, 1978; Sugaya et al., 1987; Onozuka et al., 1986). As previously reported (Clark and Brooks, 1988), Ba^{2+}-treated synaptosomes load ^3H-NE normally but do not release it under these protocols. Thus, any enhancement of ^3H-NE release would be expected to be dependent upon the mobilization of intraterminal Ca^{2+} stores necessary for the Ca^{2+}-dependent release of neurotransmitter (Llinas, 1982). Comparison of Fig. 7A and 7B shows that the structural integrity of the synaptosomes are maintained during the perfusion period regardless of the substitution of Ba^{2+} for Ca^{2+}. Upon K^+-pulsed depolarization (Fig. 7B), a characteristic Ca^{2+}-dependent peak of ^3H-NE release occurs (fractions 12-20, solid circles) but is absent in Ba^{2+} substituted buffers (open circles). Addition of a half-saturating dose of PTZ ($5x10^{-5}$M, Clark and Brooks, 1989) to synaptosomes resuspended in BaNSM (Fig. 7C) induces a release of ^3H-NE (fractions 10-21, open squares). At 10^{-5}M, deltamethrin pre-treatment produced a 66% increase in ^3H-NE release (solid triangles) over PTZ-induced release (10.5 ± 0.7 summation of fractional average differences in the presence of deltamethrin versus 6.0 ± 0.9 for PTZ-treated only). The differences in summation values of enhanced ^3H-NE release above untreated PTZ-induced values, for various concentrations of deltamethrin (from $5x10^{-11}$ to 10^{-9}M), were then subjected to a least-squares regression analysis (Fig. 8). The regression gave an r^2 value of 0.94 with an estimated ED_{50} value calculated for deltamethrin of 2.4×10^{-10}M.

Although the exact mode of action for PTZ has yet to be determined at the cellular level, two mechanisms have been proposed which directly

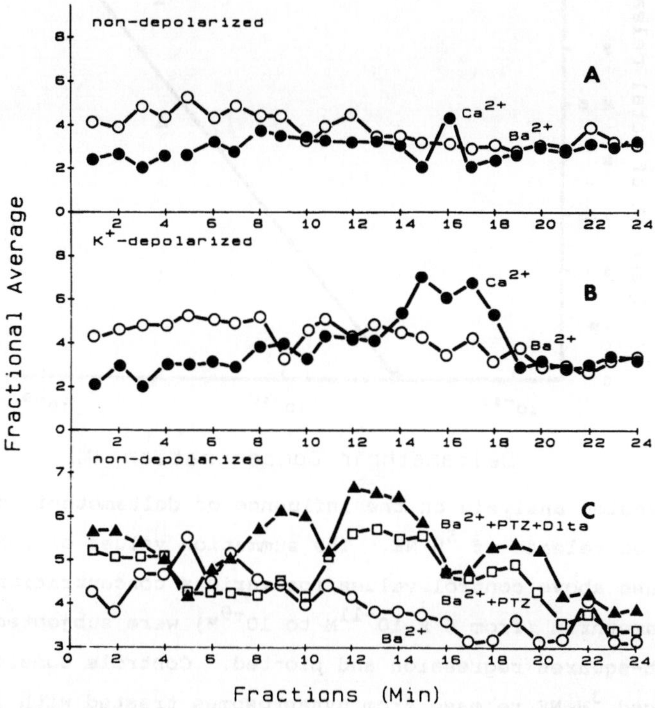

Fig. 7. Effect of Ba^{2+} and $Ca^{2}+$ on K^+-pulsed depolarized ^{3}H-NE release and the effect of deltamethrin on PTZ-induced ^{3}H-NE release: Panel A- synaptosomes loaded with ^{3}H-NE were effluxed with either NSM (Ca^{2+}, solid circles) or BaNSM (Ba^{2+}, open circles) for 24 min (24 fractions); Panel B-synaptosomes loaded with ^{3}H-NE were effluxed and pulsed depolarized for 3 min with either DM (Ca^{2+}, solid circles) or BaDM (Ba^{2+}, open circles) at min 6 (fraction 6); Panel C- Synaptosomes were pretreated (5 min) with either 10^{-5}M deltamethrin (solid triangles) or ethanol (open squares, open circles) before being loaded with ^{3}H-NE and effluxed with BaNSM. At min 6 (fraction 6), selected synaptosomes were pulsed for 3 min with $5x10^{-5}$M PTZ (open squares, solid triangles). Other synaptosomes received BaNSM for the entire 24 min collection (24 fractions, open circles).

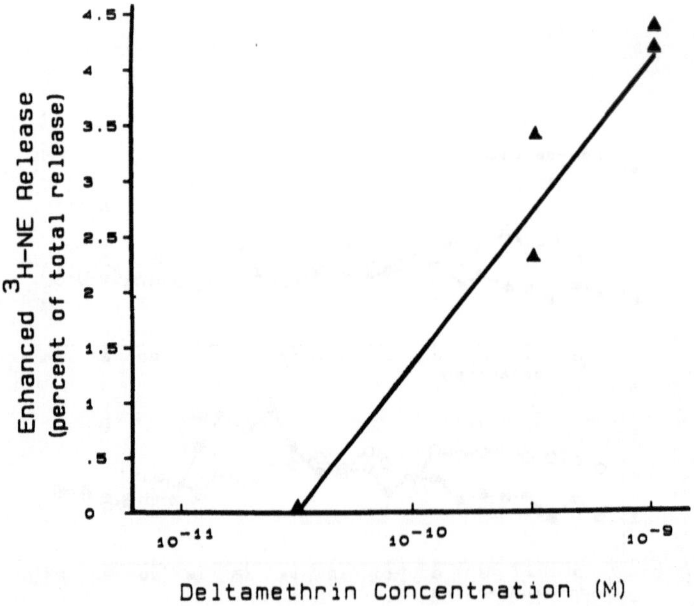

Fig. 8. Regression analysis on the influence of deltamethrin on PTZ-induced release of ^{3}H-NE. The summation values of enhanced ^{3}H-NE release above control values for various concentrations of deltamethrin (from 5 x 10^{-11}M to 10^{-9}M) were subjected to a least-squares regression and plotted. Controls consisted of PTZ-induced ^{3}H-NE release from synaptosomes treated with 1.0 ul of 95% ethanol. The regression gave an r^2 value of 0.94 with an ED_{50} value of 2.4 x 10^{-10}M. The ED_{50} value is calculated as the extrapolated value which would release a summation value of control plus 2.04%. The actual data points are represented on plot as triangles.

relate to the enhancement of monoaminergic neurotransmitter release by deltamethrin in PTZ-treated synaptosomes. The first possibility is that PTZ is involved in the release of intracellular Ca^{2+} from dense lysosome-like granules (Sugaya and Onozuka, 1978) or mitochondria (Doerner et al., 1982) during bursting activity. PTZ acts to deplete the granules of Ca^{2+} and alters their structure from dense type to lamella type. In our experiments with PTZ-induced neurotransmitter release, external Ca^{2+} was replaced by Ba^{2+} which serves as a mimic for Ca^{2+} as a charge carrier through voltage-gated calcium channels. Barium, however, does not support the phosphorylation of synapsin I (Hock and Wilson, 1984) and hence does not support Ca^{2+}-triggered neurotransmitter release from synaptosomes. Thus, any Ca^{2+}-triggered neurotransmitter release induced by PTZ is assumed to be due to the mobilization of intraterminal stores of Ca^{2+}

possibly by a mechanism similar to that described above. Enhancement of PTZ-induced neurotransmitter release by deltamethrin under these experimental restrictions may indicate an action of Type II pyrethroids at ion-selective transporters that regulate intraterminal free Ca^{2+} concentrations (e.g., Ca-pumps, Na/Ca exchange, etc.) (Clark and Matsumura, 1987). Failure to rapidly sequester or efflux Ca^{2+} liberated by the action of PTZ would result in elevated cytosolic Ca^{2+} concentrations and prolonged neurotransmitter release. A similar suppression of Ca^{2+} buffering by cypermethrin has been suggested by Seabrook et al., (1988). A second possibility is that PTZ acts at the voltage-gated sodium channel producing paraoxysmal depolarizing shifts (Matsumoto et al., 1969). This causes increased Na^{+} conductance and results in membrane depolarization which allows Ba^{2+} to influx into presynaptic nerve terminals via voltage-gated calcium channels. Both Na^{+} and Ba^{2+} are now available for exchange with intraterminal stores of Ca^{2+} (e.g., mitochondria) via Na/Ca and Ba/Ca antiporters (Blaustein and Santiago, 1977; Carafoli and Crompton, 1987; Gill et al., 1984; Clark et al., 1986). Increased cytosolic Ca^{2+} is then available to induce neurotransmitter release. In this case, enhanced neurotransmitter release caused by deltamethrin could be explained by an action at either the sodium or calcium channel or both.

DISCUSSION

Because the convulsive state is so evident in the symptomology of pyrethroid toxicity, involvement of monoaminergic systems within the CNS becomes important. Specifically, the depletion of catecholamines within the CNS causes a significant reduction in threshold levels that are necessary to induce both electroconvulsive shock (Rudzik and Johnson, 1970; Kilian and Frey, 1973) and pentylenetetrazol-induced (Maj and Vetulani, 1970) seizures. Reductions of these thresholds are apparently more highly correlated to reduction of norepinephrine than to dopamine (Quattrone et al., 1978). Recently, a variety of drugs that interfere either in synthesis or in release of norepinephrine were found to significantly potentiate pyrethroid toxicity (Staatz et al., 1982). Additionally, induction of a Type II syndrome by deltamethrin caused a substantial enhancement of plasma norepinephrine concentration in the blood of treated rats (Cremer and Seville, 1982). A similar enhancement was not apparent in cismethrin-treated rats. Interestingly, brain arterioles and capillaries in this region are innervated by adrenergic nerve fibers indicative of a central adrenergic control of cerebral blood flow (Nieuwenhuys et al., 1981). These results correlate well with those presented initially by Aldridge et al. (1978), who showed that acetylcholine was significantly depleted from rat brain after oral administration of deltamethrin. Cismethrin was greatly reduced in its ability to deplete this neurotransmitter while DDT resulted in no depletion whatsoever.

In related studies, it has also been reported that Type II alpha cyano pyrethroids were very potent in increasing the rate of miniature excitatory postsynaptic potentials in insect neuromuscular preparations. DDT and non-alpha cyano pyrethroids (Type I) were less effective in increasing this rate of potential firing (Salgado et al., 1983). Deltamethrin has also been shown to cause almost complete depletion of synaptic vesicles from presynaptic motor nerve terminals of the house fly. This indicates a presynaptic action of deltamethrin and interestingly, mitochondria of treated terminals were reported to be swollen and have vacuolated interiors (Schouest et al., 1986). It is well established that mitochondria function as a major calcium regulator within presynaptic nerve terminals (Blaustein et al., 1978).

As shown in this chapter, enhanced release of ^{3}H-NE from rat brain synaptosomes by Type II alpha cyano pyrethroids is evident only in the combined presence of external Ca^{2+} and membrane depolarization. This indicates that insecticides like deltamethrin act as use-dependent compounds (Lund and Narahashi, 1983; Ghiasuddin and Soderlund, 1985; Ruigt, 1985; Narahashi, 1985). Because release of ^{3}H-NE is highly dependent on Ca^{2+} entering the cytosol of the synaptosome, it can be assumed that deltamethrin may act on entities which transport Ca^{2+}, such as the voltage-gated sodium and calcium channels (Llinas, 1982; Mullins, 1981). In a similar study, ^{3}H-NE was released from rat brain cortex slices by electrical stimulation. This release was shown to be absolutely dependent on extracellular calcium and was partially blocked not only by TTX but also by the phenethylamine-type calcium channel antagonist, D600, and by manganese (Blaustein, 1975).

The role of both voltage-gated sodium and calcium channels in the stimulus-secretion coupling mechanism of neurotransmitter release is well established (Zucker and Lando, 1986). Voltage clamp experiments have shown that Ca^{2+} enters the neuron in two phases. The early phase is TTX-sensitive and follows the time course of fast Na^+ current via sodium channels. The latter phase of Ca^{2+} influx is not sensitive to either TTX or tetraethylammonium and, if it undergoes inactivation, it does so more slowly. It is this later phase of Ca^{2+} entry that is responsible for neurotransmitter release (Kuffler et al., 1984; Nestler and Greengard, 1983; Llinas, 1982).

The action of deltamethrin on the enhancement of monoaminergic neurotransmitter release correlated well with ^{45}Ca uptake into synaptosomes (Brooks and Clark, 1987) and the present work with TTX, D595 and PTZ indicates that other voltage-gated channels and ion-selective transporters, (e.g., calcium channel, Na/Ca exchange, etc.) may also be involved in this process in addition to the TTX-sensitive sodium channel. In doing so, deltamethrin and related Type II pyrethroids could produce the convulsive CS syndrome of poisoning by depleting specific brain regions of the CNS of monoamine neurotransmitters in a manner similar in many aspects to epileptic or PTZ-induced seizures. Virtually all of the noradrenergic pathways that have been studied at this time are efferent pathways of the locus ceruleus to cerebral cortex, cerebellum and hippocampus. The major effect of stimulating these pathways is an inhibition of spontaneous discharges (Moore and Bloom, 1979). Depletion of monoamines in the presynaptic processes of these neurons by Type II pyrethroids would then be convulsive in nature as spontaneous discharges increase. However recent studies on the action of released NE on hippocampal pyramidal cells indicate that binding of this neurotransmitter to $beta_1$ adrenergic receptors causes the slow afterhyperpolarization to be diminished (Nicoll, 1988). This apparently is due to a cAMP-dependent protein phosphorylation reaction at the Ca^{2+}-dependent K^+ channel which is responsible for the production of the slow afterhyperpolarization. Because this current plays a major role in controlling pyramidal cell excitability (Nicoll, 1988), if it is blocked or suppressed, many more action potentials will be evoked by the same depolarizing current pulse. Enhanced norepinephrine release induced by the action of deltamethrin in this region of the brain could then be convulsive by this mechanism in that pyramidal cells have been implicated as playing a key role in epileptogenesis. It is possible that the Ca^{2+} agonism elicited by deltamethrin as evidenced by enhanced ^{45}Ca uptake and ^3H-NE release is due to its shared structural similarity with D595 (Fig. 9). There are a number of instances where slight structural alterations reverse the action of a compound from antagonism to agonism. A case in point is the dihydropyridine calcium channel antagonist, nifedipine, and its structurally-related dihydropyridine agonist, CGP 28392. Whereas CGP 28392 has been shown to increase cytosolic Ca^{2+} levels and enhance cardiac contractability (positive inotropic effect), nifedipine is completely antagonistic of these actions (negative inotropic effect) (Erne et al., 1984). Apparently the conversion of the antagonistic action of nifedipine to the agonistic action of CGP 28392 is due to an increased electronegativity caused by the addition of fluorine atoms to the benzene on one end of the molecule and to increased bulkiness by the addition of a furan ring to the dihydropyridine on the other end.

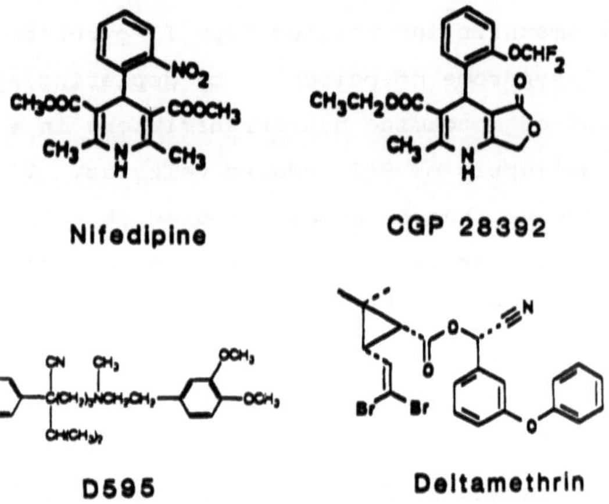

Fig. 9. Structural configurations of the calcium channel antagonists; nifedipine and D595, the Ca^{2+} channel agonist, CGP 28392, and the pyrethroid insecticide, deltamethrin.

Addition of structurally-dissimilar calcium channel antagonists, verapamil and diltiazem, did not alter the effect of CGP 28392 at the calcium antagonist receptor of the voltage-gated calcium channel indicating separate sites of action for these different classes of calcium channel antagonists (Table 4). Also unlike nifedipine, verapamil and structural analogs such as D600 and D595 are use-dependent compounds which exert an effect only after frequency-dependent depolarization (Lee and Tsien, 1983). In this study, deltamethrin acts as a use-dependent compound in the enhancement of ^{3}H-NE release and ^{45}Ca uptake which is in full agreement with other studies (Ghiasuddin and Soderlund, 1985; Ruigt, 1985) and has been shown to cause a positive inotropic effect in atrial muscle (Ruigt, 1985; Mullins, 1981). Although structurally similar to D595, deltamethrin may act as a Ca^{2+} channel agonist due to the inclusion of a strongly electronegative dibromovinyl moiety and increased bulkiness afforded by its alpha cyano 3-phenoxybenzyl alcohol component in a similar fashion as discussed for CGP 28392 above. Indeed, the overall topology of deltamethrin mimics that of D595 to a large degree and only differs markedly in the type of functional groups presented (e.g., a phenethyl-tertiary amine in D595 vs. a phenethyl-nitrile for deltamethrin). It should be noted also that the negative inotropic effect of verapamil, D595 and the other phenethylamine antagonists is stereospecific with the S(-)-enantiomer being active (Bayer et al., 1975). Although the 1R configuration is required for toxicity in cyclopropane containing-pyrethroids such as in deltamethrin (the 1S configuration is non-toxic), the S-configuration at the acidic asymmetric carbon is the toxic form for non-cyclopropane pyrethroids such as fenvalerate. This allows the two methyl groups of the isopropyl moiety to assume the same configuration presented by the gem-methyl groups attached to the C2 carbon of the cyclopropane ring in toxic 1R pyrethroids.

Table 4. Comparison of calcium channel antagonists and agonists[a] and Type II pyrethroids

Structure	Use-dependent action	Stereo-selectivity	Inotropic action	Topology	Nervous tissue selectivity	Channel subtype selectivity[b]
<u>Dihydropyridines</u>						
nifedipine	-	-	neg	-	-	L
Bay K 8644/ CGP28392	-	S(-)	pos	-	-	L
<u>Phenethylamines</u>						
Verapamil/ D595	+	S(-)	neg	+	+	L,N
<u>Type II pyrethroids</u>						
deltamethrin	+	1R	pos	+	+	(N?)
fenvalerate	+	2S	n.a.[c]	+	+	(N?)

[a]Data summarized from two recent reviews (Godfraind et al., 1986, Triggle and Janis, 1987).

[b]Channels from nerve tissues only (Boll and Lux, 1985).

[c]Not available.

The fact that Type II pyrethroids may interact with both the sodium and calcium voltage-gated channels should not be that contradictory to the existing sodium channel theory in that there appears to be a great deal of structural homology between them. As Curtis and Catterall (1986) point out, both the isolated sodium channel ionophore (Tamkun et al., 1984) and the dihydropyridine calcium antagonist receptor complex of the voltage-gated calcium channel are large membrane glycoproteins of 200-300 kd which consist of one large subunit and two smaller subunits (Catterall, 1984). Furthermore, Tanabe et al. (1987) have reported close structural and primary amino acid sequence similarities of the calcium antagonist receptor of the voltage-gated calcium channel to the voltage-gated sodium channel in support of this contention. Such overall structural similarities indicate similar requirements for rapid movement of ions across membranes via voltage-gated ionophores and may indicate similar binding regions for Type II pyrethroids. This apparent commonality in action has already been extended to the phenethylamine class of calcium channel antagonists which also have an inhibitory action at other voltage-gated channels including the sodium channel (Triggle, 1982), and Ca^{2+} transport in mitochondria (Buss et al., 1988), although at much higher dosages.

ACKNOWLEDGEMENT

This work was supported by research grant RR07048-19, NIH-BRSG and by the Massachusetts Agricultural Experimental Station, UMASS, Amherst, MA.

REFERENCES

Aldridge, W. N., Clothier, B., Forshaw, P., Johnson, M. K., Parker, V. H., Price, R. J., Skilleter, D. N., Verscholye, R. D., and Stevens, C., 1978, The effect of DDT and the pyrethroids cismethrin deltamethrin on the acetylcholine and cyclic nucleotide content of the rat brain, Biochem. Pharmacol., 27:1703.

Bayer, R., Kaufman, R., and Mannhold, R., 1975, Inotropic and electrophysiological actions of verapamil and D600 in mammalian myocardium. II. Pattern of inotropic effects of the optical isomers, Naunyn-Schmiedeberg's Arch. & Pharmacol., 290:69.

Berlin, J. R., Akera, T., Brady, T. M., and Matsumura, F., 1984, The inotropic effects of a synthetic pyrethroid decamethrin on isolated guinea pig atrial muscle, Eur. J. Pharmacol., 98:313.

Blaustein, M. P., 1975, Effects of potassium, veratridine, and scorpion venom on calcium accumulation and transmitter release by nerve terminals in vitro, J. Physiol., 247:617.

Blaustein, M. P., and Santiago, E. M., 1977, Effects of internal and external cations and ATP on sodium-calcium and calcium-calcium exchange in squid axons, Biophys. J., 20:79.

Blaustein, M. P., Ratzlaff, R. W., and Kendrick, N. K., 1978, The regulation of intracellular calcium in presynaptic nerve terminals, Ann. N.Y. Acad. Sci., 307:195.

Boll, W., and Lux, H. D., 1985, Action of organic antagonists on neuronal calcium currents, Neurosci. Letters, 56:335.

Brooks, M. W., and Clark, J. M., 1987, Enhancement of norepinephrine release from rat brain synaptosomes by alpha cyano pyrethroids, Pestic. Biochem. Physiol., 28:127.

Buss, W. C., Savage, D. D., Stepanek, J., Little, S. A., and McGuffee, L. J., 1988, Effect of calcium channel antagonists on calcium uptake and release by isolated rat cardiac mitochondria, Eur. J. Pharmacol., 152:247.

Carafoli, E., and Crompton, M., 1978, The regulation of intracellular calcium by mitochondria, Ann. N.Y. Acad. Sci., 307:269.

Catterall, W. A., 1984, The molecular basis of neuronal excitability, Science, 223:653.

Chanh, P. H., Navarro-Delmasure, C., Chamh, A. P. H., Cheav, S. L., Ziadee, F., and Samaha, F., 1984, Pharmacological effects of deltamethrin on the central nervous system, Arzneim-Forsch/Drug Res., 34:175.

Clark, J. M., and Matsumura, F., 1987, The action of two classes of pyrethroids on the inhibition of brain Na/Ca and Ca + Mg ATP hydrolyzing activities of the American cockroach, Comp. Biochem. Physiol., 86C:135.

Clark, J. M., Jones, E. L., and Matsumura, F., 1986, Is Na-Ca stimulated ATP hydrolysis found in squid retinal nerve identical with the ATP promoted aspect of Na/Ca exchange, Biochem. Biophys. Acta, 860:662.

Clark, J. M., and Brooks, M. W., 1989, Neurotoxicology of pyrethroids: single or multiple mechanisms of action?, Environ. Toxicol. Chem., 8:361.

Clark, J. M., and Matsumura, F., 1986, "Membrane Receptors and Enzymes as Targets of Insecticidal Action," Plenum Press, New York.

Cremer, J. E., and Seville, M. P., 1982, Comparative effects of two pyrethroids, deltamethrin and cismethrin, on plasma catecholamines and on blood glucose and lactate, Toxicol. Appl. Pharmacol., 66:124.

Cremer, J. E., Cunningham, V. J., and Seville, M. P., 1983, Relationships between extraction and metabolism of glucose, blood flow, and tissue blood volume in regions of rat brain, J. Cereb. Blood Flow Metabol., 3:291.

Cremer, J. E., Cunningham, V. J., Ray, D. E., and Sarna, G. S., 1980, Regional changes in brain glucose utilization in rats given a pyrethroid insecticide, Brain Res., 194:278.

Curtis, B. M., and Catterall, W. M., 1986, Reconstitution of the voltage-sensitive calcium channel purified from skeletal muscle transverse tubules, Biochem., 25:3077.

Doerner, D., Pacheco, M. F., Fowler, J. C., and Partridge, L. D., 1982, The role of calcium in pentylenetetrazol-induced bursting, Comp. Biochem. Physiol., 73C:9.

Doerner, D., Pacheco, R. M., Partridge, L. D., and Pacheco, M. F., 1984, The role of calcium in pentylenetetrazol-induced bursting, Comp. Biochem. Physiol., 79C:441.

Erne, P., Burgissen, E., Buhle, F. R., Dubach, B., Kuhnis, H., Meier, M., and Rogg, H., 1984, Enhancement of calcium influx in human platelets by CGP28392, a novel dihydropyridine, Biochem. Biophys. Res. Comm., 118:842.

Gammon, D. W., and Sander, G., 1985, Two mechanisms of pyrethroid action: electrophysiological and pharmacological evidence, Neurotoxicol., 6(2):63.

Ghiasuddin, S. M., and Soderlund, D. M., 1985, Pyrethroid insecticides: potent, stereospecific enhancers of mouse brain sodium channel activation, Pestic. Biochem. Physiol., 24:200.

Gill, D. L., Chueh, Seau-Huei, and Whitlow, C. L., 1984, Functional importance of the synaptic plasma membrane calcium pump and sodium-calcium exchanger, J. Biol. Chem. 259:10807.

Godfraind, T., Miller, R., and Wibo, M., 1986, Calcium antagonism and calcium entry blockade, Pharmacol. Rev. 38(4):321.

Gray, A. J., 1985, Pyrethroid structure-toxicity relationship in mammals, Neurotoxicol., 6(2):127.

Gray, A. J., and Richard, J., 1982, Toxicity of pyrethroids to rats after direct injection into the central nervous system, Neurotoxicol., 3:25.

Hajos, F., 1975, An improved method for the separation of synaptosomal fractions in high purity, Brain Res., 93:485.

Helmuth, D. W., Ghiasuddin, S. M., and Soderlund, D. M., 1983, Poly(ethylene glycol) preteatment reduces pyrethroid adsorption to glass surfaces, J. Agric. Food Chem., 31:1127.

Hock, D. B, and Wilson, J. E., 1984, Effects of calcium, strontium, and barium ions on phosphorylation of hippocampal proteins in vitro, J. Neurochem., 42:54.

Kilian, M., and Frey, H. H., 1973, Central monoamines and convulsive thresholds in mice and rats, Neuropharmacol., 12:681.

Kuffler, S. W., Nicholls, J. G., and Robert, A. R., 1984, "From Neuron to Brain," Sinauer Assoc. Inc., Sunderland, MA.

Lawrence, L. J., and Casida, J. E., 1982, Pyrethroid toxicology: mouse intracerebal structure-toxicity relationships, Pestic. Biochem. Physiol., 18:9.

Leake, L. D., Buckley, D. S., Ford, M. G., and Salt, D. W., 1985, Comparative effects of pyrethroids on neurones of target and non-target organisms, Neurotoxicol. 6(2):99.

Leake, L. D., Dean, J. A., and Ford, M. G., 1986, Pyrethroid action and cellular activity in invertebrate activity, in: "Neuropharmacology and Pesticide Action," M.G. Ford, G.G. Lunt, R.C. Reay and P.N.R. Usherwood, eds., Elis Horwood Ltd., Chichester, England.

Lee, K. S., and Tsien, R. W., 1983, Mechanism of calcium channel blockade by verapamil, D600, diltiazem and nitrendipine in single dialysed heart cells, Nature, 302:790.

Levi, G., and Raiteri, M., 1975, Synaptosomal transport processes, Int. Rev. Neurobiol. 19:51.

Levi, G., Gallo, V., and Raiteri, M., 1980, A reevaluation of veratridine as a tool for studying the depolarization-induced release of neurotransmitters from nerve endings, Neurochem. Res., 5:281.

Llinas, R. R., 1982, Calcium in synaptic transmitter release, Sci. Amer., 287:56.

Lund, A. E., and Narahashi, T., 1983, Kinetics of sodium channel modification as the basis for the variation in the nerve membrane effects of pyrethroids and DDT analogs, Pestic. Biochem. Physiol., 20:203.

Maj, I., and Vetulani, J., 1970, Some pharmacological properties of N,N-disubstituted dithiocarbamates and their effect on the brain catecholamine levels, Eur. J. Pharmacol., 9:183.

Matsumoto, H., and Ajmone, M. C., 1964, Cortical cellular phenomena in experimental epilepsy: interictal manifestations, Exp. Neurol., 9:286.

Matsumoto, H., Ayala, G. E., and Gumnit, R. J., 1969, Neuronal behavior and triggering mechanisms in cortical epileptic foci, J. Neurophysiol., 32:688.

Maura, G., Pittaluga, A., Ricchetti, A., and Raiteri, M., 1984, Noradrenaline uptake inhibitors do not reduce the presynaptic action of condine on ^{3}H-noradrenaline release in superfused synaptosomes, Naunyn-Schmied Arch. Pharmacol., 327:86.

Moore, R. Y., and Bloom, F. E., 1979, Central catecholamine neuron systems: anatomy and physiology of the norepinephrine and epinephrine systems, Amer. Rev. Neurosci., 2:113.

Mullins, L. J., 1981, Calcium entry upon depolarization of nerve, J. Physiol. (Paris), 77:1139.

Narahashi, T., 1985, Nerve membranes as ionic channels as the primary target of pyrethroids, Neurotoxicol., 6(2):3.

Narahashi, T., 1986, Mechanisms of action of pyrethroids on the sodium and calcium channel gating, in: "Neuropharmacology and Pesticide Action," M.G. Ford, G.G. Lunt, R.C. Reay and P.N.R. Usherwood, eds., Ellis Horwood Ltd., Chichester, England.

Narahashi, T., and Anderson, N. C., 1967, Mechanism of excitation block by the insecticide allethrin applied externally and internally to squid giant axons, Toxicol. Appl. Pharmacol., 10:529.

Nestler, E. J., and Greengard, P., 1983, Protein phosphorylation in the brain, Nature, 305:583.

Nicoll, R. A., 1988, The coupling of the neurotransmitter receptors to ion channels in the brain, Science, 241:545.

Nieuwenhuys, R., Voogd, J., and Van Juijzen, C., 1981, "Human Central Nervous System," Springer-Verlag, N.Y.

Onozuka, M., Imai, S., and Ozono, S., 1987, Involvement of pentylenetetrazole in Synapsin I phosphorylation associated with the calcium influx in synaptosomes from rat cerebral cortex, Biochem. Pharmacol., 36:1407.

Onozuka, M., Imai, S., and Sugaya, E., 1986, Pentylenetetrazol-induced bursting activity and cellular protein phosphorylation in snail neurons, Brain Res., 362:33.

Orchard, I., and Osborne, M. P., 1979, The action of insecticides on neurosecretory neurons in the stick insect, **Carausius morosus**, Pestic. Biochem. Physiol., 10:197.

Orrego, F., and Miranda, R., 1976, Electrically induced release of ^{3}H-GABA from neocortical thin slices, J. Neurochem., 26:1033.

Orrego, F., and Sanchez-Armass, S., 1981, Electrically induced release of ^{3}H-NE from rat brain cortex slices: A kinetic analysis of the dependence of extracellular calcium, Pharmacol. Res. Comm., 13:949.

Orrego, F., Jankelevich, J., Ceruti, C., and Ferrera, E., 1974, Differential effects of electrical stimulation on release of ^{3}H-noradrenaline and ^{14}C-aminoisobutyrate from brain slices, Nature, 251:55.

Quattrone, A., Crunelli, V., and Samanin, R., 1978, Seizure susceptibility and anticonvulsant activity of carbamazepine, diphenyldantoin and phenobarbitol in rats with selective depletions of brain monoamines, Neuropharmacol., 17:643.

Raiteri, M., Angelini, F., and Levi, G., 1984, A simple apparatus for studying the release of neurotransmitters from synaptosomes, Eur. J. Pharmacol., 25:411.

Ray, D. E., and Cremer, J. E., 1979, The action of decamethrin (a synthetic pyrethroid) on the rat, Pestic. Biochem. Physiol., 10:330.

Rudzik, A. D., and Johnson, G. A., 1970, Effect of amphetamine and amphetamine analogs on convulsive thresholds, in: "International Symposium on Amphetamines and Related Compounds," E. Costa and S. Garattini, eds., Raven Press, New York.

Ruigt, G. S. F., 1985, Pyrethroids, in: "Physiology, Biochemistry and Pharmacology," G. Kerkut and L. Gilbert, eds., Pergamon Press, N.Y.

Salgado, V. L., Irving, S. N., and Miller, T. A., 1983, The importance of nerve terminal depolarization in pyrethroid poisoning of insects, Pestic. Biochem. Physiol., 20:169.

Schouest Jr., L. P., Salgado, B. L., and Miller, T. A., 1986, Synaptic vesicles are depleted from motor nerve terminals of deltamethrin-treated house fly larvae, **Musca domestica**, Pestic. Biochem. Physiol., 25:381.

Seabrook, G. R., Duce, I.R. and Irving, S. N., 1988, Quantal release and pyrethroid insecticide action on the larval housefly (Musca domestica) neuromuscular junction, Pestic. Sci., 23:293.

Staatz, C. G., Bloom, A. S., and Lech, J. J., 1982, A pharmacologic study of pyrethroid neurotoxicity in mice, Pestic. Biochem. Physiol., 17:287.

Sugaya, E., and Onozuka, M., 1978, Intracellular calcium: its movement during pentylenetetrazol-induced bursting activity in snail neurons, Science, 202:1195.

Sugaya, E., Furuichi, H., Takagi, T., Kajiwara, K., and Komatsubara, J., 1987, Intracellular calcium concentration during pentylenetetrazol-induced bursting activity, Brain Res., 416:183.

Tamkun, M. M., Talvenheimo, J. A., and Catterall, W. A., 1984, The sodium channel from rat brain, J. Biol. Chem., 259:1676.

Tanabe, T., Takeshima, H., Mikami, A., Flockevzi, V., Takahashi, H., Kangawa, K., Kojima, M., Matsuo, H., Hirose, T., and Numa, S., 1987, Primary structure of the receptor for the calcium channel blocker from skeletal muscle, Nature, 328:313

Triggle, D. J., 1982, Chemical pharmacology of the calcium antagonist, in: "Calcium Regulation by Calcium Antagonists," R.G. Rahwan and D.T. Witiak, eds., ACS, Wash. D.C.

Triggle, D. J., and Janis, R. A., 1987, Calcium channel ligands, Ann. Rev. Pharmacol. Toxicol., 27:347.

Verschoyle, R. D., and Aldridge, W. N., 1980, Structure-activity relationships of some of the pyrethroids in rats, Arch. Toxicol., 45:325.

Verschoyle, R. D., and Barnes, J. M., 1972, Toxicity of natural and synthetic pyrethrins to rats, Pestic. Biochem. Physiol., 2:308.

Vijverbeg, H. P. M., and de Weille, J. R., 1985, The interaction of pyrethroids with the voltage-dependent Na channels, Neurotoxicol., 6(2):23.

West, D. P., and Fillenz, M., 1980, Storage and release of noradrenaline in hypothalamic synaptosomes, J. Neurochem., 35:1323.

Whittaker, V. P., Michaelson, I. A., and Kirkland, R. J., 1964, The poseparation of synaptic vesicles from nerve ending particles (Synaptosomes), Biochem. J., 90:293.

Zucker, R. S., and Lando, L., 1986, Mechanisms of transmitter release: voltage hypothesis and calcium hypothesis, Science, 231:574.

MECHANISM OF ACTION OF THE CYTOLYTIC TOXIN OF *Bacillus thuringiensis israelensis*

Sarjeet S. Gill, Edward Chow, Gur Jai Pal Singh, Patricia Pietrantonio, Shu-Mai Dai, Liu Shi and Leena S. Hiremath

Department of Entomology, University of California, Riverside, CA 92521

ABSTRACT

The cytolytic toxin of *Bacillus thuringiensis israelensis* and *B. t. morrisoni* are inactivated by aqueous suspensions of dioleolyl phosphatidylcholine, indicating the involvement of phospholipids in the cytotoxic action of these toxins. Using phospholipases as probes it was shown that phospholipase D does not alter erythrocyte susceptibility to lysis by the toxin, while phospholipase A_2-treated erythrocytes were less susceptible to lysis by the toxin, and these erythrocytes also bound less ^{125}I-25kDa toxin. However, phospholipase A_2-treated erythrocytes are more susceptible when treated with Triton X-100. Therefore the cytolytic toxin apparently does not act as a non-specific detergent, but rather interacts with phospholipid receptors on the cell membrane.

The 25kDa toxin of *B. t. israelensis* was also found to bind irreversibly to three cell types, human erythrocytes, and *Aedes albopictus* and *Choristoneura fumiferana* (CF1) cells. The binding of labeled toxin to cells increased both in the presence of unlabeled toxin or when the cells were first pre-incubated with unlabeled toxin. Binding kinetics data indicate the formation of toxin aggregates on the cell membrane. The toxin aggregates could be extracted with a solution of 10% Triton X-100 and separated from the monomers by means of a sucrose density gradient. Aggregates isolated from *Aedes* and CF1 cell membranes were estimated to be *ca.* 400kDa while those recovered from human red blood cells were significantly smaller. The proportion of the toxin found in aggregate form increased rapidly with the amount of toxin bound. However, the size of the aggregates remained constant.

Monoclonal antibodies raised against the denatured form of the toxin had little or no effect on the binding or aggregation of the toxin on cell membranes. However, monoclonal antibodies raised against the native form of the toxin blocked the toxin binding to cells. The epitope of one of these monoclonal antibodies was mapped to the C-terminal domain of the toxin suggesting the significance of this domain in cytolytic activity of the toxin. Binding and lytic activities were also abolished by treating the toxin with $HgCl_2$, suggesting the importance of the cysteine containing domain in toxic action.

INTRODUCTION

The biological insecticides, *Bacillus thuringiensis* and *Bacillus sphaericus*, are increasingly being used for the control of insect pests. Most subspecies of *B. thuringiensis* are effective against lepidopteran pests. However, a few subspecies of this bacterium such as *israelensis*, *morrisoni* (PG14), darmstadiensis (isolate 73-E-12-2) and *kyushuiensis*, and *B. sphaericus* are effective control agents of mosquitoes and blackflies, important vectors of human diseases (Goldberg and Margalit, 1977; deBarjac, 1978; Burges, 1982).

Both *B. thuringiensis* and *B. sphaericus* produce an intracellular parasporal protein inclusion body. The inclusion body from *B. t. israelensis* upon ingestion by susceptible insects is solubilized in the alkaline and proteolytic conditions present in the midgut giving rise to a number of toxins. The primary site of these solubilized toxins is the insect midgut epithelial cell which upon reaction with the toxin swell and lyse (Charles and deBarjac, 1983; Singh et al., 1986). Besides the insect midgut epithelial cells, a variety of other cell types are also lysed by these solubilized toxins (Thomas and Ellar, 1983a; Gill and Hornung, 1987). Some of these toxins have also been reported to disrupt functioning of the insect muscular and nervous system (Singh and Gill, 1985; Singh et al., 1986).

However, characterization of the mechanism of insecticidal action of *B. t. israelensis* toxins has been slow. This slow progress is due, in part, to the confusion until recently as to which proteins are actually involved in toxicity. A number of protein toxins are present in parasporal inclusions of *B. t. israelensis*. Of these toxins the 25kDa protein, which is a proteolytically cleaved product of the 28kDa protein present in the inclusion, is thought to be primarily responsible for the hemolytic action of *B. t. israelensis* (Armstrong et al., 1985; Cheung and Hammock, 1985; Davidson and Yamamoto, 1984; Thomas and Ellar, 1983a,b; Yamamoto et al., 1983). The other protein toxins in the parasporal inclusions, namely the 58, 128, and 130kDa proteins, have been shown to contribute to the insecticidal action of *B. t. israelensis*, but apparently not its cytolytic action (Cheung and Hammock, 1985; Visser et al., 1986). Thus it is now apparent that the in vivo toxicity of *B. t. israelensis* is not due to the action of a single toxin nor the detergent-like action of the 28kDa cytolytic protein. Studies performed to date by various investigators show that a number of toxic components are apparently involved in exerting the observed toxic effects in mosquitoes. Indeed some of these toxins appear to act synergistically with the 25kDa protein (Wu and Chang, 1985; Ibarra and Federici, 1986). Nevertheless the cytolysis of erythrocytes and mosquito cells is thought to be primarily due to the detergent-like action of the 25kDa toxin on membrane phospholipids (Thomas and Ellar, 1983a,b; Gill et al 1987a). However, the specific interaction of the 25kDa toxin with cell membranes is still not understood.

MATERIALS AND METHODS

Biochemicals and chemicals

Lipids were purchased from Avanti. Goat-anti-mouse antibodies conjugated with peroxidase and 2,2'-azino-bis(3-ethyl-benzthiazoline-6-sulfonic acid) diammonium salt (ABTS) were obtained from Boehringer Mannheim Biochemical. ^{125}I, ^{51}Cr, and ^{125}I-Goat-anti-rabbit antibodies were purchased from Amersham. Iodogen was purchased from Pierce Chemical. MEM powder, RPMI 1640 powder, penicillin and streptomycin solution were obtained from Gibco. Grace's insect medium powder was purchased from Serva. Fetal bovine serum (FBS) was obtained either from Hyclone or Irvine Scientific.

Freund's adjuvants were purchased from Calbiochem. Triton X-100 was obtained from Polysciences. Phospholipase A_2 (bee venom), phospholipase D (cabbage) were obtained from Sigma. All other chemicals and enzymes were obtained either from Sigma or Aldrich, or the best grade commercially available.

Purification of crystal proteins

B. t. subsp. *israelensis* was cultured and the 28kDa toxin converted to the 25kDa form during solubilization, and then the 25kDa toxin was purified with a DEAE column according to established procedures (Gill et al., 1987). The toxin was further purified through a Mono Q column by FPLC^R (Pharmacia) to greater than 95% homogeneity, and the purified proteins were frozen at -20°C until needed. The protein concentration was determined according to Lowry et al (1951).

Radioiodination of the 25kDa toxin

The 25kDa toxin was iodinated according to established procedures with minor modifications (Fraker and Speck, 1978). Briefly, the toxin (100μg) was incubated in ice for 10 min with 0.2 M Na_3PO_4 (5μl) and 5mCi of $Na^{125}I$ in a 12 x 75 mm glass tube previously coated with iodogen (10μg). The tube was vortexed gently and placed on ice for 10 min, and the contents then transferred to a microfuge tube containing tyrosine (20μl, 1mg/ml) to quench the reaction. Subsequently, BSA (380μl of 10% BSA) in buffered saline was added to the microfuge tube. The microfuge contents were then dialyzed against buffered saline (2 x 1L) to give a labelled toxin with a specific activity of $3\text{-}9 \times 10^6$ cpm/μg protein.

Cell culture

Aedes albopictus (C6/36) cells were obtained from D. Knudsen, Yale University and cultured in MEM medium according to established procedures (Gill and Hornung, 1987). *Choristoneura fumifera* (IPRI-CF-1, CF1) cells were obtained from S. Sohi, Canadian Forest Service, Sault Ste Marie, Ont. and cultured in Grace's insect medium supplemented with 10% FBS according to standard procedures (Gill and Hornung, 1987). Human erythrocytes were isolated from fresh whole blood drawn from volunteers by centrifugation at 1,000 x *g*. The erythrocytes were washed in PBS at least three times, resuspended to their original volume and kept at 4°C for use within two weeks.

Cytotoxicity assays

Quantitative cytotoxicity of purified proteins towards *Ae. albopictus* cells and human erythrocytes was determined by means of a ^{51}Cr release assay and a hemolytic assay as previously described (Gill and Hornung, 1987). Appropriate controls were run to measure background and maximum release from cells, and the percent cytotoxicity determined.

Phospholipid treatment of purified toxins

Stock aqueous solutions (2×10^{-4}M) of each lipid were prepared in 10mM NH_4HCO_3 and dispersed by ultrasonic treatment. The *B. t. israelensis* toxins were then separately mixed with the desired phospholipid concentration and incubated for 1 hr at 37°C. The phospholipid-treated toxins were tested for their hemolytic activity as described above.

Treatment of erythrocytes

(i) Phospholipase treatment. Aliquots (0.5ml) of human erythrocytes were washed (3X) with 5ml of 10mM glycylglycine buffer (pH 7.4) containing

100mM KCl, 50mM NaCl, 0.25mM Mg^{2+}, 0.25mM Ca^{2+} and 44mM sucrose (Lubin and Chiu, 1982). The washed erythrocytes were resuspended in glycylglycine buffer containing varying phospholipase concentrations. Following incubation at 37°C for 1 hr, the erythrocyte suspension was centrifuged (500 g x 3 min) and the supernatant analyzed for hemoglobin release due to phospholipase action. The pellet was washed (3X) in the glycylglycine buffer, and the hemolytic action of the toxin on phospholipase-treated and untreated cells determined as described above. (ii) Triton X-100 treatment. Suspensions (20μl) of phospholipase A_2- treated and untreated erythrocytes were added to polystyrene tubes each containing glycylglycine buffer (0.88ml) and buffer containing the desired concentration of Triton X-100 (100μl). Following incubation (1 hr, 37°C) the tubes were centrifuged and the supernatants analyzed for hemoglobin release.

Binding assay

Ae. albopictus cells (3 x 10^5 cells/well) or CF1 cells (2 x 10^5 cells/well) were added to wells in a 96-well tissue culture plate, the plate centrifuged at 1,000 x *g* for 5 min, and the supernatant removed by aspiration. Radiolabeled toxin at appropriate concentrations was added to each well gently, the plate then incubated at 27°C in an incubator supplied with 5% CO_2 in air. At the end of incubation, the supernatant was carefully aspirated and fresh medium (200μl) was added to each well and the plated centrifuged at 1,000 x *g* for 3 min. The plate was then washed again. The supernatant was again removed and 5% SDS in water (100μl) was added to each well and the contents transferred to a vial for counting.

Erythrocyte binding was performed in 12 x 75 mm borosilicate glass tubes treated with 1 ml of 1% bovine serum albumin solution per tube. Suspensions (50μl) of phospholipase-treated and untreated erythrocytes were added to each tube containing appropriate concentrations of labeled toxin alone or with both labeled and unlabeled toxins to give a final volume of 250μl. The mixture was then vortexed gently and incubated at 37°C for 1h. The unbound toxin was removed by centrifugation at 1,000 x *g* for 2 min and the cells washed 3x in HBSS. The pellet was then dissolved in 1% SDS for counting. Aliquots were counted to determine total radioactivity bound to erythrocytes, and also subjected to SDS-PAGE, transferred to nitrocellulose and the radioactive bands detected by autoradiography (Gill et al., 1987a).

Liposome preparation and lysis assay

All procedures were performed under a nitrogen atmosphere according to published procedures to prepare a mixture of unilamelar and oligolamelar liposomes (Szoka and Papahadjopoulos, 1978). Dioleolyl phosphatidylcholine (16 mg, 18μmol) and cholesterol (4.6 mg, 12μmol) were added to a 25 ml glass centrifuge tube. The solvent was then evaporated and the residue was dissolved in 3 ml of freshly distilled diethyl ether. Liposome buffer (1 ml, 14mM NaCl, 5mM Hepes, 0.5mM EDTA, and 0.02% NaN_3, pH 7.5) containing an appropriate amount of radiolabel (^{51}Cr, ^{86}Rb, ^{32}P-ATP or ^3H-thymidine) was then added, and the mixture was sonicated at room temperature for 5-15 min until the mixing was completed. The suspension was then transferred to a 25 ml round bottom rotary evaporatory flask and the solvent was removed under vacuum. Liposome buffer (2 ml) was added to the liposome suspension and the evaporation continued for an additional 2h. The liposome suspension was then centrifuged at 1,000 x *g* for 10 min and the supernatant containing the liposome was then dialyzed (against 3 x 800 ml liposome buffer in 24 h) and stored under nitrogen at 4°C until use.

The phosphorus content of liposomes was determined following published procedures (Barlett, 1959). A constant amount of liposome, as determined by phosphorus content, was added to glass tubes with different concentrations of 25kDa toxin. The mixture was then incubated at 37°C for 45 min. The label released from the liposomes was separated by ultrafiltration through a 0.22μm filter and then counted.

Chemical modification of the 25kDa toxin

The toxin was diluted to a final concentration of 0.5μM in Hanks' balanced salt solution (HBSS) with an appropriate concentration of the modification agents. The mixture was incubated at 37°C for either 1 or 3h. A positive control group was prepared the same way except no modification agent was added. Erythrocytes (50μl) were added to glass vials (precoated with 1% bovine serum albumin for 1h at 37°C) containing an appropriate amount of the treated toxin in 200μl. This mixture was then incubated at 37°C for 1h and centrifuged. The amount of hemoglobin released was analyzed by measuring absorbance at 650nm as described above.

For analysis of the amount of modified toxin bound to the erythrocytes, the assay was performed with glutaraldehyde-fixed erythrocytes. Erythrocytes were fixed for 15 min at 4°C with 0.25% glutaraldehyde in HBSS. The cells were then washed three times with HBSS and resuspended back to the original volume. The toxin (3.3μM) was reacted with a 15X excess of $HgCl_2$ at 37°C for 1h. The modified toxins were dialyzed overnight against a 500X volume of HBSS before they were added to the erythrocytes. The incubation procedures described above were then followed. At the end of incubation, the cell pellet obtained from centrifugation was washed three times with HBSS and then dissolved in SDS sample buffer. The sample was then analyzed by SDS-polyacrylamide gel electrophoresis (Laemmli, 1970). The protein bands were transferred onto nitrocellulose paper and probed first with rabbit anti-25kDa toxin (Gill et al., 1987a) and then with ^{125}I-goat anti-rabbit antibody.

Toxin aggregation

Cells were incubated with the toxin as described above at 27°C (insect cells) or 37°C (erythrocytes) for 30 min unless otherwise noted. The unbound toxin was then removed by washing with PBS, pH 8.0, and the cells lysed by incubating with a 5 mM sodium phosphate buffer, pH 8.0, at 4°C for 5 min. The membrane fragments of the lysed cells were recovered by centrifugation at 24,000 x *g* for 30 min. A 10% solution of Triton X-100 in PBS, pH 8.0, was then added to the pellet and extraction procedure continued for 15 min. Marker enzymes (30μl) were then added and the mixture was centrifuged in microcentrifuge tubes at 15,000 x *g* for 5 min to remove most of the unsolubilized materials. The supernatant was then loaded onto a linear 5 to 20% sucrose gradient containing 1% Triton X-100. After the sample was loaded, the tubes were centrifuged at 220,000 x *g* for 16 h at 4°C in a Beckman SW41 rotor. The gradient was then fractionated with an Isco gradient fractionator.

The marker enzymes horse radish peroxidase, acetylcholinesterase, and β-galactosidase were used as internal molecular weight markers to determine the size of the toxin aggregates. Peroxidase activity was measuring the conversion of ABTS; acetylcholinesterase activity was monitored with acetylthiocholine bromide; and β-galactosidase activity was monitored by *o*-nitrophenyl β-D-galactopyranoside.

Monoclonal antibodies

BALB/c female mice (Bantin and Kingman) were injected either three times with native 25kDa toxin at increasing dosages of 20, 40, and 80μg/mouse or with 90°C heat-denatured toxin at 50μg/mouse. The mice were monitored periodically for the production of antibodies against the toxin by screening the serum with ELISA. The mice were injected with a booster dose of 100μg of toxin in PBS 3-4 days prior to the cell fusion. Fusion of cells and the screening of positive hybridoma were performed according to established procedures with minor modifications (Galfre and Milstein, 1981). Mouse myeloma cells, P3-X63/Ag8.653 were used to fuse with the spleen cells at a 1:3 ratio with polyethylene glycol 4000. The fused cell mixture was distributed to five 96-well tissue culture plates containing mouse spleen cells (2 x 10^6 cells per well) as a feeder layer and hybridomas selected by the RPMI 1640 medium containing hypoxanthine, aminopterin, and thymidine supplemented with 15% FBS. Positive hybridomas were identified by solid phase and solution phase ELISA and cloned by limiting dilution.

RESULTS

Effect of lipids on cytotoxicity of the 25kDa protein

Incubation of the 25kDa toxin with an aqueous suspension of dioleolyl phosphatidylcholine resulted in decreased activity of this protein. A toxin:lipid molar ratio of 1:5 was required to reduce toxicity significantly, and pretreatment of the 25kDa protein at a toxin:lipid ratio of 1:50 completely neutralized the toxin. The neutralizing ability was characteristic of phosphatidylcholine with unsaturated acyl residues. In fact, unsaturation only at the syn-2 position of the phosphatidylcholine is sufficient to inhibit the hemolytic activity of the toxin (Table 1). The same lipids with saturated acyl residues did not neutralize the toxin, even up to a toxin:lipid molar ratio of 1:1000. Phosphatidylcholine was more effective at inactivating the toxins than phosphatidyl inositol (Table 1). The ability of lipids to neutralize the toxin was not observed with monooleolyl glycerol, dioleolyl glycerol or triolein, indicating that it is the phospholipid and not the triglyceride moiety that inactivates the toxins.

Table 1. Percent inhibition of hemolytic activity of the 25kDa protein of *B. thuringiensis* subsp. *israelensis* by various phospholipid dispersions.

	Toxin:Lipid Molar ratio[a]	
Lipid	1:5	1:10
Phosphatidylcholine		
Dioleolyl	28	57*
Distearoyl	7	1(NS)
β-Oleolyl, γ-stearoyl	18	29*
β-Linoleoly, γ-palmitoyl	27	48*
Phosphatidylinositol		
β-linoleoyl, γ-palmitoyl)	19	26*

[a]Control A_{540} was 0.44. Average inhibition from four assays performed in duplicate. Toxin concentration used in the assay was kept constant at 0.5μg/ml. *Significantly different from control values ($P < 0.05$). NS=nonsignificant

Similarly preincubation of the solubilized parasporal bodies from subsp. *morrisoni* with dioleolyl phosphatidylcholine at 10X the toxin concen tration reduced erythrocyte lysis to near background levels. As with subsp. *israelensis* preincubation of the solubilized toxins from subsp. *morrisoni* with distearoyl phosphatidylcholine did not have any effect on their hemolytic activity (Gill et al., 1987b).

Effect of phospholipase treatment on cytotoxicity of the 25 protein

To evaluate which membrane phosphatidylcholine moiety is involved in toxicity and whether membrane phospholipids actually participate in toxin mediated erythrocyte hemolysis, the susceptibility of erythrocytes pretreated with phospholipases was determined. Phospholipase A_2 and D were used because of their selective site of action on the phosphatidylcholine molecule. Pretreatment of erythrocytes with phospholipase D did not alter the susceptibility of these cells to the purified toxins. However, erythrocytes pretreated with phospholipase A_2 were significantly less susceptible to toxin action. This phospholipase A_2 inhibition of hemolytic activity of the 24kDa protein occurred in a dose-dependent manner from 1 to 10 units of phospholipase A_2. With 7 units of phospholipase A_2 ca. 70% inhibition of erythrocyte hemolysis by the 25kDa toxin is observed. Concentrations of phospholipase A_2 above 10 units/ml itself destabilized the erythrocyte membrane, thereby causing spontaneous hemoglobin release (Gill et al., 1987a). These phospholipase A_2-treated cells were, however, highly resistant to the lytic action of the toxin. On the other hand, phospholipase A_2-treated erythrocytes were more susceptible to subsequent treatment with Triton X-100. Compared with untreated cells, phospholipase A_2 treated erythrocytes were nearly four-fold more sensitive to the action of Triton X-100 (Gill et al, 1987a).

To determine the nature of the toxin that was bound to erythrocyte membranes, erythrocytes incubated with radiolabelled 25kDa toxin were analyzed by SDS-PAGE followed by autoradiography. The erythrocyte bound toxin was of the same molecular weight as the unbound radiolabelled toxin. In phospholipase A_2-treated cells the total toxin bound to phospholipase A_2-treated cells was considerably less than that bound to untreated cells (Gill et al, 1987a). In contrast, phospholipase D-treated cells were observed to have a similar level of binding as untreated erythrocytes.

Table 2. Increase of ^{125}I labeled 25kDa toxin binding to glutaraldehyde-fixed erythrocytes by co-incubation with a varying ratio of unlabeled toxin.

Ratio of toxin used (labeled:unlabeled)	Cpm bound
1:0	1949 ± 221
1:1	13845 ± 1186
1:10	25805 ± 658
1:100	23913 ± 3431

[a]Erythrocytes (50μl) were incubated with 0.1μg/ml of labeled toxin and various ratios of unlabeled toxin at 37°C for 1 h as described under "Materials and Methods". Values are means of three samples ± S.D.

Binding of the 25kDa toxin to cells

An unusual behavior was observed during studies performed to determine the specific binding of ^{125}I-25kDa toxin to human erythrocytes. An increase binding of the radiolabeled toxin, instead of an anticipated decrease, was observed when it was incubated with the erythrocytes in the presence of an excess of the unlabeled toxin. The increase in bound toxin varied with the amount of unlabeled toxin added, and the binding was irreversible because extensive washing of the cells for 30 min failed to remove any significant amount of the bound toxin.

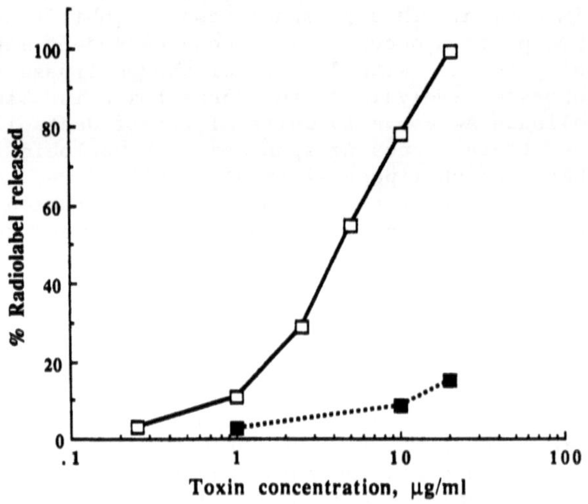

Fig. 1. Percentage release of ^{51}Cr from liposomes made of dioleolyl and distearoyl phosphatidylcholine and cholesterol. The liposomes were treated with various concentrations of 25 kDa toxin at 37°C for 45 min as described under "Materials and Methods". Values are means of three samples except for 20µg/ml where the value is average of two samples.

This phenomenon of increased binding was further studied with glutaraldehyde fixed erythrocytes. The fixed cells bound the same amount of ^{125}I-toxin as unfixed cells but could withstand a much higher toxin concentration without being lysed. At a fixed concentration of ^{125}I-toxin, the binding increased rapidly with an increasing amount of the unlabeled toxin, and finally reached a plateau when a 100-fold excess (10µg/ml) of the unlabeled toxin was utilized (Table 2). Such an increase in the binding of labeled toxin in the presence of unlabeled toxin was also observed with the *Aedes albopictus* and CF1 insect cell lines. Further, even at pH 10.5, a pH similar to that of mosquito gut pH, the increase of toxin binding in the presence of unlabeled toxin proceeded in a manner similar to that observed at the lower pH.

The role of unsaturated phospholipids in serving as binding sites for the 25kDa toxin was confirmed by the studies performed on liposomes of differing composition. These liposomes were generated with only lipids in the presence of ^{51}Cr, and the background leakage of ^{51}Cr in the absence of toxin was minimal. However, in the presence of the toxin, ^{51}Cr was rapidly released from liposomes made of dioleolyl phosphatidylcholine and cholesterol, indicating the destruction of the liposome structure by the toxin (Fig. 1). The release of ^{51}Cr proceeded nearly linearly with the log of toxin concentration and at 20μg/ml, essentially 100% of the label was released. In contrast, only 15% of the chromium label was released at that toxin concentration when the liposomes were made of distearoyl phosphatidylcholine and cholesterol.

In order to investigate the size of the pores generated by the toxin action, the radiolabeled marker within the liposome was also varied. Four chemicals, of different sizes were used: ^{86}Rb, ^{51}Cr, ^{3}H-thymidine, and ^{32}P-ATP. Among this group, ^{86}Rb had the fastest release time; with 3μg/ml, 56% of the radiolabel was released. With ATP, the largest radiolabel, only 25% of the label was released at this same toxin concentration. Both thymidine and ^{51}Cr were released at rates between ^{86}Rb and ATP. In general, the release of all markers proceeded at a concentration dependent manner.

<u>Chemical modification of the 25kDa toxin</u>

The effect of specific amino acid modifiers on toxin activity and binding on erythrocytes was also evaluated. Among the chemicals tested, 2-bromoacetamido-4-nitrophenol, 4-(chloromercuri)benzenesulfonic acid (CMBS), $CuCl_2$, 4,4'-diisothiocyano-2,2'-stilbene disulfonic acid (DIDS), 5,5'-dithiobis-(2-nitrobenzoic acid) (DTNB), N-ethylmaleimide, $FeCl_3$, $HgCl_2$, iodine, phenyl glyoxal, succinic anhydride, and $ZnCl_2$, only the mercurial compounds and DIDS caused any significant decrease in toxin activity and

Table 3. Inhibition of the hemolytic activity of the 25kDa toxin of *B. t. israelensis* by chemical modification agents.

Reagent Concentration (μM)	% inhibition[a]				
	$HgCl_2$	$HgCl_2$(3 h)	CMBS[b]	DIDS[c]	PG[d]
0.5		21			
2.5	17	68			
5	89	98			
10	98	100			
20	100				
100			4	1	
200			21	5	
500			98	58	2
1000			100	89	0

[a]Toxins (0.5μM) were reacted with the various agents for 1 h (or 3 h as indicated) at 37°C prior to the mixing with RBC as described under "Materials and Methods". Values are means of three samples. [b]CMBS - 4-chloromercuribenzenesulfonic acid. [c]DIDS - 4,4'-diisothiocyano-2,2'-stilbene disulfonic acid. [d]Phenyl glyoxal. The other compounds tried similarly caused no inhibition of hemolytic activity.

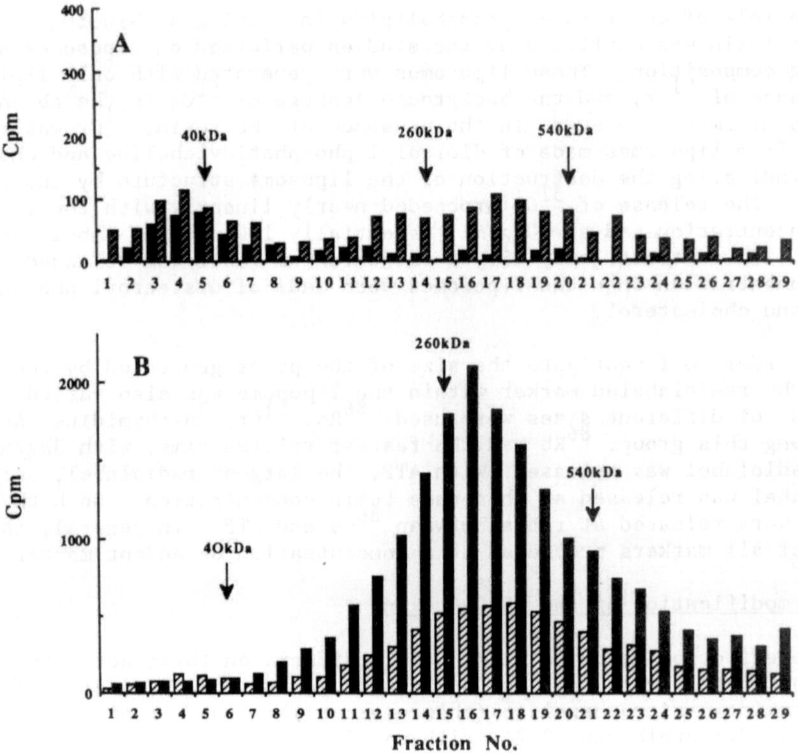

Fig. 2. Aggregation of 25kDa toxin recovered from CF1 cell membranes at varying toxin concentration. *A*. CF1 cells were incubated with ^{125}I labeled toxin at 0.25μg/ml (hatched) or 0.4μg/ml (solid) for 30 min. *B*. Cells were incubated with 0.75μg/ml (hatched) or 4.0μg/ml (solid) of ^{125}I labeled toxin at 27°C for 30 min. The cell membrane extracts were prepared as described under "Methods". Arrows indicate the positions of marker enzymes with molecular weights of 40, 260, and 540 kDa.

binding to erythrocytes. After treating the toxin with 2.5μM of $HgCl_2$ for 1h or with 0.5μM for 3h, a decrease of 17% and 21%, respectively, in the toxin's ability to lyse erythrocytes was observed. At 20μM of $HgCl_2$, the toxin's lytic action was completely abolished (Table 3). Pretreating the cells, but not the toxin, with even 100μM $HgCl_2$ had very little effect on the cells' susceptibility to toxin mediated lysis. The effect of the organic mercurial compounds and DIDS paralleled that observed with $HgCl_2$ but occurred at much higher concentrations, producing a 50% inhibition of cell lysis at concentrations of 300 and 500μM, respectively. In general, the inhibitory effect of all the effective chemicals was more prominent when the toxin was reacted for 3h, instead of 1h, with these chemicals. The other agents, tested at up to 1mM concentration, however, did not have any significant effect on toxicity.

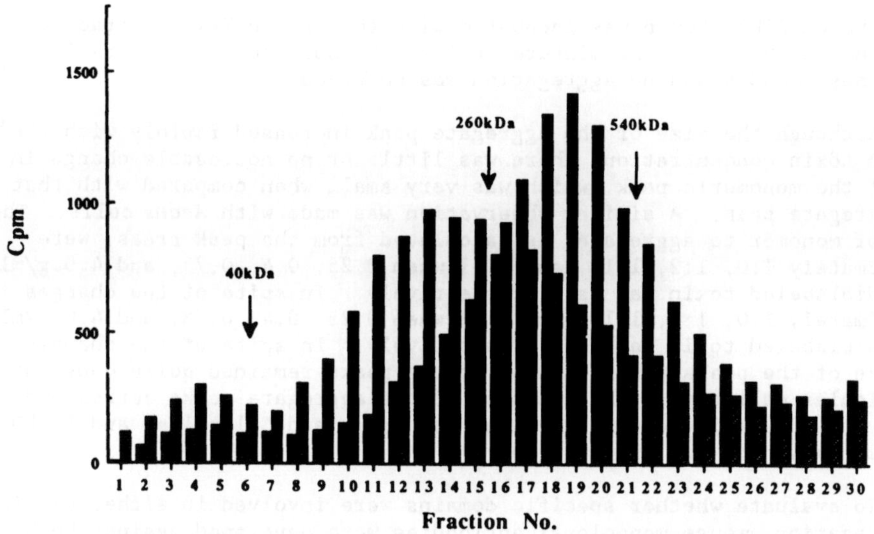

Fig 3. Aggregate forms of 25kDa toxin present on human erythrocytes and *Aedes albopictus* cell membranes at high concentrations of toxin. Human erythrocytes (hatched) were incubated with 0.5μg/ml of ^{125}I labeled toxin at 37°C for 30 min. *Ae. albopictus* (solid) cells were incubated with 2.0μg/ml of labeled toxin at 27°C for 30 min. The cell membrane extracts were prepared as described under "Methods". The slight rise of radioactivity at the end of gradient was due to insoluble materials at the bottom of the gradient tube.

The decrease in cell lysis by $HgCl2_2$-treated toxin was identified as decreased toxin binding to erythrocytes. Cell membranes recovered from erythrocytes that had been treated with unmodified toxin had approximately 10X more bound toxin in comparison to erythrocytes that were incubated with $HgCl_2$-modified toxin as determined by autoradiography.

<u>Cell membrane aggregation of the 25kDa toxin</u>

To evaluate the nature of the cell membrane bound 25kDa, CF1 cells treated with the ^{125}I-toxin were extracted with 10% Triton X-100. More than 85% of the bound toxin was extractable by this detergent. Ultracentrifugation of the extracted material on a 5 to 20% sucrose density gradient showed that the 25kDa toxin was present in its monomeric form when a low concentration (< 0.25μg/ml) of radiolabeled toxin was used and the incubation time was short (30 min) (Fig. 2).

As the concentration of the labeled toxin was raised to 0.3 or 0.4μg/ml, a second peak of 400kDa molecular weight, was observed for CF1 cell and *Aedes* cells. At 0.75 and 4.0μg/ml, this second peak became very pronounced (Fig. 2). With 0.25μg/ml of toxin but with a longer incubation time (3h), the toxin was also observed to exist as aggregates. To determine whether the toxin could form aggregates with membrane extracts, the radiolabeled 25kDa toxin was incubated with the Triton X-100 extract of cell membrane and the resulting mixture analyzed by sucrose density gradient. Under these conditions no aggregation was observed.

Although the size of the aggregate peak increased rapidly with the rise in toxin concentration, there was little or no noticeable change in the size of the monomeric peak, which was very small when compared with that of the aggregate peak. A similar observation was made with *Aedes* cells. The ratio of monomer to aggregate, as calculated from the peak areas, were approximately 1:0, 1:2, 1:13, and 1:51 when 0.25, 0.4, 0.75, and 4.0μg/ml of the radiolabeled toxin was used, respectively. In spite of the changes in
approximately 1:0, 1:2, 1:13, and 1:51 when 0.25, 0.4, 0.75, and 4.0μg/ml of the radiolabeled toxin was used, respectively. In spite of the changes in
the size of the peaks, the position of the peaks remained quite constant for a particular cell type. The location of the aggregate peaks corresponded roughly to 400kDa in molecular weight for *Aedes* and CF1 cells, and 170kDa for human erythrocytes (Figs. 2 and 3).

To evaluate whether specific domains were involved in either binding or aggregation, mouse monoclonal antibodies were developed against both heat denatured and native 25kDa toxin and tested for their effects on binding and/or aggregation. These monoclonal antibodies showed specific binding to the toxin and their binding could be competed with an excess of native toxin. However, none of these monoclonal antibodies raised against heat denatured toxin showed any significant inhibition of binding or aggregation of ^{125}I labeled toxin on cell membranes.

Mice injected with the native form of the 25kDa toxin became increasingly tolerant toward it. At the end of the injection protocol, mice could survive a dosage 10X that of the reported LC50 value (Thomas and Ellar, 1983b), indicating that the mice were producing antibodies towards the toxic domain of the 25kDa toxin. Eleven monoclonal antibodies developed utilizing splenocytes from these animals could inhibit 80-97% of labeled toxin binding to cells and protect them from lysis by the toxin. Preliminary findings from mapping studies on one of these antibodies, 10B4, indicated that the C-terminal fragment from amino acid #95 contained the epitope for this antibody. Another monoclonal antibody, 10B9, that only recognized the

Table 4. Inhibition of ^{125}I labeled 25kDa toxin binding to CF1 cells by monoclonal antibody 10B9.

MAb Concentration, (μg/ml)	Cpm bound[a]	% Inhibition
0	10812 ± 931	
0.25	11032 ± 1502	-2
0.5	10589 ± 692	2
1.0	9297 ± 539	14
5.0	6400 ± 142	41
20	603 ± 33	94
80	495 ± 37	95

[a]Toxin (1.0μg/ml) was incubated with a varying concentration of 10B9 MAb at 27°C for 1h. CF1 cells were then incubated with the mixture at 27°C for 45 min. Values are means of three samples ± S.D.

Table 5. Binding of ^{125}I labeled 25kDa toxin to Aedes albopictus and CF1 cells at varying concentrations of labeled toxin.

Toxin concentration, ($\mu g/ml$)	Cpm bound per well	
	CF1	*Aedes albopictus*
0.25	88 ± 46	76 ± 17
0.5	-	789 ± 432
0.6	593 ± 177	-
0.75	-	2214 ± 68
1.0	3945 ± 647	3390 ± 140
2.0	15346 ± 723	9880 ± 432
4.0	34224 ± 378	19328 ± 451
6.0	50129 ± 3745	27986 ± 740

[a]The incubations were performed at 27°C for 30 min. For *Ae. albopictus* and CF1 (3×10^5 cells/well) and (2×10^5 cells/well) were used, respectively. Values are means of four samples ± S.D.

native form of the 25kDa toxin was purified and shown to inhibit 95% of toxin binding (Table 4).

Although the measurement of specific binding to cells was not possible, an attempt was made to determine the net amount of toxin bound under various conditions. The binding of toxin to erythrocytes, CF1 cells, and *Aedes* cells all appeared to increase linearly and rapidly with an increasing concentration of labeled toxin (Table 5). No saturation of binding sites was observed in the 30 min assay performed. An increase in binding was also observed when the *Aedes* and CF1 cells were washed after incubation with unlabeled toxin prior to the addition of labeled toxin (data not shown). The increased binding depended very much on the concentration of the unlabeled toxin used. However, concentrations equal to or greater than 2.0$\mu g/ml$ apparently led to a decrease in binding of ^{125}I-toxin added subsequently.

When the toxin concentrations were held constant while the time of incubation was varied, the net amount of binding to *Aedes* and CF1 cells also showed a more rapid phase of binding after an initial lag period. The amount of toxin bound increased slowly when the membrane toxin concentration was low. But when the membrane concentration reached a critical threshold level, the rate of binding increased 2 to 3-fold.

DISCUSSION

Lipid inactivation of the cytolytic toxins of *B. t. israelensis* as previously shown by Thomas and Ellar (1983a) suggests a role for phospholipids in the action of these toxins. Our results with purified toxins confirm previous findings performed with crude toxins (Thomas and Ellar 1983a) that the acyl moiety of phospholipids is crucial in neutralizing toxin action. However, the concentration of lipid required to inhibit the toxin action is an order of magnitude less than previously reported.

The mechanism by which phospholipids effect the hemolytic activity of *B. thuringiensis* subsp. *israelensis* is not known. The results presented here on the effect of phospholipases on erythrocyte susceptibility to toxin action show that membrane phospholipids play a crucial role in the manifestation of cytotoxicity. The failure of the toxin to lyse phospholipase A_2-treated erythrocytes strengthens the role of fatty acyl residues in the hemolytic action of the toxin, because phospholipase A_2 hydrolyzes the fatty acid at the syn-2 position of phospholipids, which in most eukaryotic cell membranes is usually unsaturated (Dennis, 1983). The role of this unsaturated fatty acid of membrane phospholipids in toxin-induced hemolysis is not known, but potentially it may be involved in the binding and/or insertion of the toxin into cell membranes. Our observations that phospholipase A_2-treated erythrocytes bind significantly less toxin than untreated cells provide additional support for the role of membrane phospholipids in toxin action. However, the observation that phospholipase D did not alter erythrocyte susceptibility to the toxin indicates that the lipid-head group may play a lesser role in the hemolytic action of this toxin.

Membrane lipids have been implicated in the cytolytic action of a number of bacterial toxins. In some cases the phospholipids are involved in the toxin action (Bangham et al., 1965; Duncan and Buckingham, 1981), whereas cholesterol mediates the action of some other bacterial proteins (Cowell and Bernheimer, 1978). Based on the ability of phospholipids to neutralize the toxins and the observation that other bacterial toxins with affinity for phospholipids disrupt cell membrane in a detergent-like action, it has been suggested (Thomas and Ellar, 1983a) that *B. thuringiensis* subsp. *israelensis* toxins disrupt cells in a similar fashion. However the lytic action of these toxins towards erythrocytes cannot be explained solely on the basis of a non-specific "detergent-like action" hypothesis because phospholipase A_2 treatment of erythrocytes increases the susceptibility of these cells to the lytic action of non-ionic detergents like Triton X-100, whereas the same treatment has an opposite effect on the susceptibility of erythrocytes to the toxin (Gill et al., 1987a).

The binding of the 25kDa cytolytic toxin of BTI to both insect and mammalian cell membranes is an irreversible process. This binding phenomenon does not reflect that of a typical ligand-receptor interaction; instead an increase in the binding of the radiolabeled toxin is observed in the presence of an increasing amount of unlabeled toxin. This interaction of the toxin with the cell membranes, however, is similar to that observed with another bacterial toxin, the staphylococcal α-toxin (Phimister and Freer, 1984).

The lack of saturation of cell binding observed with increasing concentrations of ^{125}I-toxin (up to 6.0μg/ml for *Aedes* and CF1 cells) is consistent with the assumption that unsaturated phospholipids are the primary sites for toxin binding. These phospholipids are widespread in nature and are especially abundant in insects of the order Diptera (Fast, 1966). Unsaturated phospholipids, which make up approximately 40% of the total membrane lipids for *Aedes* or *Culex* cells (Luukkonen et al, 1973; Jenkin et al, 1976), can account for the large amount of toxin bound at a concentration of 6.0μg/ml.

The ability of the 25kDa toxin to lyse liposomes made of dioleoyl phosphatidylcholine confirms the importance of unsaturated phospholipids as toxin binding sites. Similar results with liposomes were reported previously although the release of radiolabel from the liposomes reported here is much more complete (nearly 100%) than the 40% release from multilamellar liposomes reported (Drobniewski and Ellar, 1987). The differences observed between the previous study and those reported here is probably a result of the quality of the liposome preparation. However, the liposome model does not appear to be a true reflection of what happens with cells because these liposomes are only damaged at concentrations 10X of that required to lyse cells (Gill et al., 1987a; Gill and Hornung, 1987). Thus it is possible that a mixture of lipids or a combination of lipids and proteins on the cell membrane may serve as even better receptors as observed with the diphtheria toxin (Olsnes et al, 1985). Alternatively, the toxin may be unable to penetrate the layered structure of the oligolamelar liposomes, which is part of the mixture produced, as previously suggested (Drobniewski and Ellar, 1987).

Liposomes made of saturated lipids such as dimyristoyl phosphatidylcholine (Drobniewski and Ellar, 1987) or distearoyl phosphatidylcholine are also lysed by the toxin but at an even higher range of concentrations. Thus it appears that the nature of fatty acid chains rather than the phosphate head group is the primary determinant for toxin binding. A similar conclusion was obtained previously using phospholipases as probes for toxin-membrane interaction (Gill et al., 1987a). Nevertheless liposomes derived from saturated phosphatidylcholine do interact with the toxin, albeit at high toxin concentrations.

The appearance of aggregation is dependent on the surface concentration of bound toxins. At low toxin concentration the membrane bound toxins show no aggregate formation. But when a higher toxin level is used the concentration of membrane bound toxin increases and toxin aggregates are observed. It appears that aggregates form when a critical membrane concentration of toxin is reached. Thus when the concentration of bound toxins is high the toxins exists predominantly in the aggregate form. Such a finding parallels that of erythrocytes bound staphylococcal α-toxins, which exist mostly in an aggregate form (Hugo et al., 1986). It has also been reported for staphylococcal α-toxin that bound toxins remain in monomeric form at low temperature but will spontaneously assemble into a dodecameric form when the temperature is raised (Reichwein et al., 1987). It is possible that aggregate formation for the 25kDa toxin is also temperature dependent.

The dependence of aggregate formation on membrane toxin concentration as noted above indicates that toxins interact with each other and suggests that there is more than one 25kDa toxin in each aggregate. Therefore, if the whole aggregate is composed solely of the 25kDa toxin, the aggregate has to consist of sixteen toxin molecules for *Aedes* or CF1 cells. We have no evidence, however, that the aggregate is composed solely of 25kDa toxin or 25kDa toxin along with other membrane macromolecules. It has been reported that certain lipids (*e.g.*, phosphatidylserine) are tightly associated with membrane protein (van Zoelen et al., 1977), and thus it is likely that these lipid-protein complexes contribute toward the large molecular size of the toxin aggregate. The slightly smaller size of aggregate observed in erythrocytes may be caused by differences in cell membrane proteins involved in toxin aggregate formation.

The mechanism by which unlabeled toxins increase the binding of the labeled toxin has not been determined but it apparently is occurring on the cell surface rather than occurring in the solution prior to the binding of toxin to cells. It appears that formation of toxin aggregates results in the increased binding of toxin with higher toxin concentrations. Thus additional binding sites are actually generated by the bound toxin. A cooperative effect alone, as reported for diphtheria toxin (Alving et al, 1980), cannot explain the dramatic increase of toxin binding observed here. It is also likely that the formation of toxin aggregate is the mechanism by which the binding of labeled toxin increases after an initial lag period. After this initial lag period, the rate of binding increased with time.

Giant aggregates of streptolysin O isolated from cell membranes have been reported to require no cholesterol for stability although cholesterol is the receptor for the toxin's initial binding (Bhakdi et al., 1985). The streptolysin O molecules are capable of attaching to each other even in the absence of cholesterol, the receptor molecule. If this model is also operative for the 25kDa toxin, it is possible that the 25kDa toxin can bind directly to the toxin aggregate on the cell membrane in addition to binding to cell membrane lipids. Consequently, more binding sites are generated on the cell surface as aggregates are formed and the rate of binding may increase as a result.

The contribution of aggregation to toxicity has not been studied in our experiments. But the fact that aggregates of the toxin were recovered by sucrose density gradient from all three cell types used suggests that aggregation may play a crucial role in toxicity. The time dependence of the lytic effect of the toxin at 0.3μg/ml (Gill and Hornung, 1987) when it is incubated with the cells for several hours parallels the process of formation of aggregate observed in our experiment when similar conditions are used. This aggregation of the 25kDa toxin apparently contributes to toxicity, and this contribution may be mediated by two different means. First, since aggregation of toxins may lead to a larger amount of binding, toxic action may be increased. Alternatively, functional pores may not be generated until the aggregates are formed. In the case of streptolysin O, the formation of aggregate has been found to be critical for toxic effects (Hugo et al, 1986). But in the case of hemolysin produced by *E. coli*, aggregates are never formed while pores are generated as individual toxins bind to the membrane (Bhakdi et al, 1986).

The functional pore size of the pores generated by the 25kDa toxin on live cells has been estimated to be 2nm in diameter (Knowles and Ellar, 1987). Under the 10μg/ml of toxin concentration used in that experiment, the cell membrane bound toxin should exist as aggregates rather than as monomers. Thus it appears that the 2 nm pore size reported for the 25kDa toxin is that of an aggregate. This pore size of the 25kDa toxin aggregate is therefore similar to that of the staphylococcal α-toxin and *E. coli* aerolysin (Fussle et al, 1981; Howard and Buckley, 1982) but quite different from the >30nm pores generated by streptolysin O (Buckingham and Duncan, 1983). It is likely that only when the aggregate is of an enormous size do pores of this large diameter occur. For staphylococcal α-toxin and streptolysin O, pore formation apparently results from toxin insertion through the cell membrane (Fussle et al, 1981; Bhakdi et al, 1985). It is possible, therefore, that a similar toxin insertion into the cell membrane accounts for the irreversibility of the binding observed here for the 25kDa toxin.

The critical point in the interaction of the 25kDa toxin with the cell membrane is the initial binding process. It appears that a specific toxin domain is crucial for this binding phenomenon. Among the various chemicals used to modify the different amino acids of the toxin, $HgCl_2$ is the most effective. A similar observation was made for carboxypeptidase Y, for which $HgCl_2$ was also the most effective compound in reacting with the single cysteine in that protein (Bai and Hayashi, 1979). The modification of the 25kDa toxin appears to be a very specific reaction for the cysteine moiety for two reasons: 1) The effect can be observed when an approximately equimolar ratio of toxin and $HgCl_2$ was used. 2) The small amount of $HgCl_2$ left after dialysis is too low to cause any significant damage to the cell membrane lipids to alter their affinity for the toxin. Although other sulfhydryl reagents such as DTNB and CMBS are only marginally effective, this may be potentially due to the inaccessibility of the cysteine at this hydrophobic site. In contrast to the decrease in cytotoxic effects of the 25kDa toxin by sulfhydryl reagents, Pfannenstiel and coworkers (1985) reported that sulfhydryl reagents ($HgCl_2$ and 2-hydroxyethyldisulfide) actually increased the mosquitocidal activity of solubilized crystal proteins from *B. t. israelensis*. They attributed this increase to the inherent toxicity of these sulfhydryl reagents to the larvae.

There are only two cysteines in the native 28kDa toxin molecule (Waalwijk et al, 1985; Ward and Ellar, 1986). Cys-7 is removed during the proteolytic conversion to the 25kDa toxin form and thus it is not important for in vitro cytotoxicity (Gill et al., 1987a). The other cysteine at position 190 is located in the middle of a major hydrophobic region from amino acid 170 to 210 (Ward and Ellar, 1986). According to a recently proposed three dimensional structural model, this cysteine is located at the coil/turn loop region adjacent to helix 3 and 4 (McPherson et al, 1987; Ward et al, 1988). Therefore it appears that this domain is important for binding and toxicity. This domain apparently also contains the epitope for the monoclonal antibody that inhibits binding thus confirming the importance of this site to toxicity.

Residues outside of the helix 3 and 4 structure can also be replaced by site-directed mutagenesis. It has been reported that replacement of certain basic (arginine and lysine) and acidic (aspartic and glutamate acid) amino acids with alanine will change the toxic properties of the toxin (Ward et al, 1988). Some of the effects observed have been attributed to the alteration in the packaging of the toxin into crystal form. We have found that amino acids such as arginine and lysine, which are modified by chemicals used in our experiment, do not appear to play a crucial role in the binding site of the toxin.

It is not unequivocal from our experiment that the binding and lysis steps are two distinct steps carried out by separate domains of the toxin as observed with the staphylococcus α-toxin (Blomqvist and Sjogren, 1988), although it appears that they are. Theoretically binding must precede toxic action. Since the great decrease in the toxin's binding affinity by $HgCl_2$ can account totally for the observed decrease in lysis, possibly only the binding domain is affected by $HgCl_2$.

Identification of the site important for binding and/or aggregate formation has also been attempted by monoclonal antibodies. Many monoclonal antibodies against the 25kDa toxins have no effect on the binding or lytic process of the toxin, but some monoclonal antibodies raised against the native form of the 25kDa toxin can significantly decrease the binding of the toxins. When these antibodies are co-incubated with the 25kDa toxin and cells, the lytic effect is greatly diminished. However, it has not been determined conclusively if all these antibodies affect only the binding site or perhaps some of them are directed against domains important for aggregation. It has been suggested previously that aggregation may facilitate binding. Consequently, a blockage of aggregation will also be reflected as a decrease in overall binding. Further, the monoclonal antibodies developed are directed against several different epitopes since they do not all recognize denatured toxins and some of them can recognize (though only weakly) toxins bound on the cell surface. Presumably, some of these monoclonal antibodies may be directed to the binding site while others are directed to the aggregation site(s). Experiments are in progress to define the specific epitopes of these monoclonal antibodies.

In summary, our experiments demonstrate that the 25kDa toxin binds to cell membranes irreversibly in a concentration dependent manner. When a critical cell membrane concentration of the toxin is reached, the bound toxins form aggregates of 400kDa in size on insect cell membranes. A specific toxin domain involved in the binding of the toxin to cells has been identified. This domain, which can be destroyed by $HgCl_2$ or inactivated by specific monoclonal antibodies, incorporate the amino acids 154 - 203, which form the helix 3/4 and a coil/turn loop region.

ACKNOWLEDGEMENTS

This work was support in part by NIH ES03298, and the Mosquito Control Program and Biotechnology Program of the University of California.

REFERENCES

Alving, C. R., Iglewski, B. H., Urban, K. A., Moss, J., Richards, R. L., and Sadoff, J. C, 1980, Binding of diphththeria toxin to phospholipids in liposomes, Proc. Natl. Acad. Sci., USA, 77:1986-1990.

Armstrong, J. L., Rohrmann, G. F., and Beaudreau G. S., 1985, Delta-endotoxin of *B. thuringiensis* subsp. *israelensis*, J. Bacteriol., 161:39-46.

Bai, Y., and Hayashi, R., 1979, Properties of the single group of carboxypeptidase Y. Effects of alkyl and aromatic mercurials on activities toward various synthetic substrates, J. Biol. Chem., 254:8473-8479.

Bangham, A. D., Standish, M. M., and Weissmann, G., 1965, The action of steroids and Streptolysin S on the permeability of phospholipid structures to cations, J. Mol. Biol., 13:253-259.

Barlett, G. R., 1959, Phosphorus assay in column chromatography, J. Biol. Chem., 234:466-468.

Bhakdi, S., Tranum-Jensen, J., and Sziegoleit, A., 1985, Mechanism of membrane damage by streptolysin-O, Infect. Immun., 47:52-60

Bhakdi, S., Mackman, N., Nicaud, J., and Holland, I., 1986, *Escherichia coli* hemolysin may damage target cell membranes by generating transmembrane pores, Infect. Immun., 52:63-69.

Blomqvist, L., and Sjogren, A., 1988, A staphlococcal α-toxin fragment, Toxicon, 26:265-273

Buckingham, L., and Duncan, J. L., 1983, Approximate dimensions of membrane lesions produced by streptolysin S and streptolysin O, Biochim. Biophys. Acta, 729:115-122

Burges, D. D., 1982, Control of insects by bacteria, Parasitology, 84:79-117.

Cassidy, P., and Harshman, S., 1979, Characterization of detergent-solubilized iodine-125-labeled α-toxin bound to rabbit erythrocytes and mouse diaphargm muscle, Biochemistry, 18:232-236

Charles, J. K., and deBarjac, H., 1983, Action des cristeaux de *B. thuringiensis* var. *israelensis* sur l'intestin moyen des larves de Aedes aegypti L., en microscopie electronique, Ann. Microbiol. (Paris), 134A:197-218.

Cheung, P. Y. K., and Hammock, B. D., 1985, Separation of three biologically distinct activities from the parasporal crystal of *Bacillus thuringiensis* var. *israelensis*, Current Microbiol., 12:121-126.

Cowell, J. L., and Bernheimer, A. W., 1978, Role of cholesterol in the action of cereolysin on membranes, Arch. Biochem. Biophys., 190:603-610.

Davidson, E. W., and Yamamoto, T., 1984, Isolation and assay of the toxic component of *B. thuringiensis* var. *israelensis*, Current Microbiol., 11:171-174.

De Barjac, H., 1978, Une nouvelle variete de Bacillus thuringiensis tres toxique pour les moustiques: B. thuringiensis var. israelensis serotype 14, C. R. Acad. Sci. Paris Ser. D, 286, 797-800.

Dennis, E. A. 1983. Phospholipases. In Enzymes Vol. XVI, P. D. Boyer, ed., p. 307-353, Academic Press, N.Y.

Drobniewski, F. A., and Ellar, D. J., 1987, Toxin-membrane interactions of *Bacillus thuringiensis* δ-endotoxins, Biochem. Soc. Trans., 16:39-40

Duncan, J. L., and Buckingham, L., 1981, Effects of Streptolysin S on liposomes: Influence of membrane lipid composition of toxin action, Biochim. Biophys. Acta, 648:6-12.

Fast, P. G., 1966, A compartive study of the phospholipids and fatty acids of some insects. Lipids 1:209-215

Fraker, P. J., and Speck, J. C., Jr., 1978, Protein and cell membrane iodinations with a sparingly soluble chloroamide, 1,3,4,6-tetrachloro-3a,6a-diphenylglycoluril, Biochem. Biophy. Res. Commun., 80:849-857.

Fussle, R., Bhakdi, S., Sziegoleit, A., Tranum-Jensen, J., Kranz, T., and Wellensiek, H. J., 1981, On the mechanism of membrane damage by *Staphylococcus aureus* α-toxin, J. Cell Biol., 91:83-94.

Galfre, G., and Milstein, C., 1981, Preparation of monoclonal antibodies: strategies and procedures, Methods Enzymol., 73:3-46

Gill, S. S., Singh, G. J. P., and Hornung, J. M., 1987a, Cell membrane interaction of *Bacillus thuringiensis* subsp. *israelensis* cytolytic toxins, Infect. Immun., 55:1300-1308

Gill, S. S., Hornung, J. M., Ibarra, J. E., Singh, G. J. P., and Federici, B. A., 1987b, Cytolytic avtivity and immunological similarity of the *Bacillus thuringiensis* subsp. *israelensis* and *Bacillus thuringiensis* subsp. *morrisoni* isolate PG-14. Appl. Environ. Microbiol., 53:1251-1256.

Gill, S. S., and Hornung, J. M., 1987, Cytolytic activity of Bacillus thuringiensis proteins to insect and mammalian cell lines, J. Inverterbr. Pathol., 50:16-25.

Goldberg, L. J., and Margalit, J., 1977, A bacterial spore demonstrating rapid larvicidal activity against *Anopheles sergentii*, *Uranotaenia unguiculata*, *Culex univatattus*, *Aedes aegypti* and *Culex pipiens*, Mosquito News, 37:355-358

Howard, S. P., and Buckley, J. T., 1982, Membrane glycoprotein receptor and hole-forming properties of a cytolytic protein toxin, Biochemistry, 21:1662-1667

Hugo, F., Reichwein, J., Arvand, M., Kramer, S., and Bhakdi, S., 1986, Use of a monoclonal antibody to determine the mode of transmembrane pore formation by streptolysin O, Infect. Immun., 54:641-645

Ibarra, J. E. and Federici, B. A., 1986, Isolation of a relatively non-toxic 65-kilodalton protein inclusion from parasporal body of *Bacillus thuringiensis* subsp. *israelensis*, J. Bacteriol., 165:527-533.

Jenkin, H. M., McMeans, E., Anderson, L. E., and Yang, T. K., 1976, Phospholipid composition of *Culex quinquefasciatus* and *Culex tritaeniorhynchus* cells in logarithmic and stationary growth phases, Lipids, 11:697-704

Knowles, B. H., and Ellar, D. J., 1987, Colloid-osmotic lysis is a general feature of the mechanism of action of *Bacillus thuringiensis* δ-endotoxins with different insect specificity, Biochim. Biophys. Acta, 924:509-518

Laemmli, U. K., 1970, Cleavage of structural proteins during the assembly of the head of the Bacteriophage T4, Nature, 227:680-685.

Lowry, O. H., Rosebrough, J. J., Farr, A. L., and Randall, R. J., 1951, Protein measurement with the Folin phenol reagent, J. Biol. Chem., 193:265-275.

Lubin, B., and Chiu, D., 1982, Membrane phospholipid organization in pathologic human erythrocytes, Prog. Clin. Biol. Res., 97:137-150.

Luukkonen, A., Brummer-Korvenkontio, M., and Renkonen, O., 1973, Lipids of cultured mosquito cells (*Aedes albopictus*). Comparision with cultured mammalian fibroblasts (BHK cells), Biochim. Biophys. Acta, 326:256-261

McPherson, A., Jurnak, F., Singh, G. J. P., and Gill, S. S., 1987, Preliminary x-ray diffraction analysis of crystals of *Bacillus thuringiensis* toxin, a cell membrane disrupting protein, J. Mol. Biol., 195:755-757

Olsnes, S., Carvajal, E., Sundan, A., and Sandvig, K., 1985, Evidence that membrane phospholipids and protein are required for binding of diphtheria toxin in Vero cells, Biochim. Biophys. Acta, 846:334-341

Phimister, G. M., and Freer, J. H., 1984, Binding of ^{125}I-alpha toxin of *Staphylcoccus aureus* ro erythrocytes, J. Med. Microbiol., 18:197-204

Pfannenstiel, M. A., Couche, G. A., Muthukumar, G., and Nickerson, K. W., 1985, Stability of larvicidal activity of *Bacillus thuringiensis* subsp. *israelensis*: Amino acid modification and denaturants, Appl. Environ. Microbiol., 50:1196-1199.

Reichwein, J., Hugo, F., Roth, M., Sinner, A., and Bhakdi, S., 1987, Quantitative analysis of the binding and oligomerization of staphylococcal alpha-toxin in target erythrocyte membranes, Infect. Immun., 55:2940-2944

Singh, G.J.P., and Gill S. S, 1985, Myotoxic and neurotoxic activity of *Bacillus thuringiensis* var *israelensis* crystal toxin, Pestic. Biochem. Physiol., 24, 406-414.

Singh, G. J. P., Schouest, L. P., Jr. and Gill, S. S., 1986, Action of *Bacillus thuringiensis* subsp. *israelensis* δ-endotoxin on the ultrastructure of the housefly neuromuscular system in vitro, J. Invertebr. Pathol., 47:155-166.

Szoka, F., Jr., and Papahadjopoulos, D., 1978, Procedure for preparation of liposomes with large internal aqueous space and high capture by reverse-phase evaporation, Proc. Natl. Acad. Sci., USA, 75:4194-4198

Thomas, W. E., and Ellar, D. J., 1983a, Mechanisms of action of *B. thuringiensis* var. *israelensis* insecticidal δ-endotoxin, FEBS Lett., 154:362-368.

Thomas, W. E., and Ellar, D. J., 1983b, *Bacillus thuringiensis* var. *israelensis* crystal δ-endotoxin: Effects on insect and mammalian cells in vitro and in vivo, J. Cell Sci., 60:181- 197.

Van Zoelen, E. J. J., Zwaal, R. F. A., Reuvers, F. A. M., Demel, R. A., and Van Deenen, L. L. M., 1977, Evidence for the preferential interaction of glycophorin with negatively charged phospholipids, Biochim. Biophys. Acta, 464:484-492

Visser, B., van Workum, M., Dullemans, A., and Waalwijk, C., 1986, The mosquitocidal activity of *B. thuringiensis* var. *israelensis* is associated with Mr 230,000 and 130,000 crystal proteins, FEMS Microbiol. Lett., 30:211-214.

Waalwijk, C., Dullemans, A. M., Van Workman, M. E. S., and Visser, B., 1985, Molecular cloning and nucleotide sequence of Mr 28000 crystal protein gene of *Bacillus thuringiensis* subsp. *israelensis*, Nucleic Acids Res., 13:8207-8217.

Ward, E. S., and Ellar, D. J., 1986, *Bacillus thuringiensis* var. *israelensis* δ-endotoxin. Nucleotide sequence and characterization of the transcripts in *Bacillus thuringiensis* and *Escherichia coli*, J. Mol. Biol., 191:1-11.

Ward, E. S., Ellar, D. J., and Chilcott, C. N., 1988, Single amino acid changes in the *Bacillus thuringiensis* var. *israelensis* δ-endotoxin affect the toxicity and expression of the protein, J. Mol. Biol., 202:527-535

Wu, D., and Chang, F. N., 1985, Synergism in mosquitocidal activity of 26 and 65kDa proteins from *Bacillus thuringiensis* subsp. *israelensis* crystal, FEBS Lett., 190:232-236.

Yamamoto, T., Iizuka, T., and Aronson. J. N., 1983, Mosquitocidal protein of *Bacillus thuringiensis* subsp. *israelensis*: Identification and partial isolation of the protein, Current Microbiol., 9:279-284.

SYNAPTIC TOXINS FROM ARACHNID VENOMS: PROBES FOR NEW INSECTICIDE TARGETS

Michael E. Adams, Vytautas P. Bindokas, and Eliahu Zlotkin[1]

Division of Toxicology and Physiology
Department of Entomology
University of California
Riverside, CA 92521

and

[1]Department of Zoology
Institute of Life Sciences
Hebrew University of Jerusulam
91904 Jerusalem, Israel

ABSTRACT

Ion channel toxins of remarkable potency and selectivity have evolved in venomous arachnid predators of insects. Biochemical and physiological studies of spider and scorpion venoms reveal three classes of toxins defined by their specificity for ion channels in insect excitable membranes. The first class consists of acylpolyamine antagonists of glutamate-activated receptor channels in muscle. Insecticidal polypeptide toxins comprise a second class which alter the properties of voltage-activated sodium channels, causing either excitatory or depressant effects. Some of these toxins appear to be highly specific for insect sodium channels. The third class of venom toxins suppress neurotransmitter release by specific antagonism of voltage-activated calcium channels. Arachnid venom toxins thus offer pharmacological and biochemical probes for the characterization of ion channels as novel points of attack in the insect nervous system for new insecticides.

INTRODUCTION

The majority of agrochemicals used in insect control today are neurotoxins. Their insecticidal actions generally result from disruption of impulse conduction along nerve processes or interference with chemical signaling at synaptic junctions. While such insecticides are highly effective and continue to be a mainstay in pest control, their use is complicated by problems of resistance development, non-target toxicity and environmental impacts, all of which make the discovery of new, more sophisticated control strategies clearly desirable.

Ideal insecticides would be tailored as much as possible to physiological targets in specific to insects. The task of designing novel agents with high selectivity can be facilitated by expanding our knowledge of insect neural processes and the means by which toxins, both natural and synthetic, disrupt insect systems. What essential physiological and biochemical differences exist between insect systems and those of non-target organisms? We consider this question using the insect neuromuscular junction as a model synaptic system.

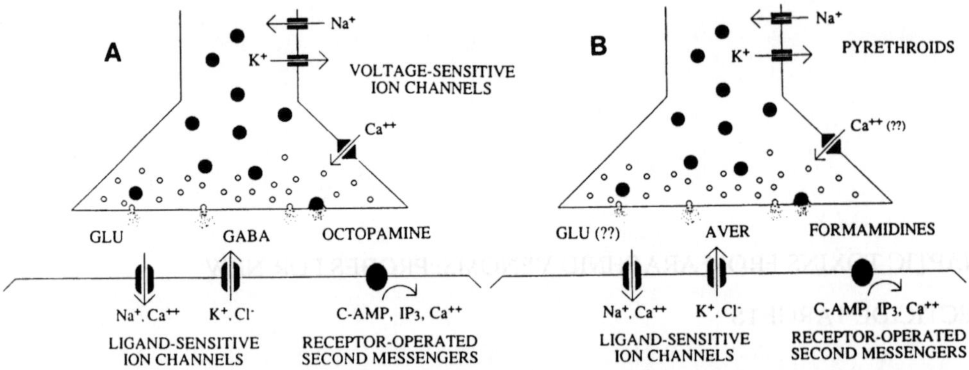

Fig. 1. (A) Illustration of a generalized insect neuromuscular junction depicting ion channels, neurotransmitters and receptor types. Voltage-sensitive sodium and potassium channels propagate action potentials to nerve endings, thus activating voltage-sensitive calcium channels involved in regulating neurotransmitter release. Release of L-glutamate (GLU), or GABA from small, electron translucent vesicles activates postsynaptic cation or anion channels to cause synaptic potentials. Biogenic amines (e.g., octopamine) and neuropeptides (e.g., proctolin) released from large neurosecretory granules interact with postsynaptic receptors coupled to intracellular second messengers. (B) Insecticides such as the pyrethroids modify voltage-activated sodium channels to produce dramatic changes in nervous system excitability; no insecticides are yet known to affect potassium or calcium channels. Insecticidal avermectins activate the GABA channel. Formamidine insecticides interact with octopamine receptors. Insecticides mimicking glutamate or neuropeptides have not yet been described.

Insect Synapses as Insecticide Targets

Striking differences in neuromuscular mechanisms exist between insects and vertebrates. Insect skeletal neuromuscular transmission is both excitatory glutamatergic, and inhibitory GABAergic (Fig. 1A; Pitman, 1985). This contrasts with exclusively excitatory cholinergic transmission at the vertebrate neuromuscular junction. Interestingly, a relatively new class of naturally-occurring insecticides with high selectivity, the avermectins, are believed to act on the insect GABA system (Fig. 1B), and consequently much interest is directed toward this site of action for new insecticide development. As yet, no known insecticide is targeted to glutamatergic mechanisms, although venomous predators of insects appear to contain high concentrations of paralytic toxins affecting the glutamate-activated receptor channel. These toxins may prove to be valuable tools in the characterization of glutamate receptor channels in the insect nervous system, and are described in subsequent sections of this paper (see also Usherwood, this volume). A third component of insect synaptic transmission, that of octopaminergic modulation (O'Shea and Evans, 1979), constitutes another potential target for insecticidal action and may be the target site of formamidine insecticides currently used in insect control.

Voltage-sensitive ion channels mediating neurotransmitter release are primary sites of action for certain neuroactive insecticides. Voltage-sensitive sodium channels serve as the primary site of action for pyrethroid and DDT-type insecticides (Narahashi, 1986; Fig. 1B). The pyrethroids are synthetic derivatives of naturally-occurring pyrethrins of chrysanthemum flowers.

Venom Toxins as Probes of Synaptic Processes

We are analyzing synaptic toxins occurring in the venoms of insect natural enemies. Toxins from orb weaver and funnel web spiders as well as scorpions paralyze insects by interacting with either ligand-sensitive or voltage-sensitive ion channels. While these toxins are not themselves envisioned as future insecticides, they may serve as valuable biochemical and pharmacological probes for processes vital to insect synaptic transmission. Detailed investigations of the structure and physiological actions of the arachnid venom toxins are generating new knowledge about the insect nervous system and may be of assistance in the characterization of new insecticide targets.

MATERIALS AND METHODS

Purification

Whole venoms from orb weaver spiders (*Argiope aurantia*, *Araneus gemma*, *Neoscona arabesca*) and a funnel web spider (*Agelenopsis aperta*) were purchased from Spider Pharm, Black Canyon City, AZ. Venoms were dissolved in 1% aqueous trifluoroacetic acid (TFA) and fractionated by reversed-phase liquid chromatography (RPLC) using a Perkin Elmer 410 solvent delivery system. Toxins were purified using C_4, C_8 or C_{18} bonded silica columns (Brownlee or Vydac) and various combinations of the following three solvent programs: 1. acetonitrile/water gradient in constant 0.1% TFA, 2. acetonitrile/water gradient in constant 0.1% heptafluorobutyric acid (HFBA), and 3. 1-propanol/water gradient in constant 1.0% TFA. UV_{280} or UV_{220} absorbance peaks of purified toxins were integrated and quantified by amino acid analysis. Further details regarding purification methods may be found in recent publications (Adams et al., 1989; Skinner et al., 1989).

Paralysis assays

In vivo bioassays for paralysis were performed on adult *Musca domestica* house flies, NAIDM strain. Adult females at three days post-eclosion were injected intra-abdominally with crude venoms or RPLC-purified fractions dissolved in insect saline (in mM: NaCl, 140; KCl, 5; $CaCl_2$, 5; $MgCl_2$, 1; $NaHCO_3$, 4; HEPES, 5; pH = 7.2). Paralysis was scored as loss of righting response at one hour post-injection; mortality was scored at 24 hours post-injection. ED_{50} and LD_{50} values (Table 1) were determined by probit analysis (Finney, 1971) according to the method of Raymond (1985), using 10-30 flies per dose. Further details appear elsewhere (Adams et al., 1987; 1989).

Neuromuscular assays

The effects of whole venoms and purified toxins were investigated on longitudinal ventrolateral muscles 6A and 7A (Irving and Miller, 1980) of pre-pupal *Musca domestica*. Muscles 6A and 7A are rectangular (300 x 700 µm), segmentally-repeated single cell pairs. Excitatory junctional potentials (EJPs) were neurally-evoked using suction electrodes and recorded intracellularly using standard techniques. Ionophoretic glutamate potentials were obtained by passing 10 nA, 10 msec negative current pulses through microelectrodes filled with 0.5 M L-glutamate (sodium salt) adjusted to pH 8.0; electrode impedances measured 70-100 megohms. Toxins were dissolved in physiological saline (same composition as above, except that $CaCl_2$ was adjusted to 0.75 mM to maximize the amplitude of ionophoretic glutamate potentials) and applied by continuous superfusion via a 400 µm pipette directed at the muscle. Quantal size and content were determined by loose-patch current recordings from neuromuscular terminals using methods similar to those of Dudel (1981). These and other methods used in the pre-pupal house fly preparation are described in detail in Adams et al. (1989) and Bindokas and Adams (1989).

POSTSYNAPTIC ANTAGONISTS: ACYLPOLYAMINE TOXINS

Chemical Characterization

Insect neuromuscular transmission is mediated by the excitatory neurotransmitter, L-glutamic acid (Usherwood, 1981). Until only a few years ago, postsynaptic antagonists of glutamatergic transmission were unavailable. This state of affairs changed rapidly with reports that orb weaver spider venoms contain potent postsynaptic antagonists (Fig. 2) with selectivity for glutamatergic synapses (Kawai et al., 1982; Abe et al., 1983; Bateman et al., 1984; Usherwood et al., 1984; Magazanik et al, 1987). Structure elucidation has followed quickly, beginning with "argiopine" (AR_{636}), an acylpolyamine toxin from *Argiope lobata* (Grishin et al., 1986). Over the past three years, acylpolyamines of similar size and structural motif have been purified and characterized from venoms of Eurasian and North American orb weaver genera, including *Argiope*, *Nephila*, and *Araneus* (Grishin et al, 1986; Aramaki et al., 1986; Adams et al., 1987; Budd et al., 1988; Toki et al., 1988) and from the funnel web spider *Agelenopsis aperta* (Adams et al., 1989; Skinner et al., 1989).

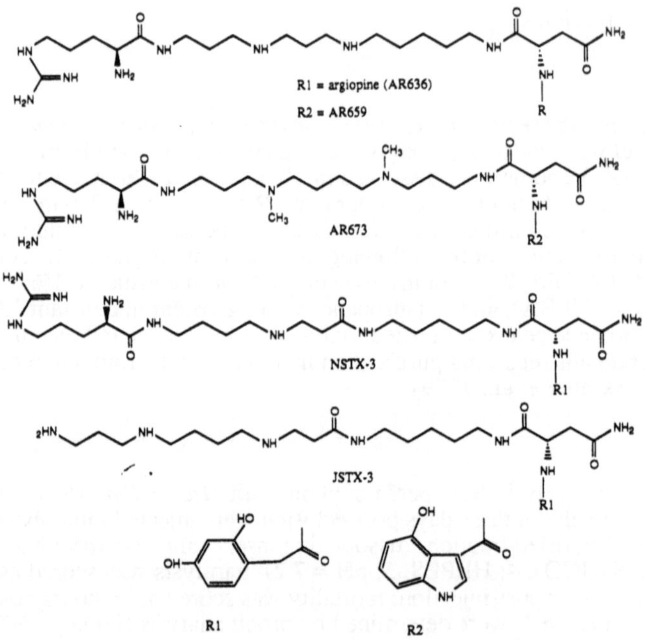

Fig. 2. Structures of acylpolyamine toxins from orb weaver spider venoms. Argiopine and argiotoxins (AR_{636}, AR_{659}, AR_{673}) occurring in *Argiope* venoms feature N-terminal arginine attached to a polyamine of variable structure, which in turn is attached to asparagine coupled to either 2,4-dihydroxyphenylacetic acid (R1) or 4-hydroxyindole-3-acetic acid (R2). Two additional toxins from *Nephila* species contain exclusively 2,4-dihydroxyphenylacetic acid coupled to putreanylcadaverine.

Three "argiotoxins" (AR_{636}, AR_{659}, and AR_{673}) from the North American species, *Argiope aurantia*, exhibit the basic structural motif of the acylpolyamine antagonists from orb weaver spiders (Fig. 2). N-terminal arginine is coupled to one of two polyamines: 1,13-diamino-4,8-diazatridecane (AR_{636}, AR_{659}) or 1,12-diamino-4,9-dimethyl-4,9-diazadodecane (AR_{673}). In each instance, the opposite end of the polyamine is attached via an second amide linkage to the carboxyl group of asparagine, whose amino group is coupled by a third amide bond to either 2,4-dihydroxyphenylacetic acid (AR_{636}) or to 4-hydroxyindoleacetic acid (AR_{659}, AR_{673}). Structural assignments of AR_{636} and AR_{659} were verified by chemical synthesis and bioassay. Molar concentrations of AR_{636}, AR_{659}, and AR_{673} in crude *Argiope aurantia* venom are approximately 20 mM each, or 12-14 mg/ml.

AR_{636} originally was described as argiopine, a constituent of *Argiope lobata* venom (Grishin et al., 1986) and is probably identical to argiotoxin 636 from *Argiope trifasciata* venom (Budd et al, 1988). The polyamine toxins NSTX-3 and JSTX-3 from the New Guinea spiders *Nephila clavata* and *Nephila maculata*, respectively, are similar to argiopine. They contain 2,4-dihydroxyphenylacetic acid coupled to asparagine (Fig. 2), but differ in the structure of the polyamine group, which is described as putreanylcadaverine (Aramaki et al., 1986). JSTX-3 is substituted with an aminopropyl group in place of N-terminal arginine. Additional *Nephila* toxin structures containing unsubstituted indoleacetic acid and further variations on the polyamine constituent have recently appeared (Toki et al., 1988).

Polyamine toxins of a similar chemical nature occur also in venom of the funnel web spider, *Agelenopsis aperta* (Fig. 3). Five of these "α-agatoxins" have been described: AG_{488}, AG_{489}, AG_{504}, AG_{505}, and AG_{452} (Adams et al, 1989; Skinner et al, 1989). Like the argiotoxins, the α-agatoxins are present at millimolar concentrations (3-32 mM) in crude venom (Table 1). The most abundant α-agatoxin is AG_{489}, which is present at a concentration of 32 mM. The α-agatoxins contain polyamine and aromatic moieties characteristic of the orb weaver toxins, but lack conventional amino acids (Skinner et al., 1989; Adams et al., 1989). UV absorbance characteristics of the α-agatoxins suggest the presence of both substituted and unsubstituted indolic moeities (Adams et al., 1989).

Biological Actions of Acylpolyamine Toxins

The argiotoxins produce a temporary, flaccid paralysis when injected into house flies, cockroaches, locusts or moths. The ED_{50} values (effective dose to paralyze 50% of treated flies at 1 hour post-injection) for AR_{636}, AR_{659} and AR_{673} in house flies of 20 mg body weight are in the range of 0.8-4.5 pmol/mg (0.5-3.0 µg/g), AR_{659} being the most active (Table 1). Recovery usually occurs within 2-3 hours of injection at these doses. We estimated house fly hemolymph volume to be about 30 µl, suggesting that *in vivo* concentrations produced by ED_{50} doses are in the range of 0.5-3.0 µM. This concentration range is in reasonable correspondence to that producing 50% block of neurally evoked excitatory junctional potentials (EJPs) in pre-pupal house fly muscle (0.2-0.5 µM; Table 1 in Adams et al., 1987).

The orb weaver acylpolyamine toxins antagonize insect neuromuscular transmission blocking the postsynaptic glutamate-activated receptor channel. The mechanism of action has been

Table 1. Abundances (mM), molecular weights (M_r) and *in vivo* biological activities (ED_{50} or LD_{50}) of *Argiope aurantia* and *Agelenopsis aperta* spider venom toxins against the house fly, *Musca domestica* and the tobacco hornworm, *Manduca sexta.* Values for argiotoxins (AR_{xxx}) are from Adams *et al*., (1987). Relative abundances and biological activities for α- (AG_{xxx}), µ- (µ-Aga-I-VI) and an ω-agatoxin (ω-Aga-IA) against *M. domestica* are from Adams et al., (1989). All values for *M. sexta* are from Skinner et al., (1989). ED_{50} values are the injected dose necessary to cause 50% paralysis at one hour post-injection. LD_{50} values are 50% lethal injected dose at 24 hours post-injection.

Toxin	Abundance (mM)	M_r	ED_{50} (µg/g) *M. domestica*	*M. sexta*
AG_{489}	32.0	489	35.6 ± 5.4	54 ± 61
AG_{505}	5.3	505	14.8 ± 2.0	80 ± 21
AG_{488}	8.7	488	95.6 ± 10.7	124 ± 56
AG_{504}	3.0	504	69.0 ± 15.6	>74
AR_{659}	20.0	659	0.52 ± 0.06	62 ± 20
AR_{636}	20.0	636	2.1 ± 0.2	---
AR_{673}	20.0	673	3.0 ± 0.2	---
			LD_{50} (µg/g)	
µ-Aga I	0.4	4264	0.32 ± 0.02	28 ± 7
µ-Aga II	0.8	4137	0.57 ± 0.30	75 ± 27
µ-Aga III	0.6	4188	0.24 ± 0.25	28 ± 12
µ-Aga IV	1.3	4199	0.12 ± 0.01	40 ± 9
µ-Aga V	1.3	4199	0.33 ± 0.08	48 ± 11
µ-Aga VI	1.6	4159	0.62 ± 0.02	38 ± 12
			ED_{50} (µg/g)	
ω-Aga IA	0.10	~7500	2.2 ± 0.6	

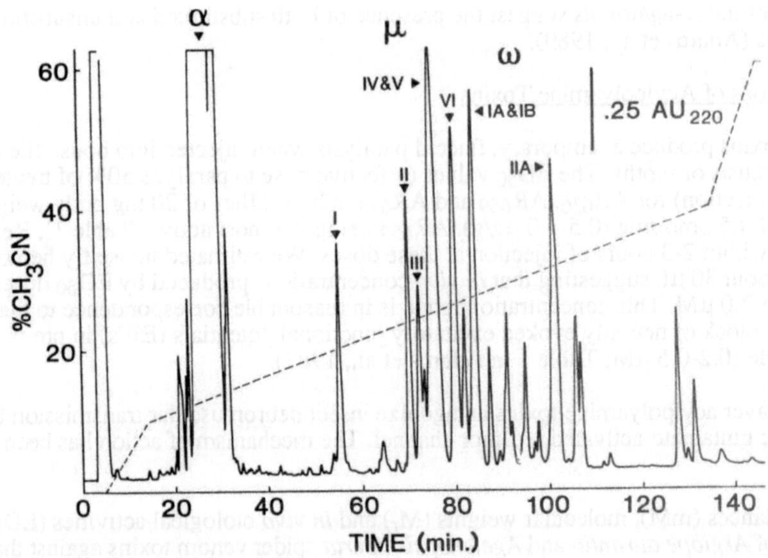

Fig. 3. Reversed-phase liquid chromatography of whole *Agelenopsis aperta* venom, showing elution profiles of three classes of ion channel-specific toxins, the α- , μ- and ω-agatoxins. The venom is fractionated using a Brownlee Aquapore C_8 column and a linear gradient of acetonitrile (CH_3CN)/water in constant 0.1% trifluoroacetic acid at a flow rate of 1.0 ml/min. Multiple acylpolyamine α-agatoxins are condensed into the peak labelled "α". Two types of presynaptically active polypeptide toxins appear subsequently: The μ-agatoxins (μ-Aga-I, II, III, IV, V, VI) elute between 50-80 min, followed by the ω-agatoxins (ω-Aga-IA, IB, and IIA), appearing between 80-100 min.

described in greatest detail for argiopine (AR_{636}). Argiopine antagonism at the blowfly and locust neuromuscular junctions is use-dependent and appears to result from non-competitive or un-competitive binding to the receptor channel complex. Voltage clamp and noise analyses suggest that the toxin exerts a voltage-dependent block of the channel in its open state (Bateman et al., 1984; Magazanik et al., 1987). Argiopine also inhibits cholinergic transmission at the frog neuromuscular junction, but is some 50-fold less potent as compared with the blowfly neuromuscular junction. On the strength of these data, argiopine appears to be primarily a cation channel blocker with some selectivity for glutamatergic transmission. Studies on the effects of argiopine on extrajunctional glutamate channels in locust muscle have provided further evidence for open channel block and a possible allosteric influence of the toxin on the glutamate binding site (Kerry et al., 1988 and see Usherwood, this volume). In our hands, the argiotoxins (AR_{636}, AR_{659}, AR_{673}) block house fly neuromuscular transmission in identical fashion (Adams et al., 1987; Adams, 1988), suggesting that the indolic and phenylacetic acid toxins antagonize transmission in similar fashion.

The α-agatoxins of *Agelenopsis aperta*, like the argiotoxins, produce reversible paralysis in house flies (Table 1) that is correlated with use-dependent block of neuromuscular transmission (Fig. 4). EJPs in pre-pupal house fly muscle are reduced by 50% following treatment with 0.8 μM AG_{489} under conditions of continuous nerve stimulation. Synaptic potentials decrease almost immediately upon introduction of toxin into the bath (Fig. 4A). However, when toxin is applied in the absence of stimulation, (Fig. 4B) no significant block of evoked EJPs is produced until stimuli are resumed. This shows that synaptic activation is required for the toxin to act. In a third experiment (Fig. 4C), the muscle preparation is exposed to AG_{489} in the absence of nerve stimulation, but in the presence of the neurotransmitter, L-glutamate. Under these conditions, EJPs are substantially blocked in the absence of nerve stimulation. Thus, block of EJPs is observed when the transmitter is supplied either from endogenous stores by nerve stimulation or by bath application.

The presence of glutamate in the bath also slows the rate of recovery from synaptic block by

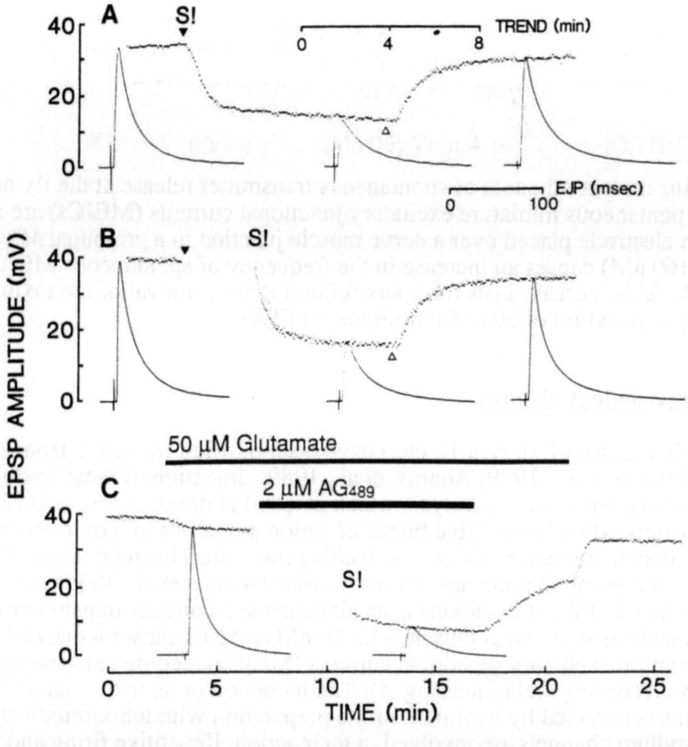

Fig. 4. Use-dependence of AG489 antagonism at the pre-pupal fly neuromuscular junction. A) Application of 0.8 µM AG489 (filled arrow) applied with nerve stimulation (S!) immediately reduces the amplitude of the neurally-evoked EJP; recovery is rapid upon removal of the toxin (open arrow). B) AG489 is applied in identical manner (filled arrow), except that stimulation is stopped during the first 4 min of exposure to the toxin. Upon resumption of stimulation (S!), EJPs of normal amplitude are evoked initially, but they rapidly decrease. This suggests that antagonism by the toxin requires synaptic activation. C) AG489 is applied in the absence of nerve stimulation (wide black bar), but in the presence of 50 µM L-glutamic acid (narrow black bar). When stimuli are resumed following a 4 minute pre-exposure to the toxin (S!), EJP amplitudes already have been suppressed. Note that during the first phase of toxin washout (following end of wide black bar), the rate of recovery of EJP amplitude is slower than following the removal of glutamate (end of narrow black bar). A) and B) are modified from Adams et al. (1989).

AG489. As shown in Fig. 4C, the initial phase of recovery following removal of the toxin is slower than the subsequent phase during which glutamate is removed from the bath. Thus, the presence of glutamate is necessary for synaptic block to occur, and as well slows the rate of recovery. This result is intriguing, since *Agelenopsis* venom contains additional toxins which increase glutamate release from nerve terminals. As we describe in the next section, polypeptide toxins in *Agelenopsis* venom which increase transmitter release enhance the paralytic actions of AG489.

PRESYNAPTIC POLYPEPTIDE TOXINS

Further biochemical analysis of *Agelenopsis aperta* venom has revealed two classes of synaptic toxins in addition to the α-agatoxins. These polypeptides affect voltage-sensitive ion channels in presynaptic axons and nerve terminals. The first of these, the µ-agatoxins, interact with sodium channels in a manner similar to scorpion toxins from *Androctonus* and *Leiurus* species. The second class of toxins, the ω-agatoxins, are somewhat larger polypeptides which antagonize neuronal calcium channels.

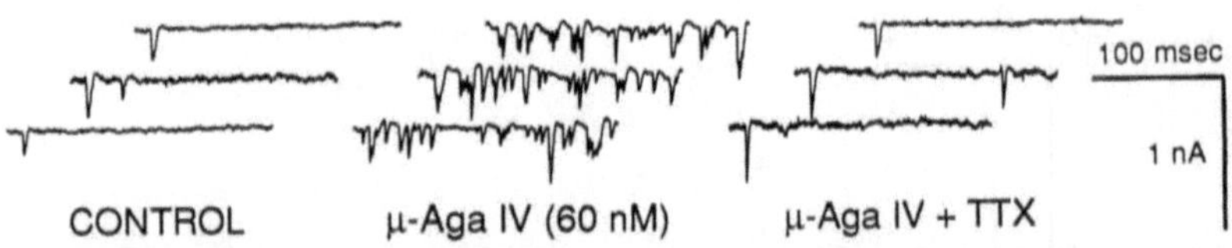

Fig. 5. μ-agatoxins increase the rate of spontaneous transmitter release at the fly neuromuscular junction. Spontaneous miniature excitatory junctional currents (MEJCs) are recorded with a loose patch electrode placed over a nerve muscle junction in a pre-pupal *Musca domestica*. μ-Aga-IV (60 nM) causes an increase in the frequency of spontaneous MEJCs over the CONTROL. This increase lasts for hours following the removal of the toxin, but it can be reversed by application of 50 nM tetrodotoxin (TTX).

μ-Agatoxins: Presynaptic Activators

Six μ-agatoxins (μ-Aga-I, μ-Aga-II, etc.) have been purified by RPLC from *Agelenopsis* venom (Fig. 3; Skinner et al., 1989; Adams, et al., 1989). Injection of these toxins into adult house flies produces a convulsive paralysis which is lethal at doses as low as 0.12 μg/g (Table 1). Paralysis is correlated with repetitive bursts of action potentials in motor neurons which persist long after toxins are removed from the bathing medium. This repetitive activity originates presynaptically in the terminal branches of motor axons (Adams et al., 1989). An additional and quite remarkable effect of the μ-agatoxins is an elevation spontaneous transmitter release (Fig. 5). For example, treatment of insect muscle with 20 nM μ-Aga-I causes a marked increase in the frequency of miniature excitatory junctional currents (MEJCs) recorded at synaptic sites using a loose patch clamp recording technique (Fig. 5). The induction of an increased MEJC frequency by the μ-agatoxins is reversed by treatment of the preparation with tetrodotoxin (Fig. 5), suggesting that sodium channels are involved in their action. Repetitive firing and increased spontaneous transmitter release also are observed following treatment of insect nerve muscle preparations with scorpion toxins (Zlotkin et al., 1983; 1985) and pyrethroid insecticides (Salgado et al, 1983), both of which have been shown by voltage clamp analysis to alter the properties of voltage-sensitive sodium channels.

Sequence analysis of the six μ-agatoxins reveals that they contain 36-38 amino acids, including 8 half-cystines (Skinner et al., 1989). The amino acid sequences of the μ-agatoxins are highly conserved, particularly with respect to placement of the cysteine residues, indicating that disulfide bridging may be crucial to biological activity. The μ-agatoxins occur in the whole venom at concentrations of approximately 1 μM, or about 10-30 fold less than the α-agatoxins (Table 1).

Joint action of α- and μ-agatoxins produces enhanced paralysis in house flies

Since the α- and μ-agatoxins occur together in *A. aperta* venom, we investigated the possibility that the two types of toxin may act synergistically in paralysis assays. The potential for interaction between these two types of toxin was suggested by the observation that the μ-agatoxins increase the release of the neurotransmitter L-glutamate from nerve terminals (Fig. 5). Furthermore, the presence of glutamate is necessary for α-agatoxin-induced block to occur and as well slows recovery of synaptic potentials following removal of the toxin (Fig. 4).

The degree and time course of paralysis produced by injection of adult house flies with AG_{489} and μ-Aga-I alone and in combination were measured (Fig. 6). Injection of AG_{489} at the ED_{50} dose (35.2 μg/g; Table 1) produced paralysis within 30 minutes followed by recovery over the ensuing 2 hours. Injection of μ-Aga-I at 1.5 x LD_{50} (0.48 μg/g) caused lethal paralysis, but symptoms developed over a relatively slow time course of 4-5 hours. When each toxin was co-injected at the same dose, 100% paralysis occurred within 30 minutes, with little or no recovery ensuing over the subsequent 24 hour period. In particular, recovery from AG_{489} paralysis between 2 and 5 hours was almost completely suppressed. This lack of recovery could result from increased transmitter release caused by co-injection of μ-Aga-I, an hypothesis that

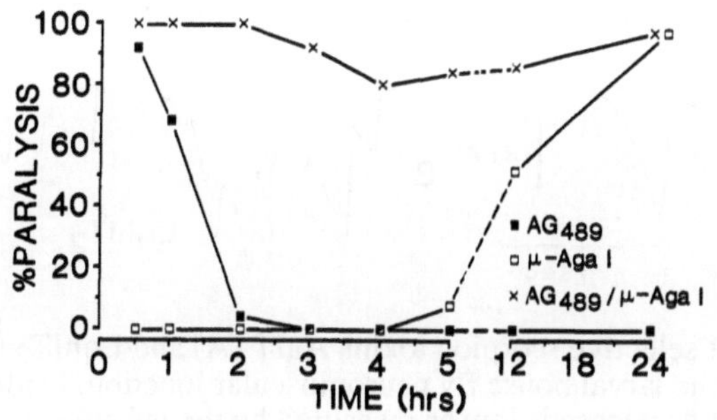

Fig. 6. Joint action of α- and μ-agatoxins in a house fly paralysis assay. Injection of AG_{489} at the ED_{50} dose (35.2 ug/g; Table 1) produced paralysis within 30 minutes followed by recovery over the ensuing 2 hours. Injection of μ-Aga-I at 1.5 x LD_{50} (0.48 μg/g) led to lethal paralysis characterized by slow onset over a period of 4-5 hours. When each toxin was co-injected at the doses used previously, 100% paralysis was observed within 30 minutes, with little or no recovery occurring over the subsequent 24 hour period. In particular, recovery from AG_{489} paralysis between 2 and 5 hours was almost completely suppressed.

is supported by *in vitro* data (Fig. 4C), showing that the presence of glutamate dramatically slowed the rate of recovery from AG_{489} block.

Effects of Insect-Selective Scorpion Toxins on House Fly Neuromuscular Activity

The active ingredients of scorpion venoms are polypeptide sodium channel toxins which show selectivity for insects, crustaceans or mammals (Zlotkin, 1983). The insect selective toxins AaIT (*Androctonus australis)* and $LqhIT_2$ (*Leiurus quinquestriatus hebraeus*) induce opposite symptoms when injected into blowfly larvae; AaIT produces tetanic paralysis, while LqhITproduces flaccid paralysis. Based on these symptoms, AaIT and $LqhIT_2$ are referred to as "excitatory" and "depressant" toxins, respectively. Using the house fly neuromuscular junction preparation, we compared the physiological actions of AaIT and LqhIT with those of the μ-agatoxins.

Exposure of house fly pre-pupae to 50 nM AaIT leads to persistant repetitive activity in motor neurons (Fig. 7A). Although this excitatory activity resembles that produced by the μ-agatoxins, AaIT does not produce the elevation of MEJC frequency observed after μ-Aga-IV treatment (Fig. 5). $LqhIT_2$ causes only minimal repetitive activity; its effect is dominated by an increased level of spontaneous transmitter release, measured as an increase in MEJC frequency (Fig. 7B). As in the case of the μ-agatoxins, the effects of both AaIT and $LqhIT_2$, that is repetitive firing and increased transmitter release, were reversed by treatment of the nerve-muscle preparation with TTX (not shown). The physiological actions of the μ-agatoxins and the scorpion toxins AaIT and $LqhIT_2$ are attributable to their actions on a single target site: the voltage-sensitive sodium channel. However, their relative effects on the processes of channel activation and inactivation determine whether excitatory or depressant symptoms predominate. The effects of scorpion toxins on sodium channel function have been analyzed in detail by voltage clamp techniques using the cockroach giant axon preparation. The excitatory toxins AaIT and $LqqIT_1$ cause the following two actions on sodium conductance: 1. a small depolarization resulting from increased sodium permeability at resting membrane potential without suppression of activatable peak sodium conductance, and 2. a slowing of voltage-dependent sodium channel inactivation (Pelhate and Zlotkin, 1981; Zlotkin et al., 1985). Both of these actions lead to repetitive action potentials. By contrast, the dominant effects of the depressant toxin $LqqIT_2$ are: 1. a large, sustained depolarization of membrane potential and suppression of peak sodium channel activation

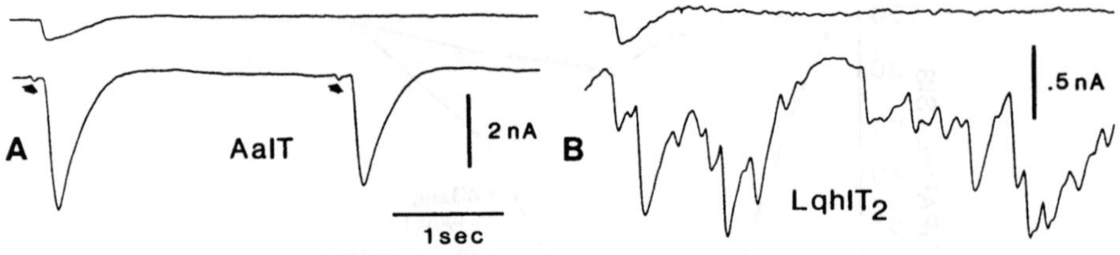

Fig. 7. Effects of the insect selective scorpion toxins AaIT (A) and $LqhIT_2$ (B) on synaptic currents recorded at the larval house fly neuromuscular junction. Under normal conditions, spontaneous transmitter release is low as measured by the infrequent occurrence of MEJCs (top traces in A and B). However, upon treatment with the insect selective scorpion toxin, AaIT (A, lower trace) spontaneous EJCs preceeded by nerve action potentials (arrows) appear. This repetitive activity is correlated with convulsive activity in injected whole insects. In contrast, the dominant effect of 6 nM $LqhIT_2$ is an increase in spontaneous transmitter release with very little repetitive activity.

(Zlotkin et al., 1985). The suppression of peak sodium channel activation produce blockade of action potentials in giant axons and would be expected to suppress synaptic transmission in the same manner; hence the depressant action of $LqqIT_2$ and probably $LqhIT_2$ may be hypothesized to involve the suppression of action potentials in motor axons. Using these studies as reference points, we hypothesize that the elevation in MEJC frequency produced by the μ-agatoxins and $LqhIT_2$ results from increased permeability of sodium channels at the resting membrane potential. However, since the μ-agatoxins do not show obvious depressant actions, they may not produce the strong suppression of peak sodium conductance that characterizes $LqqIT_2$, and probably $LqhIT_2$.

PRESYNAPTIC ANTAGONISTS: ω-AGATOXINS

The ω-agatoxins are the third and most recent class of polypeptide toxins to be described from *Agelenopsis* venom. They were recognized in physiological assays to antagonize neuromuscular transmission by a presynaptic action (Bindokas and Adams, 1989), in contrast to the α-agatoxins, which are postsynaptic antagonists (Adams et al., 1989). We have shown that the ω-agatoxins block presynaptic, voltage-activated calcium channels.

ω-agatoxins suppress neurally-evoked EJPs in house fly muscle without affecting the ionophoretic glutamate potential (Fig. 8A). This suggests that this class of toxin does not interfere with the interaction of glutamate-activated postsynaptic response and may instead reduce the amount of transmitter released from presynaptic stores. This hypothesis is supported by data showing that amplitudes of evoked EJCs but not MEJCs are suppressed by ω-Aga-IA (Bindokas and Adams, 1989). Such reduction in the ratio of evoked EJC to MEJC, or quantum content, could be explained by direct antagonism of either sodium or calcium channels. The second distinguishing characteristic of ω-Aga-IA is its irreversible action; 18 nM toxin irreversibly suppresses neurally evoked EJPs within 12 minutes (Fig. 8A).

The results of several experiments indicate that ω-Aga-IA blocks neuronal calcium channels without affecting sodium channels (Bindokas and Adams, 1989). For example, following blockade of EJPs by ω-Aga-IA, depolarization of nerve terminals generates sodium action potentials. In experiments designed to examine the functioning of calcium channels more directly, sodium action potentials were eliminated by treatment of the preparation with tetrodotoxin. Under this condition, EJPs may be elicited by direct nerve terminal depolarization to activate presynaptic calcium channels. Such EJPs are suppressed by both cobalt and ω-Aga-IA. These data suggest that ω-Aga-IA blocks calcium channels in nerve terminals.

To more directly test for effects on neuronal calcium currents, we examined the effect of ω-Aga-IA on action potentials in dorsal unpaired median (DUM) neurons of the locust metathoracic

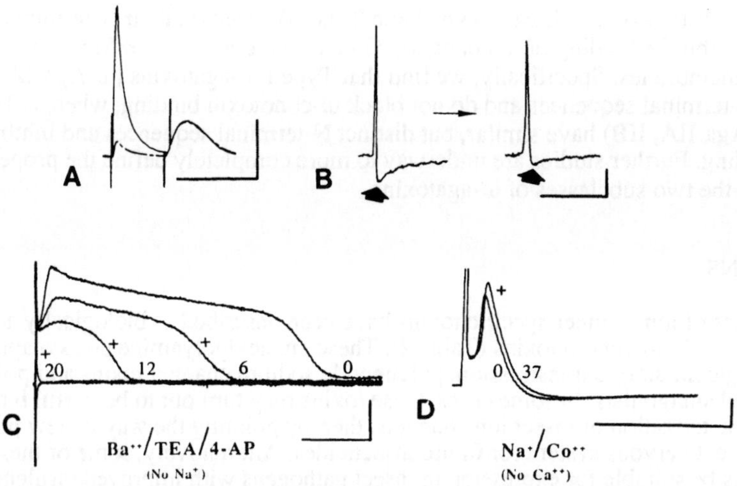

Fig. 8. ω-Aga-IA blocks neuronal calcium channels in insects. (A) Neurally-evoked EJPs in house fly muscle are blocked by the toxin, whereas iontophoretic glutamate potentials (IP) are unaffected. The upper EJP trace (solid line) is the pre-treatment control; the lower trace (broken line) shows the reduced EJP 12 minutes after exposure to 18 nM ω-Aga-IA. The EJP amplitude remains blocked after up to 3 hours of washing the preparation in toxin-free saline. All waveforms are the average of 10 individual sweeps. (B) ω-Aga-IA blocks the calcium, but not the sodium component of action potentials (APs) recorded from locust dorsal unpaired median (DUM) neuron somata. Under normal saline conditions, spike amplitude and hyperpolarizing after potential (arrows) of spontaneously occurring APs are diminished after treatment with 36 nM ω-Aga-IA. (C) DUM APs in saline containing tetraethylammonium (TEA), 4-aminopyridine (4-AP), no sodium, and barium substituting for calcium are blocked by 36 nM ω-Aga-IA. Records (+) show progressive block at 6, 12, and 20 min after treatment with toxin. Little more than the passive electric response remains (lower record is the response to hyperpolarizing current). (D) Toxin treatment (+) does not block the pure sodium potential recorded under conditions in which calcium is replaced by cobalt. Records show mean (n = 20) pretreatment (0) and 37 min postreattment potentials. Scale bars: (A) 10 mV, 100 msec; (B), (C): 20 mV, 100 msec; (D): 20 mV, 50 msec (adapted from Bindokas and Adams, 1989).

ganglion (Bindokas and Adams, 1989). Action potentials in DUM neuron somata result from a combination of sodium, calcium and potassium currents (Goodman and Spitzer, 1981). Application of ω-Aga-IA at 36 nM (Fig. 8B) produces a slight reduction in the peak amplitude of normal DUM action potential and also reduces the amount of undershoot. Further analysis was carried out under conditions in which the calcium and sodium components of the action potential were studied in isolation. To monitor calcium responses, barium was used as a charge carrier in sodium-free saline containing tetraethylammonium and 4-aminopyridine to block potassium channels. These potentials were completely blocked by ω-Aga-IA (Fig. 8C). In contrast, pure sodium potentials were not affected (Fig. 8D). We conclude that the effect of ω-Aga-IA involves a direct effect on neuronal calcium channels.

Three ω-agatoxins have been partially characterized. Gel electrophoresis analysis suggests molecular weights of 7.5, 8.2 and 10 kDa for ω-Aga-IA, IB and IIA, respectively. Amino acid analyses indicate that that, like the μ-agatoxins and scorpion toxins, the ω-agatoxins contain multiple disulfide bridges. Preliminary sequence information shows that the positions of several cysteine residues in μ- and ω-agatoxins are significantly conserved. The ω-agatoxins occur at about 10-50-fold lower abundance than the μ-agatoxins in whole venom (ca. 1-100 μM; 0.1-0.6% total protein).

The ω-agatoxins may provide useful biochemical probes for characterization of calcium channel binding sites in diverse animal phyla. For example, ω-Aga-IA also blocks transmission at the frog neuromuscular junction (Bindokas and Adams, 1989). Recent data indicate that ω-agatoxins occur as two sub-classes (Type I and Type II) based on their N-terminal sequences and ability to inhibit the binding of ω-conotoxin GVIA (Olivera et al., 1985) to chick synaptosomal membranes. Specifically, we find that Type I ω-agatoxins (ω-Aga-IA, IB, IC) have similar N-terminal sequences and do not block ω-conotoxin binding, whereas Type II ω-agatoxins (ω-Aga-IIA, IIB) have similar, but distinct N-terminal sequences and inhibit ω-conotoxin binding. Further studies are underway to more completely define the properties and specificities of the two subclasses of ω-agatoxins.

CONCLUSIONS

Three classes of ion channel-specific toxins have been described as biologically active constituents of arachnid venom toxins (Table 2). These are acylpolyamine postsynaptic antagonists of glutamatergic transmission, polypeptide sodium channel toxins and polypeptide calcium channel antagonists. In some cases, these toxins may turn out to be useful biochemical probes for characterization of insect ion channels, thereby pointing the way to new points of attack in the insect nervous system for future insecticides. Alternatively, some of these toxins may themselves be suitable for expression in insect pathogens with improved virulence.

Acylpolyamine antagonists of glutamatergic transmission appear to be a common constituent of the venoms of the spider families Araneidae (*Argiope, Araneus, Neoscona, Nephila*)and Agelenidae (*Agelenopsis*). The use-dependence and voltage-sensitivity of antagonism by these toxins support their designation as cation channel blockers, but they nevertheless show significant preference for glutamate-activated channels in insect muscle (Magazanik et al., 1987). Although it seems unlikely that the polyamine toxins will themselves be developed as insecticides, they are already being put to use in the molecular characterization of glutamate-activated receptor channels in insects. These ongoing studies will help to resolve whether agents focused on glutamatergic transmission in insects can serve as a new class of insecticides.

Sodium channel toxins occurring in both funnel web spider and scorpion venoms have irreversible effects on neuronal sodium channels and show impressive insecticidal potencies. Whereas the μ-agatoxins and AaIT modify sodium channels to induce convulsive paralysis, $LqhIT_2$ produces a depressant action. It is worth mentioning the obvious analogy between venom sodium channel toxins and pyrethroid insecticides (Narahashi, 1986). Voltage clamp analyses of pyrethroid effects on squid and crayfish giant axons show a tendency for Type I insecticides, like AaIT and μ-agatoxins, to slow sodium channel inactivation with minimal

Table 2. Three classes of channel-specific toxins from orb weaver and funnel web spider venoms

Class of Toxin	Source	Physiological Target
I. Postsynaptic Antagonists	*Argiope aurantia* ($AR_{636, 673, 659}$) *Araneus gemma* ($AN_{622, 743, 758}$) *Neoscona arabesca* ($NE_{673, 659}$) *Agelenopsis aperta* ($AG_{488, 489, 504, 505}$)	Postsynaptic, glutamatergic receptor channel
II. Presynaptic Activators	*Agelenopsis aperta* (μ-agatoxins I, II, III, IV, V, VI)	Presynaptic, voltage-sensitive sodium channel
III. Presynaptic Antagonists	*Agelenopsis aperta* (ω-agatoxins IA, IB, IIA, IIB)	Presynaptic, voltage-sensitive calcium channel

suppression of activation. Similarly, the Type II insecticides strongly depolarize and suppress peak sodium activation as has been shown for the scorpion toxin $LqqIT_2$ (Zlotkin, 1985). It may be feasible to employ radiolabelled scorpion toxins to more clearly characterize Type I and Type II pyrethroid binding to insect sodium channels.

It is fascinating that insecticidal sodium channel toxins have evolved in spiders, scorpions, and chrysanthemum flowers. Can arachnid venom sodium channel toxins lend additional insight into the development of new, selective insecticides? It has already been shown that different scorpion toxins can distinguish between insect, crustacean and mammalian sodium channels (Zlotkin, 1983). Although such selectivity has not been demonstrated for the μ-agatoxins, we have preliminary information suggesting that vertebrate sodium channels are relatively insensitive to their effects. Further studies of the interactions between arachnid toxins and insect sodium channels therefore may provide insights into points of selectivity for new insect-specific insecticides.

The ω-agatoxins of *Agelenopsis aperta* venom appear to constitute a new class of neuronal calcium channel antagonists. Our information thus far suggests that, unlike arachnid sodium channel toxins, the ω-agatoxins are biologically active over a broad range of animal phyla. Thus, ω-Aga-IA blocks neuromuscular transmission in both insect and frog, and also suppresses calcium action potentials in insect neurons. In addition, Type II ω-agatoxins block the binding of ω-conotoxin GVIA in vertebrate brain. This lack of ω-agatoxin selectivity for insects over other animal groups contrasts toxins from *Hololena curta* (Bowers et al., 1987) and *Plectreurys tristis* (Branton et al., 1987) spider venoms. Toxins from these venoms cause presynaptic antagonism at the insect neuromuscular junction but appear to be inactive against vertebrate calcium channels. Further characterization of spider venom calcium channel antagonists hopefully will lead to explanations for their differing spectrum of activities and help to define aspects of calcium channel structure and function that may make them potential targets for novel insecticides.

REFERENCES

Abe, T., Kawai, N. and Miwa, A., 1983, Effects of a spider toxin on the glutaminergic synapse of lobster muscle, J. Physiol., 339: 243.

Adams, M. E., 1988, Synaptic Ion Channel Toxins From Spider Venoms, in: "Neurotox '88. Molecular Basis of Toxic Action", G.G. Lunt, ed., pp. 49-59, Elsevier, Amsterdam.

Adams, M. E., Carney, R. L., Enderlin, F. E., Fu, E. T., Jarema, M. A., Li, J. P., Miller, C. A., Schooley, D. A., Shapiro, M. J., and Venema, V. J., 1987, Structures and biological activities of three synaptic antagonists from orb weaver spider venom, Biochem. Biophys. Res. Comm., 148: 678.

Adams, M. E., E. E., Herold, and Venema, V. J., 1989, Two classes of channel-specific toxins from funnel web spider venom, J. Comp. Physiol. A, 164: 333.

Aramaki, Y., Yasuhara, T., Higashijima, T., Yoshioka, M., Miwa, A., Kawai, N., and Nakajima, T., 1986, Chemical characterization of spider toxin, JSTX and NSTX, Proc. Japan Acad. 62B, 359.

Bateman, A., Boden, P. , Dell, A., Duce, I. R., Quicke, D. L. J, and Usherwood, P. N. R., 1985, Postsynaptic block of a glutamatergic synapse by low molecular weight fractions of spider venom. Brain Res., 339: 237.

Bindokas, V. P. and Adams, M. E., 1989, ω-Aga-I: A presynaptic calcium channel antagonist from venom of the funnel web spider, *Agelenopsis aperta*., J. Neurobiol., 20: 171.

Bowers, C. W., Phillips, H. S., Lee, P., Jan, Y. N., and Jan, L. Y., 1987, Identification and purification of an irreversible presynaptic neurotoxin from the venom of the spider, *Hololena curta*, Proc. Natl. Acad. Sci., 84: 3506.

Branton, W. D., Kolton, L., Jan, Y. N., Jan, L. Y., 1987, Neurotoxins from *Plectreurys* spider venom are potent presynaptic blockers in *Drosophila*, J. Neurosci., 7: 4195.

Budd, T., Clinton, P., Dell, A., Duce, I. R., Johnson, S. J., Quicke, D. L. J., Taylor, G. W., Usherwood, P. N. R., and Usoh, G., 1988, Isolation and characterization of glutamate receptor antagonists from venoms of orb-web spiders, Brain Res., 448: 30.

Dudel, J., 1981, The effect of reduced calcium on quantal unit current and release at the crayfish neuromuscular junction, Pflugers Arch., 391: 35.

Evans, P. D., 1984, A modulatory octopaminergic neuron increases cyclic nucleotide levels in locust skeletal muscle, J. Physiol. (Lond), 348: 307.

Finney D. J., 1971, "Probit Analysis", Cambridge University Press, Cambridge.

Goodman, C. S. and Spitzer, N. C., 1981, The mature electrical properties of identified neurones in grasshopper embryos, J. Physiol., 313: 369.

Grishin, E. V., Volkova, T. M., Arseniev, A. S., Reshetova, O. S., Onoprienko, V. V., Magazanic, L. G., Antonov, S. M., and Fedorova, I. M., 1986, Structure-functional characterization of argiopine- an ion channel blocker from the venom of spider, *Argiope lobata*, Bioorg. Khim., 12: 110 (In Russian).

Irving, S. N. and Miller, T. A., 1980, Ionic differences in fast and slow neuromuscular transmission in body wall muscles of *Musca domestica*, J. Comp. Physiol., 135: 291.

Kawai, N., Niwa, A., and Abe, T., 1982, Spider venom contains specific receptor blocker of glutaminergic synapses, Brain Res., 247: 169.

Kerry, C. J., R. L. Ramsay, M. S. P. Sansom, and Usherwood, P. N. R., 1988, Single channel studies of non-competitive antagonism of a quisqualate-sensitive glutamate receptor by argiotoxin$_{636}$- a fraction isolated from orb-web spider venom, Brain Res., 459: 312.

Magazanik, L. G., Antonov, S. M., Fedorova, I. M., Volkova, T. M., Grishin, E. V., 1987, Argiopin - a naturally occurring blocker of glutamate-sensitive synaptic channels, in: "Receptors and Ion Channels", Y.A. Ovchinnikov and F. Hucho, eds., pp. 305-312, de Gruyter, New York.

Narahashi, T., 1986, Mechanisms of action of pyrethroids on sodium and calcium channel gating, in: "Neuropharmacology and Pesticide Action", M.G. Ford, G.G. Lunt, R.C. Reay and P.N.R. Usherwood, eds, pp. 35-60, Ellis Horwood Ltd., Chichester, England.

Olivera, B. M., Gray, W. R., Zeikus, R., McIntosh, J. M., Varga, J., Rivier, J., de Santos, V., and Cruz, L. J., 1985, Peptide neurotoxins from fish-hunting cone snails, Science, 230: 1338.

O'Shea, M. and Evans, P.D., 1979, Potentiation of neuromuscular transmission by an octopaminergic neurone in the locust, J. Exp. Biol., 79: 169-190.

Pelhate, M. and Zlotkin, E., 1981, Voltage dependent slowing of the turn off of Na^+ current in the cockroach giant axon induced by the scorpion venom "insect toxin", J. Physiol. (Lond.), 319: 30.

Pitman, R.M., 1985, Nervous System, in: "Comprehensive Insect Physiology, Biochemistry and Pharmacology", G. A. Kerkut and L. I. Gilbert, eds, pp. 5-54, Pergamon Press, New York.

Raymond, M., 1985, Presentation d'un programme Basic d'analyse log-probit pour micro-ordinateur. Cah. O.R.S.T.O.M., Ser. Entomol. Med. & Parasitol., 23: 117.

Salgado, V. L., Irving, S. N., and Miller, T. A., 1983, Depolarization of motor nerve terminals by pyrethroids in susceptible and *kdr*-resistant house flies, Pest. Biochem. Physiol., 20: 100.

Skinner, W. S., Adams, M. E., Quistad, G. B., Kataoka, H., Cesarin, B. J., Enderlin, F. E., and Schooley, D. A., 1989, Purification and characterization of two classes of neurotoxins from the funnel web spider, *Agelenopsis aperta*. J. Biol. Chem., 264: 2150-2155.

Toki, T., Yasuhara, T., Aramaki, Y., Kawai, N., and Nakajima, T., 1988, A new type of spider toxin, Nephilatoxin, in the venom of the joro spider, *Nephila clavata*, Biomed. Res., 9: 75.

Usherwood, P. N. R., 1981, Glutamate synapses and receptors on insect muscle, in: "Glutamate as a Neurotransmitter", G. Di Chiara and G. L. Gessa, eds., pp. 183-193, Raven Press, New York.

Usherwood P. N. R., Duce I. R., and Boden P., 1984, Slowly-reversible block of glutamate receptor-channels by venoms of the spiders, *Argiope trifasciata* and *Araneus gemma*, J. Physiol. (Paris), 79: 241.

Zlotkin, E., 1983, Insect selective toxins derived from scorpion venoms: an approach to insect neuropharmacology, Insect Biochem., 13: 219.

Zlotkin, E., Kadouri, D., Gordon, D., Pelhate, M., Martin, M. F., Rochat, H., 1985, An excitatory and a depressant insect toxin from scorpion venom both affect sodium conductance and possess a common binding site, Arch. Biochem. Biophys., 240: 877.

STUDIES ON THE MOLECULAR PATHOGENESIS OF ORGANOPHOSPHORUS COMPOUND-INDUCED DELAYED NEUROTOXICITY (OPIDN)

Mohamed B. Abou-Donia and Daniel M. Lapadula

Duke University Medical Center
Department of Pharmacology
Durham, North Carolina

ABSTRACT

Tri-o-cresyl phosphate (TOCP) produces organophosphorus compound-induced delayed neurotoxicity (OPIDN) in humans and in sensitive animal species. After a delay period of 6 to 14 days, neurologic dysfunctions are manifested as ataxia and bilateral paralysis. Early lesions are aggregates of neurotubules and neurofilaments which later condense. Neurotoxic doses of TOCP resulted in an increased activity of Ca^{2+}-calmodulin kinase II (CaM kinase II). This resulted in enhanced CaM kinase II-dependent autophosphorylation as well as phosphorylation of cytoskeletal proteins, i.e., α- and β-tubulin, MAP-2, and neurofilament triplet proteins that interfere with their assembly. Also, phosphorylation of MAP-2 reduces its interaction with tubulin and diminishes its ability to promote tubulin assembly into microtubules. Aggregation and accumulation of such structures disrupt axonal transport and lead to the accumulation of mitochondria and/or endoplasmic reticulum in the distal parts of the axons with subsequent release of Ca^{2+} into the axoplasm. This disrupts axonal membrane mechanisms for intracellular/extracellular ionic gradient, which results in focal internodal swelling and degeneration that spreads somatofugally to involve the entire distal axon.

INTRODUCTION

Many organophosphorus compounds are manufactured. Compounds with anticholinesterase activity are used as insecticides. Others have industrial uses. In 1981 more than 396.5 million pounds of phosphorus-containing compounds were produced in the United States (Chemical Economics Handbook, 1983). This number includes over 50,000 pesticides used as insecticides, acaricides, fungicides, or cotton defoliants (Matsumura, 1976). Among industrial chemicals is triphenyl phosphite with a reported production in the public file of the Toxic Substances Control Act (TSCA) inventory for 1977 that ranged from 10.1 million to 51 million pounds. Studies from other laboratories, as well as ours, have shown that this compound is capable of producing neurotoxicity in experimental animals (Veronesi, et al., 1986; Carrington, et al., 1988).

Organophosphorus compounds are used as pesticides, flame retardants, plasticizers, stabilizers/antioxidants, antiwear and antifriction additives in numerous synthetic lubricants, and chemical intermediates for synthesis of pharmaceuticals and pesticides (Chemical Economics Handbook, 1983). The skin is the most important port of entry for organophosphorus compounds into the body since these chemicals are rapidly absorbed through the intact skin because of their high lipid solubility (Abou-Donia, 1979; 1983). Workers could be exposed to organophosphorus compounds during their manufacturing and processing and during their use as stabilizers/antioxidants, flame retardants, and lubricants. Additional occupational exposures may occur in field workers when organophosphorus compounds are used as pesticides. During the manufacturing of this compound , workers engaged in sampling, cleaning, and replacing filters, and packaging may be exposed by dermal contact or inhalation.

Although the immediate hazard associated with organophosphorus compounds seems to be related to their ability to inhibit acetylcholinesterase, several of these compounds also are capable of producing delayed neurotoxicity (Abou-Donia, 1985). Organophosphorus compound-induced delayed neurotoxicity (OPIDN) is characterized by a delay of 6-14 days before onset of ataxia and paralysis (Smith, et al., 1930; Abou-Donia, 1981). Histopathologic lesions are seen as Wallerian-type degeneration of the axon and myelin in the distal parts of the longest tracts in both central and peripheral nervous systems (Cavanagh, 1964).

Tri-o-cresyl phosphate (TOCP), a plasticizer used in lacquers and varnishes, was first shown to cause OPIDN in humans and sensitive animal species (Smith, et al., 1930; Abou-Donia, et al., 1986). OPIDN was observed first at the end of the nineteenth century. Since then, estimated 40,000 human cases of delayed neurotoxicity have been attributed to OPIDN (Abou-Donia, 1981); this number seems greatly underestimated. Misdiagnosis of OPIDN in field workers exposed to organophosphorus compounds is well documented (Xintaras, et al., 1978). Occupational exposure to several of these compounds has caused delayed neurotoxicity in workers. These chemicals are leptophos (Xintaras, et al., 1978), mipafox (Bidstrup, et al., 1953), methamidophos (SenanYake, et al., 1982), trichlorphon (Shiraishi, et al., 1977), trichlornate (Jedrezjowska, et al., 1980), EPN (Xintaras and Burg, 1980), and chlorpyrifos (Lotti and Morretto, 1986). Other organophosphorus compounds shown to produce OPIDN in experimental animals are DEF (Abou-Donia, et al., 1979a), merphos (Abou-Donia, et al., 1980b), S-Seven (Abou-Donia, et al., 1979b), haloxon Malone, 1964), coumaphos (Abou-Donia, et al., 1982), cyanofenphos (Abou-Donia and Graham, 1979), leptophos (Abou-Donia, et al., 1974), and desbromoleptophos (Abou-Donia, et al., 1980a).

MOLECULAR MECHANISMS OF OPIDN

Although OPIDN has been recognized since 1899, its molecular mechanisms are not yet defined (Cavanagh, 1964). Rapid progress in the field of neuropathology, mainly due to the development of electron microscopic techniques, has substantially expanded our knowledge of the pathological changes in OPIDN. The early ultrastructural changes are aggregation, accumulation, and partial condensation of neurofilaments and neurotubules accompanied by the proliferation of smooth endoplasmic reticulum (Bischoff, 1967; 1970). The common characteristics of delayed neurotoxic organophosphorus compounds are that they a) contain a phosphorus atom and b) are direct or indirect inhibitors of esterases (Abou-Donia, 1981).

Inhibition of Esterases

Early attempts to study the mechanisms of action of OPIDN recognized that neurotoxic organophosphorus esters exert their action by phosphorylating a protein in the nervous system. Brain acetylcholinesterase (Bloch and Hottinger, 1943) and pseudocholinesterase (Earl and Thompson, 1952) were hypothesized as the targets for OPIDN. Both enzymes were subsequently ruled out because of inconsistency between the potential of some organophosphorus compounds to produce OPIDN and their ability to inhibit these enzymes (Aldridge and Barnes, 1966).

In vivo and *in vitro* studies have demonstrated that a small portion (about 6%) of the total phenylvalerate-hydrolyzing activity in hen brain is sensitive to inhibition by delayed neurotoxic organophosphorus compounds such as mipafox but not by non-delayed neurotoxic esters such as paraoxon (Johnson, 1969). This enzymatic activity known as neurotoxic esterase or neuropathy target esterase (NTE) has to undergo "aging" by organophosphorus compounds to produce OPIDN (Johnson, 1975). For an organophosphorus ester to have the potential to produce OPIDN, it must cause at least 70% inhibition of hen brain NTE 24 hours after the administration of an LD_{50} dose. NTE has the following characteristics: (1) it is a membrane bound protein (Carrington and Abou-Donia, 1985a), (2) it has a molecular weight of 155-178 kDa (Carrington and Abou-Donia, 1985a), (3) it has a target size, as determined by radiation inactivation, of 105 kDa (Carrington, et al., 1985), (4) it is transported in the sciatic nerve of hens at a fast axonal transport rate of about 300 mm/day (Carrington and Abou-Donia, 1985b), (5) it is reversibly inhibited by paraoxon (Carrington and Abou-Donia, 1985c), and (6) studies on the kinetics of substrate hydrolysis and inhibition by mipafox of NTE indicated that the higher apparent K_i values that occur with low concentrations of mipafox are attributed to the formation of a Michaelis complex at high concentrations (Carrington and Abou-Donia, 1986), rather than to the formation of two NTE isoenzymes as has been suggested (Chemnitius, et al., 1983; Abou-Donia, et al., 1984).

Although 20 years have passed since NTE was proposed as the initial biochemical target for OPIDN, NTE theory has not furthered our understanding of the mechanisms of this syndrome. Evidence of this theory are correlative, with no hypothesis as to how the pathological damage is initiated or progresses. Also, NTE has no known biochemical or physiological functions. Thus, despite the numerous reports on the assaying of the enzyme in various tissues of many species and the correlation of its inhibition and aging by organophosphorus compounds with their ability to produce OPIDN, no real advance was made to explain its involvement in the mechanisms of OPIDN.

EFFECT ON KINASE-MEDIATED PROTEIN PHOSPHORYLATION

Hypothesis

To elucidate the molecular mechanisms of OPIDN, we have been investigating the hypothesis that delayed neurotoxic organophosphorus compounds may interfere with endogenous kinase-mediated protein phosphorylation of a neuronal enzyme or cytoskeletal proteins, thereby adversely affecting the regulation of normal neuronal processes and resulting in axonal degeneration. Our hypothesis was based on the histopathological picture produced by delayed neurotoxic organophosphorus compounds that showed the involvement of cytoskeletal proteins in OPIDN. Bischoff (Bischoff, 1967; 1970) reported that the earliest ultrastructural alteration in OPIDN was an aggregation and accumulation of neurofilaments and neurotubules together with partial condensation of these proteins accompanied by proliferation of smooth endoplasmic reticulum (SER). In later stages, these cytoskeletal proteins become condensed with the proliferated SER to form disordered solid masses that are finally replaced by a granular and electron dense webbing before they eventually destruct.

Effect of TOCP on Kinase-dependent Protein Phosphorylation

A preliminary study was carried out to examine the effect of a 750 mg/kg single oral dose of TOCP on brain protein phosphorylation after the development of delayed neurotoxicity in chickens (Abou-Donia, et al., 1984; Patton, et al., 1983; 1985a; 1985b; Patton, et al., 1986). Synaptosomal (P_2) membrane and synaptosomal cytosolic proteins were phosphorylated *in vitro* using [γ-^{32}P]ATP as a phosphoryl group donor. Phosphorylated proteins were analyzed using sodium dodecyl sulfate polyacrylamide gel electrophoresis (SDS-PAGE). Protein phosphorylation was detected by autoradiography and quantified by microdensitometry. TOCP treatment increased *in vitro* kinase-mediated protein phosphorylation which was Ca^{2+} and calmodulin-dependent of both cytosolic proteins, 65 kDa (149%) and 55 kDa (196%), and membrane protein 20 kDa (146%). Without further confirmation, the identity of these proteins was suggested as the α- and β-tubulin for the 55 kDa doublet protein and myelin basic protein for the 20 kDa protein band (Patton, et al., 1983; Abou-Donia, et al., 1984).

Correlation between TOCP Enhanced Kinase-dependent Protein Phosphorylation and OPIDN

In order to determine whether this effect was specific for OPIDN, we examined its ability to fulfill the criteria for OPIDN (Abou-Donia, 1981). The results demonstrated that TOCP-induced enhanced kinase-dependent protein phosphorylation fits the following criteria for OPIDN:

1. Clinical Condition. It correlated well with the onset and progress of clinical signs of OPIDN.
2. Test Chemical. Only delayed neurotoxic organophosphorus compounds, e.g., TOCP, DFP, and mipafox produced this effect; non-neurotoxic paraoxon or tri-*p*-cresyl phosphate did not.
3. Species Selectivity. Sprague Dawley rats neither developed OPIDN nor exhibited this effect.
4. Sex. Both male and female chickens exhibited this effect.
5. Age Sensitivity. Chicks neither developed OPIDN nor showed this effect.
6. Protection with PMSF. Phenyl methylsulfonyl fluoride *in vivo* protected against OPIDN and did not change kinase-dependent protein phosphorylation.

Enzyme-catalyzed Phosphoryl Transfers

Phosphorylation mediated by kinases is a post-translational modification of proteins that has been shown to be involved in synthesis of DNA/RNA, regulation of neurotransmitter release, modulation of ion channels, regulation of glycogen metabolism, and control of eukaryotic gene expression (Kenyon and Garcia, 1987). The function of kinases is the transfer of the γ-phosphoryl group ($-PO_3^{-2}$ not a phosphate group $-OPO_3^{2-}$) of a nucleoside triphosphate, usually ATP, to a particular acceptor protein. Opposing the action of kinases are two other enzymes that catalyze phosphoryl transfers: 1) Adenosine triphosphatase (ATPase) which catalyzes the hydrolysis of ATP to ADP and phosphate and consequently, determines its availability for kinases. 2) Phosphoprotein phosphatase which catalyzes the removal of the phosphoryl group from the phosphoprotein yielding the protein and phosphate. These reactions are illustrated on the following page:

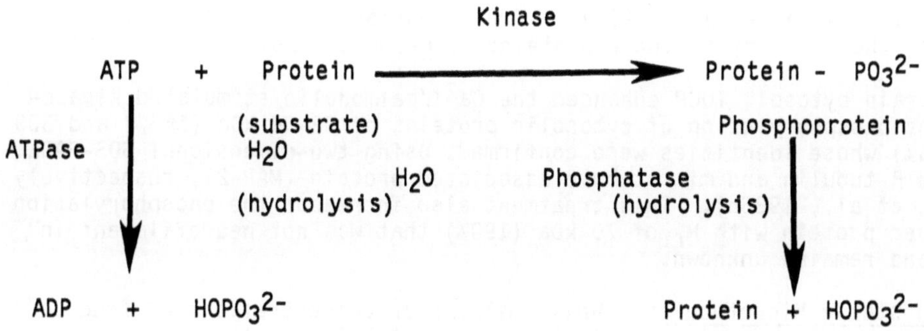

Subsequently, we have investigated the role of these reactions on TOCP increased phosphorylation of neural proteins. These studies have eliminated two of these possibilities for TOCP-enhanced kinase-dependent protein phosphorylation: 1) The possibility that ATPase was inactivated by TOCP, thus making more ATP available for protein phosphorylation, was ruled out by examining the effect of varying ATP concentrations (Pattton, et al., 1986). The finding that there was more than a 5-fold increase in the phosphorylation of the 55 kDa doublet protein from the TOCP-treated compared to the control hens suggested that ATPase activity was not a limiting factor in TOCP-enhanced protein phosphorylation. 2) The other possibility that phosphoprotein phosphatase activity increased by TOCP treatment resulting in more in vivo dephosphorylation and more in vitro phosphorylation was also eliminated by performing a pulse-chase experiment (Patton, et al., 1986). In this experiment, the maximum ^{32}P (from [γ-^{32}P]ATP) incorporated in the 55 kDa band from TOCP treated hens was 5 times greater than from the control hens. However, when the unused "hot" [γ-^{32}P]ATP was "chased" (diluted) after 1.5 minutes of incubation, with a 1000-fold excess of "cold" ATP, rapid decreases were noted in the ^{32}P labeling of both the control and the treated 55 K band with the same initial half-life of 6 min for 10 min followed by a more gradual decline thereafter. These results indicated that phosphatase activity, as determined by the rate of protein dephosphorylation, was the same in the TOCP-treated hens and the control groups. This experiment eliminated the possibility that TOCP-enhanced kinase-dependent protein phosphorylation may have resulted from an increased in vivo dephosphorylation by phosphatase induction, stimulation, or release from lysosomes, leading to increased availability of phosphorylation sites assayed in vitro.

Identification of Proteins Affected by In Vivo Treatment with TOCP

The effect of a single 750 mg/kg oral dose of TOCP on the endogenous kinase-dependent protein phosphorylation of spinal cord neurofilaments and brain microtubule preparations was studied in chickens after they developed OPIDN. Protein phosphorylation with [γ-^{32}P]ATP was analyzed by one-dimensional and two-dimensional SDS-PAGE.

Neurofilaments. TOCP stimulated the kinase-dependent phosphorylation of spinal cord proteins of 70 kDa (119%) and 160 kDa (129%) in a Mg^{2+}-dependent fashion in treated hens when compared to control hens (Suwita, et al., 1986). Addition of Ca^{2+} and calmodulin further enhanced the phosphorylation of these 70 kDa (563%) and 160 kDa (221%) proteins as well as 210 kDa (196%). Two-dimensional electrophoresis confirmed the identity of these proteins as the neurofilament triplet proteins.

α- and β-Tubulin. TOCP enhanced the Ca^{2+}/calmodulin stimulated kinase-dependent phosphorylation of spinal cord proteins of 52 kDa (160%) and 59 kDa (140%) (Suwita, et al., 1986a). Two-dimensional electrophoresis confirmed the identity of these proteins as α-and β-tubulin, respectively.

In brain cytosol, TOCP enhanced the Ca^{2+}/calmodulin stimulated kinase-dependent phosphorylation of cytosolic proteins of 52-59 kDa (155%) and 300 kDa (166%) whose identities were confirmed, using two-dimensional SDS-PAGE, as α-and β-tubulin and microtubule associated protein (MAP-2), respectively (Suwita, et al., 1986b). This treatment also increased the phosphorylation of another protein with M_r of 70 kDa (199%) that was not neurofilament in origin and remains unknown.

Cold-stable Microtubules. While millimolar concentrations of free Ca^{2+} induces rapid disassembly of cold-labile microtubules, cold-stable microtubules are not affected (Suwita, et al., 1986a). In contrast, cold-stable microtubules are disassembled rapidly in the presence of Ca^{2+}/calmodulin, at a sub-stoichiometric concentration, and ATP. Since TOCP also enhanced Ca^{2+}/calmodulin stimulated kinase-dependent phosphorylation of tubulin and MAP-2, its effect on the two physical states of microtubules was investigated.

In crude chicken brain cytosol, TOCP treatment enhanced the Ca^{2+}/calmodulin-dependent kinase-modulated phosphorylation of α-tubulin (160%), β-tubulin (140%), and MAP-2 (145%) (Suwita, et al., 1986a). In cold-stable microtubule preparation, TOCP increased the phosphorylation of α-tubulin (165%), β-tubulin (155%) and MAP-2 (133%). In the cold-labile microtubule preparation, TOCP only enhanced the phosphorylation of MAP-2 (328%). There was a significant increase in phosphorylation of an unknown protein with M_r of 70 kDa in the brain cytosol and in the cold-stable microtubule fractions.

Phosphopeptide Mapping of Brain Cold-stable Microtubules and Spinal Cord Neurofilaments

Since ATPase inhibition and phosphatase activation were ruled out as explanations for TOCP-induced endogenous phosphorylation of cytoskeletal proteins, two other possibilities may be considered: 1) an increased number of phosphorylation sites on the protein substrate, or 2) an increased protein kinase activity/availability.

The effect of TOCP on the number of phosphorylation sites of hen brain cold-stable microtubules and spinal cord neurofilaments was investigated by peptide mapping and autoradiography (Suwita, et al., 1986a). Corresponding individual ^{32}P-labeled polypeptide spots from two-dimensional electrophoresis were excised and digested with *Staphylococcus aureus* V8 protease. The proteolysis product was subjected to SDS-PAGE and then to autoradiography. The results showed identical protein patterns in both TOCP-treated and control animals, indicating that phosphorylation sites of hen brain cold-stable α- and β-tubulin and spinal cord neurofilament triplet proteins were not changed by TOCP treatment. This experiment also showed enhanced ^{32}P incorporation into cytoskeletal proteins in TOCP-treated animals compared to control animals. Since the results of the peptide mapping experiment eliminated the possibility that TOCP did not change the substrate, the other possibilities that remain to be investigated are whether the activity or the amount of kinase was increased after TOCP treatment (Suwita, et al., 1986a).

Increased Ca^{2+}/Calmodulin-dependent Protein Kinase Phosphorylation of Cytoskeletal Proteins in Sciatic Nerve from Hens Treated with TOCP

The proximal and distal parts of sciatic nerves from control and TOCP-treated hens, at 1, 6, 14, and 21 days after treatment, were cut into

6mm segments, homogenized, and subjected to phosphorylation, SDS electrophoresis, and autoradiography (Suwita, Lapadula, and Abou-Donia, unpublished data). Ca^{2+}/calmodulin-dependent protein kinase phosphorylation of the three neurofilament proteins was significantly increased in TOCP-treated hens, particularly at 1 and 6 days after dosing, in both proximal and distal portion of the nerve. Increased phosphorylation of these cytoskeletal proteins one day after treatment suggests a direct effect on the kinase since it is unlikely that increased synthesis of new kinase would occur in the sciatic nerve.

The Involvement of Ca^{2+}/Calmodulin-dependent Protein Kinase Type II (CaM Kinase II)

Ca^{2+}/calmodulin-dependent protein kinase II (CaM kinase II) normally mediates ATP phosphorylation of neuronal cytoskeletal proteins including α- and β-tubulin, MAP-2, neurofilament triplet proteins, and synapsin I, as well as tyrosine hydroxylase, tryptophan hydroxylase, and myelin basic protein (Lin, et al., 1987). There are several pieces of evidence suggesting that CaM Kinase II may be the enzyme involved in the increased protein phosphorylation with TOCP: 1) Our results showed that in the cold stable microtubule fraction, the kinase activity was enriched at least 10-fold and that MAP-2 phosphorylation increased 20-fold. The kinase activity has subsequently been shown to be CaM kinase II (Suwita, et al., 1986a). These results suggest that the same enzyme is involved in the enhanced phosphorylation of cytoskeletal protein in TOCP-treated animals. 2) In the cold-labile microtubule fraction, tubulin was very lightly phosphorylated, while MAP-2 phosphorylation was greatly enhanced in TOCP-treated animals (Suwita, et al., 1985a). Since MAP-2 has been reported to be a better substrate than tubulin for CaM kinase II, our result that MAP-2 was more phosphorylated than tubulin is a further evidence that CaM kinase is involved in TOCP effect. 3) Since CaM kinase II co-purifies with cold-stable microtubules, our finding that kinase activity was diminished in the cold-labile fraction further suggests the involvement of CaM kinase II (Suwita, et al., 1986a).

CaM Kinase II

CaM kinase II consists of a 10-subunit holoenzyme of 550-650 kDa (Lou, et al., 1986). Each subunit consists of a major polypeptide (α) of M_r 50 kDa and of minor polypeptides (β_1 and β_2) of M_r 58/60 kDa, present in a 4:1 ratio. β_1 and β_2 polypeptides have very similar peptide mapping that are very distinct from that of the α.

Each polypeptide is autophosphorylated in a Ca^{2+}/calmodulin-dependent manner, incorporating from 1 to 3 mol of P per polypeptide. The enzyme has an isoelectric point (PI) of 7.0, and the following distinct sites: a) autophosphorylation, b) calmodulin-binding, c) Ca^{2+}-binding, and d) polypeptide-binding sites at which the enzyme associates with cytoskeletal proteins. CaM Kinase II co-purifies with cytoskeletal proteins and has been suggested to be associated directly with either microtubules and neurofilaments or MAP-2. CaM Kinase II regulates the phosphorylation state of MAP-2 and the dynamics of microtubule assembly/disassembly. Phosphorylation of MAP-2 diminishes its ability to promote tubulin assembly into microtubules by reducing the interaction between MAP-2 and tubulin.

Working Hypothesis

The molecular mechanisms of OPIDN (Fig. 1) are outlined in the following assumptions and their probable consequences (Abou-Donia, et al., 1988).

Assumption 1: Delayed neurotoxic organophosphorus compounds affect CaM kinase II either a) directly by phosphorylating the enzyme or b) indirectly, e.g., by inhibiting a protease.

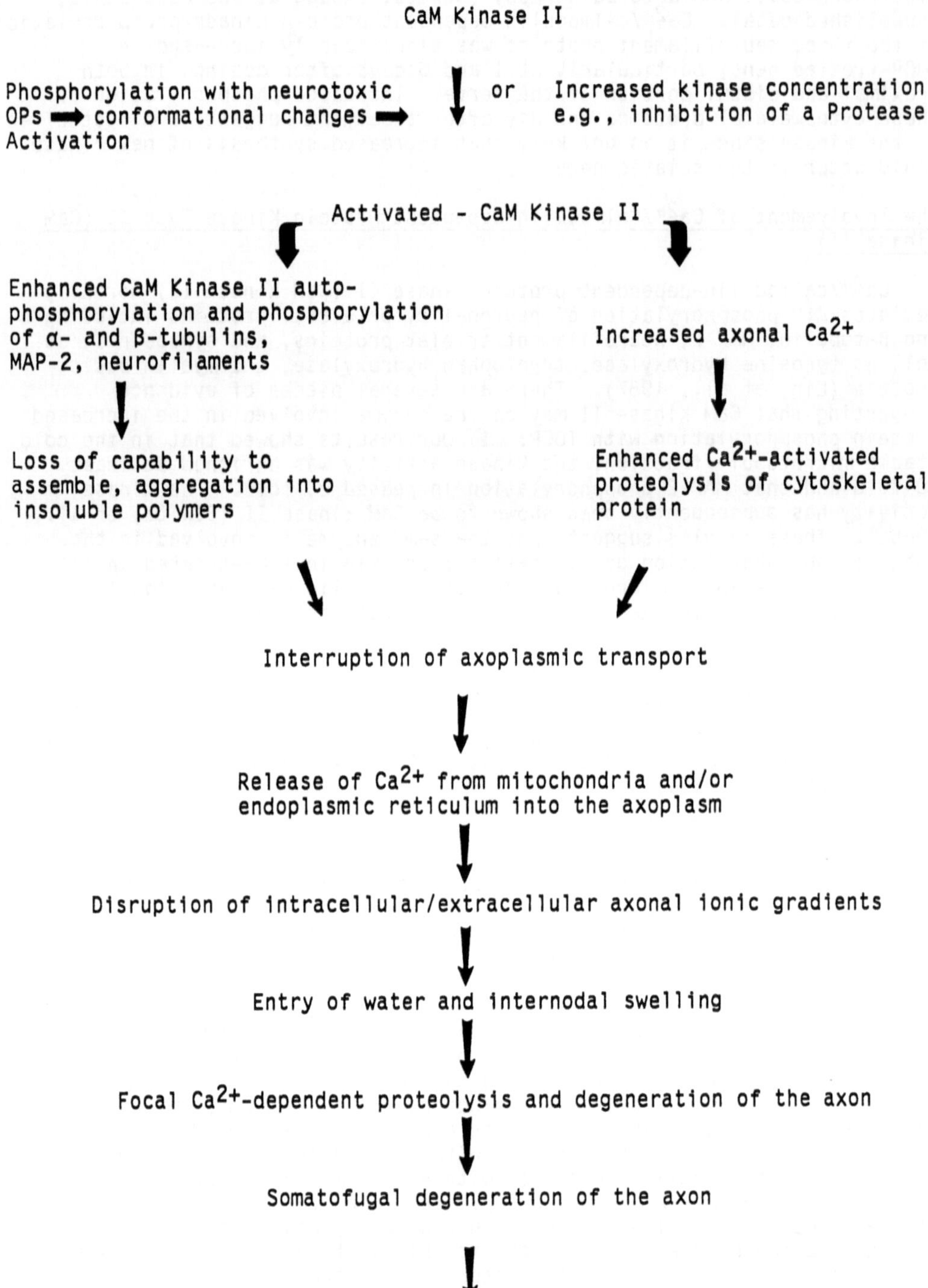

Fig. 1. A proposed scheme of the molecular events leading to OPIDN.

Consequences. Either a) an increase in the activity of CaM kinase II due to conformational changes of phosphorylated enzyme, e.g., change in affinity of their kinase for substrates, or b) an increase in the amount of the enzyme resulting from the inhibition of the protease. These events would lead to increased phosphorylation of cytoskeletal proteins.

Assumption 2: Increased CaM kinase II activity results in enhanced autophosphorylation and the phosphorylation of cytoskeletal proteins.

Consequences. a) Increased phosphorylation of MAP-2 diminishes the polymerization of tubules and promotes the depolymerization of cold-labile microtuble. b) Increased phosphorylation of tubulin results in aggregated polymers. c) Increased autophosphorylation of CaM kinase II leads to increased axonal Ca^{2+}.

Assumption 3: Enhanced phosphorylation of cytoskeletal proteins leads to their inability to assemble into polymers; instead, they aggregate into solid masses; also, they may undergo Ca^{2+}-activated proteolysis.

Consequences. Axonal transport is impaired which results in the accumulation of mitochondria at the distal portion of the axon.

Assumption 4: Ca^{2+} is released from broken down mitochondria or endoplasmic reticulum.

Consequences. Overloading and disruption of intracellular/extracellular ionic gradients.

Assumption 5: Entry of water into the axon.

Consequences. Focal internodal swelling and Ca^{2+}-activated proteolysis followed by focal degeneration that spreads somatofugally to involve the entire axon.

The events presented in this hypothesis are consistent with an emerging view that CaM kinase II may play a key role in the mechanisms of OPIDN (Abou-Donia, et al., 1988). These changes are in agreement with early ultrastructural alterations induced by delayed neurotoxic organophosphorus esters and seen as accumulation of microtubules and neurofilaments followed by their condensation to more solid masses. They are also in harmony with the report that *in vitro* endogenous Ca^{2+}/calmodulin-dependent phosphorylation of tubulin resulted in its transformation into twisted filamentous polymers distinct from microtubules.

Acknowledgement

This work was supported in part by Grants ES 02717 and ES 05154 from the National Institute of Environmental Health Sciences and OH 00823 and OH 02003 from the National Institute for Occupational Safety and Health of Centers for Disease Control.

References

Abou-Donia, M.B., 1979, Pharmacokinetics and metabolism of a topically applied dose of O-4-bromo-2,5-dichlorophenyl O-methyl phenylphosphonothioate in hens, Toxicol. Appl. Pharmacol. 51:311.

Abou-Donia, M.B., 1981, Organophosphorous ester-induced delayed neurotoxicity, Annu. Rev. Pharmacol. Toxicol. 21:511.

Abou-Donia, M.B., 1983, Toxicokinetics and metabolism of delayed neurotoxic organophosphorus esters, Neurotoxicology 4:89.
Abou-Donia, M.B., 1985, Biochemical Toxicology of Organophosphorus Compounds, in "Neurotoxicology," K. Blum and L. Manzo, eds., Marcel Dekker, New York.
Abou-Donia, M.B. and Graham, D.G., 1979, Delayed neurotoxicity of a single dose of O-ethyl O-4-cyanophenyl phenylphosphonothioate in the hen, Neurotoxicology 2:449.
Abou-Donia, M.B., Graham, D.G., Abdo, K.M., and Komeil, A.A., 1979a, Delayed neurotoxic, late acute and cholinergic effects of S,S,S-tributyl phosphorotrithioate (DEF) in hens, Toxicology 14:229.
Abou-Donia, M.B., Graham, D.G., Ashry, M.A., and Timmons, P.R., 1980a, Delayed neurotoxicity of leptophos and related compounds: Differential effects of subchronic oral administration of pure, technical grade, and degradation products on the hen, Toxicol. Appl. Pharmacol. 53:150.
Abou-Donia, M.B., Graham, D.G., and Komeil, A.A., 1979b, Delayed neurotoxicity of O-ethyl O-2,4-dichlorophenyl phenylphosphonothioate: effects of a single oral dose on hens, Toxicol. Appl. Pharmacol. 49:293.
Abou-Donia, M.B., Graham, D.G., Timmons, P.R., and Reichert, B., 1980b, Late acute, delayed neurotoxic, and cholinergic effects of S,S,S-tributyl phosphoro trithioate (Merphos) in hens, Toxicol. Appl. Pharmacol. 53:439.
Abou-Donia, M.B., Lapadula, D.M., and Suwita, E., 1988, Cytoskeletal proteins as targets for organophosphorus compound and aliphatic hexacarbon-induced neurotoxicity, Toxicology 49:469.
Abou-Donia, M.B., Makkawy, H.A., and Graham, D.G., 1982, Coumaphos: delayed neurotoxic effect following dermal administration in hens, J. Toxicol. Environ. Health 10:87.
Abou-Donia, M.B., Othman, M.A., Tantawy, G., Khalil, A.Z., and Shawer, M.G., 1974, Neurotoxic effect of leptophos, Experimentia (Basel) 30:63.
Abou-Donia, M.B., Patton, S.E., and Lapadula, D.M., 1984, Possible role of endogenous protein phosphorylation in organophosphorus compound-induced delayed neurotoxicity. Proceedings of the Symposium on "Cellular and Molecular Neurotoxicity," T. Narahashi, ed., Raven Press, New York.
Abou-Donia, M.B., Trofatter, L.P., Graham, D.G., and Lapadula, D.M., 1986, Electromyographic, neuropathologic, and functional correlates in the cat as the result of tri-o-cresyl phosphate delayed neurotoxicity, Toxicol. Appl. Pharmacol. 83:126.
Aldridge, W.N. and Barnes, J.M., 1966, Further observations on the neurotoxicity of organophophorus compounds, Biochem. Pharmacol. 15:541.
Bidstrup P.L., Bonnell, J.A., and Beckett, A.G., 1953, Paralysis following poisoning by a new organic phosphorous insecticide (mipafox), Br. Med. J. 1:1068.
Bischoff, A., 1967, The ultrastructure of tri-ortho-cresyl phosphate poisoning, I. Studies on myelin and axonal alterations in the sciatic nerve, Acta. Neuropath. 9:158.
Bischoff, A., 1970, Ultrastructure of tri-ortho-cresyl poisoning, II. Studies on spinal cord alterations, Acta. Neuropath. 15:142.
Bloch, H. and Hottinger, A., 1943, Uber die spezifitat der cholinesteraserhemmung durch tri-o-kresyl phosphat, Z. Vitaminfursch. 13:90.
Carrington, C.D. and Abou-Donia, M.B., 1985a, Characterization of [^{3}H]di-isopropyl phosphorofluoridate-binding proteins in hen brain. Rates of phosphorylation and sensitivity to neurotoxic and non-neurotoxic organophosphorus compounds, Biochem. J. 228:537.
Carrington, C.D. and Abou-Donia, M.B., 1985b, Axoplasmic transport and turn-around of neurotoxic esterase in hen sciatic nerve, J. Neurochem. 44:616.
Carrington, C.D. and Abou-Donia, M.B., 1985c, Paraoxon reversibly inhibits neurotoxic esterase, Toxicol. Appl. Pharmacol. 79:175.

Carrington, C.D. and Abou-Donia, M.B., 1986, Kinetics of substrate hydrolysis and inhibition by mipafox of paraoxon-preinhibited hen brain esterase activity, Biochem. J. 236:503.
Carrington, C.D., Brown, R.G., and Abou-Donia, M.B., 1988, Histopathologic assessment of triphenyl phosphite neurotoxicity in the hen, Neurotoxicology 9:223.
Carrington, C.D., Fluke, D.J., and Abou-Donia, M.B., 1985, Target size of neurotoxic esterase and acetylcholinesterase as determined by radiation inactivation, Biochem. J. 231:789.
Cavanagh, J.B., 1964, Peripheral nerve changes in ortho-cresyl phosphate poisoning in the cat, J. Pathol. Bacteriol. 87:365.
Chemical Economics Handbook, 1983, Lubricating oil additives. Menlo Park, Menlo Park, CA: SRI International.
Chemnitius, J.M., Haselmeyer, K.H., and Zech, R., 1983, Neurotoxic esterase: Identification of two isozymes in hen brain, Arch. Toxicol. 53:235.
Earl, C.J., and Thompson, R.H.S., 1952, The inhibitory action of tri-ortho-cresyl phosphate and cholinesterases, Br. J. Pharmacol. 7:261.
Jedrezjowska, H., Rowinska-Marcinska, K., and Hoppe, B.C., 1980, Neuropathy due to phytosol (agritox): report of a case, Acta Neuropathol. 49:163.
Johnson, M.K., 1969, A phosphorylation site in brain and the delayed neurotoxic effect of some organophosphorus compounds. Biochem. J. III:487.
Johnson, M.K., 1975, The delayed neuropathy caused by some organophosphorus esters: Mechanism and challenge, CRC Crit. Rev. Toxicol. 3:289.
Kenyon, G.L., and Garcia, G.A., 1987, Design of kinase inhibitors, Med. Res. Rev. 7:389.
Lin, C.R., Kapifoff, M.S., Durgerian, S., Tatemoto, K., Russo, A.F., Hanson, P., Schulman, H., and Rosenfeld, M.G., 1987, Molecular cloning of a brain specific calcium/calmodulin-dependent protein kinase, Proc. Nat. Acad. Sci. (USA) 84:5962.
Lotti, M., and Morretton, A., 1986, Inhibition of lymphocyte neuropathy target esterase predicts the development of organophosphate polyneuropathy in man, Hum. Toxicol. 5:114.
Lou, L.L., Lloyd, S.J., and Schulman, H., 1986, Activation of the multifunctional Ca^{2+}/calmodulin-dependent protein kinase by autophosphorylation: ATP modulation production of an autonomous enzyme, Proc. Natl. Acad. Sci. (U.S.A.) 83:9497.
Malone, J.C., 1964, Toxicity of Haloxon, Res. Vet. Sci. 5:17.
Matsumura, F., 1976, "Toxicology of Pesticides," Plenum Press, New York.
Patton, S.E., Lapadula, D.M., and Abou-Donia, M.B., 1985a, Partial characterization of endogenous phosphorylation conditions for hen brain cytosolic and membrane proteins, Brain Research 328:1.
Patton, S.E., Lapadula, D.M., and Abou-Donia, M.B., 1985b, Comparison of endogenous phosphorylation of hen and rat spinal cord proteins and partial characterization of optimal phosphorylation conditions for hen spinal cord, Neurochem. Int. 1:111.
Patton, S.E., Lapadula, D.M., and Abou-Donia, M.B., 1986, The relationship of tri-o-cresyl phosphate-induced delayed neurotoxicity to enhancement of in vitro phosphorylation of hen brain and spinal cord proteins, J. Pharmacol. Exp. Therap. 239:597.
Patton, S.E., Lapadula, D.M., O'Callaghan, J.P., Miller, D.B., and Abou-Donia, M.B., 1985, Changes in in vitro brain and spinal cord protein phosphorylation after a single oral administration of tri-o-cresyl phosphate in hens, J. Neurochem. 45:1567.
Patton, S.E., O'Callaghan, J.P., Miller, D.B., and Abou-Donia, M.B., 1983, The effect of oral administration of tri-o-cresyl phosphate on in

vitro phosphorylation of membrane and cytosolic proteins from chicken brain, J. Neurochem. 41:897.
SenanYake, N., and Johnson, M.K., 1982, Acute polyneuropathy following poisoning by new organophosphate insecticide: a preliminary report, New Eng. J. Med. 306:155.
Shiraishi, S., Goto, I., Yamashita, Y., Onishi, A., and Nagao, H., 1977, Dipterex polyneuropathy, J. Neurol. Med. (Japan) 6:34.
Smith, M.I., Elvove, E., Frazier, W.H., 1930, The pharmacological action of certain phenol esters, with special reference to the etiology of so called ginger paralysis, Pub. Health Rep. 45:2509.
Suwita, E., Lapadula, D.M., and Abou-Donia, M.B., 1986a, Calcium and calmodulin enhanced in vitro phosphorylation of hen brain cold-stable microtubules and spinal cord neurofilament triplet proteins folloing a single oral dose of tri-o-cresyl phosphate, Proc. Nat. Acad. Sci. 83:6174.
Suwita, E., Lapadula, D.M., and Abou-Donia, M.B., 1986b, Calcium and calmodulin stimulated in vitro phosphorylation of rooster brain tubulin and MAP-2 following a single oral dose of tri-o-cresyl phosphate, Brain Res. 374:199.
Veronesi, B., Padilla, S., and Newland, D., 1986, Biochemical and neuropathological assessment of triphenyl phosphite in rats, Toxicol. Appl. Pharmacol. 83:203.
Xintaras, C., and Burg, J.R., 1980, Screening and prevention of human neurotoxic outbreaks: issues and problems, in "Clinical and Experimental Neurotoxicology," P.S. Spencer and H.H. Schaumburg, eds., Williams and Wilkins, Co., Baltimore, MD.
Xintaras, C., Burg, J.R., Tanaka, S., Lee, S.T., Johnson, B.L., and Cottrill, J., 1978, "NIOSH Health Survey of Velsicol Pesticides Workers, Occupational Exposure to Leptophos and Other Chemicals, Occupational Exposure to Leptophos and Other Chemicals," NIOSH, Cincinnati.

PHYSIOLOGICAL COMPENSATION FOR TOXIC ACTIONS OF

ORGANOPHOSPHATE INSECTICIDES

Janice E. Chambers and Howard W. Chambers

Departments of Biological Sciences and Entomology
Mississippi State University
Mississippi State, Mississippi 39762

ABSTRACT

The neurochemical changes which could serve as physiological compensatory mechanisms during recovery from a high level acute intoxication with an organophosphate anticholinesterase were studied concurrently with behavioral deficits resulting from the intoxication. Paraoxon, the active metabolite of the common insecticide parathion, was administered to male rats at lethal (antidoted with atropine) and high sub-lethal levels. Brain acetylcholinesterase, the target enzyme, was inhibited about 90% on the day of treatment, and recovered slowly over the next several days. Rats displayed some relatively small and transient behavioral deficits 1 and 2 days after treatment with paraoxon in an active avoidance protocol. Of four neurochemical parameters in the brain which could have helped compensate for the cholinergic hyperactivity induced by the acetylcholinesterase inhibition (high affinity choline uptake, choline acetyltransferase, acetylcholine release and muscarinic acetylcholine receptors), none showed definitive changes at the time points monitored. Thus, unlike the response to chronic exposures to organophosphate anticholineterases, an acute intoxication, even at high dose levels, does not seem to result in the employment of cholinergic compensatory mechanisms despite persistent acetylcholinesterase inhibition.

INTRODUCTION

Because of the high acute toxicity of some of the organophosphate (OP) insecticides, they pose a threat of accidental poisoning and are responsible for numerous poisonings annually (Hayes, 1969). The OP insecticides are widely used in agriculture, and therefore, have a great potential for occupational exposures to manufacturers, formulators, pesticide applicators and field workers. There is also a great potential for accidental poisonings to farmers, careless consumers and frequently children. Oddly, they have even been the vehicle used for some suicides. The acute oral LD_{50} to rats of parathion, a widely used OP insecticide, is 3-10 mg/kg (Gaines and Linder, 1986). The OP insecticides or their activated metabolites (the phosphate or oxon forms) are potent anticholinesterases; this action, inhibition of acetylcholinesterase (AChE), the target enzyme, leads to an accumulation of the neurotransmitter acetylcholine (ACh), and therefore, hyperactivity within cholinergic pathways. Since cholinergic synapses are prominent in the central nervous system, and both the somatic and autonomic portions of the peripheral nervous system, an intoxication with an anticholinesterase can lead to wide- spread perturbation throughout the organism. If the dose of the OP is lethal, the usual cause of death is respiratory collapse, primarily from bronchoconstriction, increased bronchiolar mucus, spasm of the respiratory muscles and inhibition of the respiratory control center in the brain. However, frequently the administration of proper therapy, usually consisting of atropine plus the oxime AChE reactivator 2-PAM along with artificial respiration, will allow poisoned individuals to survive a lethal dose. Atropine is an antagonist of muscarinic cholinergic receptors in both the central and peripheral nervous systems and therefore, antagonizes the action of the accumulated ACh which leads to the lethal symptoms. The oxime 2-PAM is capable of enhancing dephosphorylation of the inhibited AChE, so that catalytic activity of the AChE is restored.

Because of the slow reactivation of the OP-inhibited AChE, a severe acute poisoning can lead to massive AChE inhibition which is extremely persistent, lasting for several days. Since AChE inhibited by some OP's can age (i.e., one of the alkoxy groups can hydrolyze leaving the remaining phosphorylated AChE charged and therefore resistant to reactivation) (Berends et al., 1959), the inhibited AChE can become permanently deactivated, so the time course of inhibition would be even more prolonged since de novo synthesis of AChE would have to occur to restore original levels of enzyme. The aged AChE is refractory not only to spontaneous reactivation, but also oxime-mediated reactivation. Thus, inhibition by an OP anticholinesterase can lead to a very persistent lesion within the cholinergic system. Even though the initial symptomology (such as tremors) subsides, basically during the first day, the primary lesion (i.e., inhibition of AChE) persists for several more days. It is logical to assume that some compensatory mechanism may come into play during this time of prolonged AChE inhibition following an acute poisoning with an OP insecticide to attempt to return cholinergic pathway activity to normal levels.

Four possible compensatory mechanisms, which could reduce the amount of ACh available for action or to reduce the action of ACh, are:
1) decreased high affinity choline uptake into presynaptic nerve terminals;
2) decreased synthesis of ACh by choline acetyltransferase (ChAT);
3) reduced release of ACh in response to a nerve impulse; and
4) down-regulation of post-synaptic ACh receptors. These four mechanisms are illustrated schematically in Fig. 1. An acute severe poisoning with an OP insecticide might resemble physiologically a chronic exposure because of the persistent inhibition caused by OP anticholinesterases and therefore a prolonged altered cholinergic state could result. Therefore, during a severe acute poisoning in order to return cholinergic pathways to a near normal level of activity, mechanisms may be employed which are similar to those employed during chronic OP exposures to compensate for the hypercholinergic activity.

Numerous studies have demonstrated that chronic exposures to OP anti-AChE's result in a reduction in both central and peripheral muscarinic receptors, as measured by [^{3}H]quinuclidinyl benzilate (QNB) binding to membranes (Aas et al., 1987; Churchill et al., 1984 a and b; Gazit et al., 1979; Lim et al., 1986; Schiller, 1979; Smolen et al., 1986; Uchida et al., 1979; Yamada et al., 1983). Most of this work has been with the very potent AChE inhibitors, diisopropylfluorophosphate (DFP) or soman. These changes in receptor density in response to drug alteration of cholinergic pathway activity could serve to compensate for the drug's effect and return the system to a more normal level of activity, by reducing the amount of physiological response to the ACh present. Additionally, there are reports

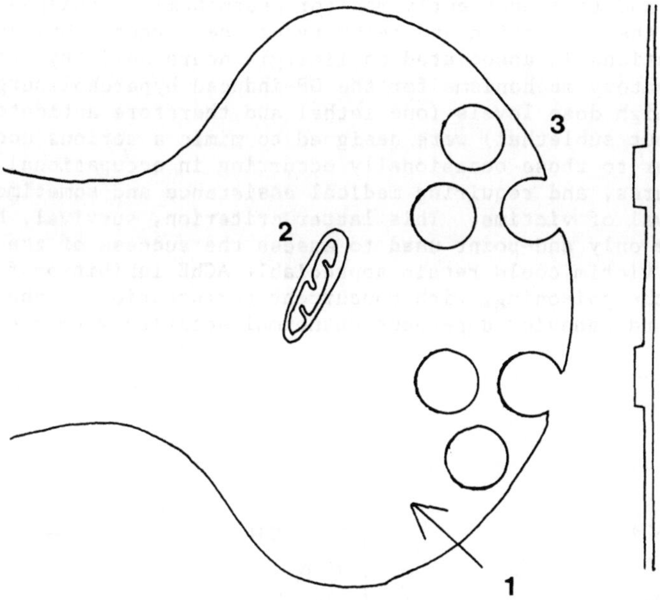

Fig. 1. Schematic diagram of a cholinergic synapse illustrating four possible sites for physiological compensation for the hypercholinergic activity resulting from high level organophosphate anticholinesterase intoxication. 1: sodium dependent high affinity choline uptake; 2: mitochondrial choline acetyltransferase activity; 3: release of acetylcholine from synaptic vesicles; 4: muscarinic acetylcholine receptors.

of reduced high affinity choline uptake into nerve terminals following chronic or acute OP exposures (Atweh et al., 1975; Yamada et al., 1983). This transport system is the rate-limiting step in the process of ACh formation; its reduction, therefore, would decrease the amount of ACh available for release.

Since the cortical and striatal cholinergic systems are implicated in behavior (Beninger et al., 1986a and b; Bermudez-Rattoni et al., 1986; Gonzalez and Ellinwood, 1984; Rauch and Raskin, 1984; Sanberg et al., 1978; Sandberg et al., 1984), behavioral changes could certainly be an expected result of OP poisoning. In fact, effects on operant conditioning, avoidance learning, maze learning, and spatial learning have been reported in experimental animals including primates following OP intoxication (Bignami et al., 1975; Bignami and Giardini, 1983; Brezenoff et al., 1985; Desi, 1983; Gardner et al., 1984; Geller et al., 1984; Haggerty et al., 1986; Lehotzky, 1982; Lerer et al., 1984; McMillan, 1982; Raffaele et al., 1987; Upchurch and Wehner, 1987; Woolley et al., 1979). Most of these studies have used relatively low doses in a chronic dosing protocol, while little has been reported on the behavioral effects from a high dose acute poisoning protocol which resembled a severe accidental poisoning. Most of the behavioral effects of OP compounds reported have been deleterious. We have reported a limited deleterious effect of such an acute dosing protocol on shock avoidance performance in rats (Chambers et al., 1988).

The major thrust of our experiments has been to determine what alterations in learned behavior result following a severe acute poisoning by an OP and what neurochemical compensatory mechanisms might correlate with these. Our experimental design was to administer a single very high dose of an organophosphate and to subsequently monitor the animal for memory deficits and to concurrently monitor neurochemical changes in the brain including the inhibition and recovery of the target AChE, as well as potential alterations in associated cholinergic neurochemistry which might serve as compensatory mechanisms for the OP-induced hypercholinergic activity. The high dose levels (one lethal and therefore antidoted, and the other high but sublethal) were designed to mimic a serious accidental poisoning similar to those occasionally occurring in occupational and household exposures, and requiring medical assistance and sometimes therapy to assure survival of victims. This latter criterion, survival, has usually been the only end-point used to assess the success of the medical treatment. The victim could retain appreciable AChE inhibition for several days following the poisoning, with concurrent perturbation of cholinergic neurochemistry and behavior dependent on normal activity within cholinergic pathways.

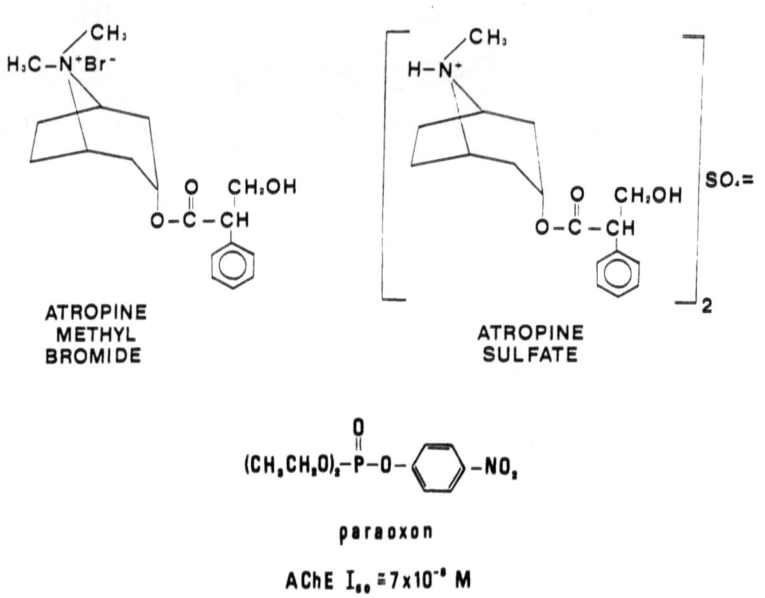

Fig. 2. Structural formulas of the two muscarinic receptor antagonists (atropine sulfate and atropine methyl bromide) and the organophosphate anticholinesterase (paraoxon) used.

The protocol for most experiments has been to administer intraperitoneally a high sublethal dose (1.3 mg/kg) or a lethal dose (2 mg/kg) of paraoxon, the potent anticholinesterase metabolite of the common OP insecticide parathion. Survival of the lethal dose was assured by concurrent administration of a relatively high level of either a centrally active (atropine sulfate, AS; 90 mg/kg) or a non-centrally active (atropine methyl bromide, AMB; 100 mg/kg) form of atropine, a muscarinic receptor antagonist. Structures are given in Fig. 2. Neurochemical parameters and behavioral responses were monitored for up to four days following the intoxication.

PRIMARY LESION

Acetylcholinesterase, the target enzyme for the OP insecticides or their metabolites, is a widely distributed enzyme throughout the central and peripheral nervous systems. It hydrolyzes and thereby inactivates the important neurotransmitter ACh into the products acetate and choline. AChE was monitored in three critical brain regions: the cerebral cortex, the ultimate control center for behavior and physiological functions; the corpus striatum, important in postural control and possessing a very high level of cholinergic activity; and the medulla oblongata, site of the respiratory control center whose inhibition may contribute heavily to the respiratory failure occurring during a fatal intoxication. AChE activity was monitored spectrophotometrically in whole homogenates by a modification of the Ellman et al. (1961) technique, which uses acetylthiocholine as the substrate and 5,5'-dithiobis(nitrobenzoic acid) as the chromogen (Chambers et al., 1988; Chambers and Chambers, 1989). Per cent inhibition in homogenates from paraoxon-treated rats was calculated by comparisons of specific activities to those of corresponding controls.

The lethal and high sub-lethal doses selected for these experiments yielded a very high level of AChE inhibition, about 90% within 2 hours after intoxication (Chambers and Chambers, 1989). The high sub-lethal dose yielded only slightly less inhibition than the antidoted lethal dose. Enzyme activity recovered with time such that on 1 day after treatment there was about 70% inhibition, whereas at 4 days there was only about 40% inhibition. AChE activity in the medulla oblongata recovered slightly more rapidly than did that in the corpus striatum or cerebral cortex. Concurrently with this decrease in AChE inhibition over the 4 day period was an increase in the amount of aging, as indicated by inhibition which persisted in the *in vitro* presence of the oxime reactivator trimedoxime bromide (TMB-4), from no aging on the day of treatment to complete aging by day 4. The relationship between AChE inhibition and aging with time for the medulla oblongata are illustrated in Fig. 3; the other two brain parts monitored, cerebral cortex and corpus striatum, demonstrated very similar trends, although overall AChE specific activity in the corpus striatum was about 4.6 and 6.5 fold higher than that in the medulla oblongata and the cerebral cortex, respectively. As noted above, recovery of the inhibited AChE was slightly faster in the medulla oblongata than in the other two brain parts. Thus, the brain AChE inhibition produced in these experiments was high initially and was persistent. Additionally, the relatively rapid aging of the phosphorylated AChE would prevent any further reactivation of inhibited AChE molecules, and recovery of enzyme activity to pre-treatment levels would have to rely on *de novo* enzyme synthesis. Therefore, the treatment protocol used induced a severe and persistent biochemical lesion in the cholinergic system, and one would expect the cholinergic pathways to be perturbed for several days.

BEHAVIOR

The cholinergic system is strongly implicated in learned behavior (Beninger et al., 1986; Walsh et al., 1984; Wenk, et al., 1987), so it is a potential target for OP insecticide toxicity. Behavioral deficits have been reported to occur after exposures to OP anticholinesterases (Brezenoff et al., 1985; Geller et al., 1984; Upchurch and Wehner, 1987). Most of the reports represent effects observed after chronic exposures.

We have investigated the effect of a high level acute exposure to paraoxon on the retention of a negatively reinforced learned task (Chambers and Chambers, 1988; Chambers et al., 1988). We have trained rats prior to treatment in an active avoidance protocol, similar to that described by

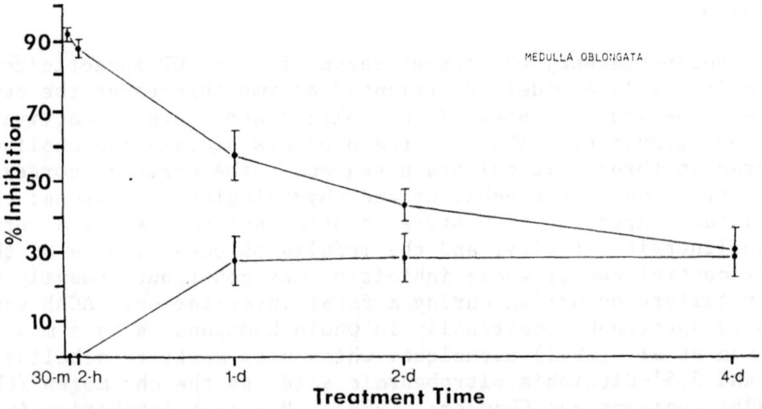

Fig. 3. Per cent of acetylcholinesterase inhibited (top line) and inhibited and aged (bottom line) in the medulla oblongata of rats receiving a lethal dose (2 mg/kg) paraoxon plus antidote (90 mg/kg atropine sulfate or 100 mg/kg atropine methyl bromide, combined data). Data represent mean ± S.E.M.

Brush (1962) and Davis and Hatoum (1987). Rats were trained to change sides of a shuttle box at the sound of a buzzer to avoid or escape from a mild footshock. When their performances were stabilized, they were injected with paraoxon and/or atropine, and were tested for retention of the learned behavior at one and two days after treatment. Avoidance and escape percentages and latencies were recorded. A repeated measures analysis of variance was conducted on the resultant data comparing pre-treatment performance ("day 0") to performance at 1 and 2 days after treatment.

Since the first test of behavioral retention occurred 24 hours after treatment, the acute symptomology, such as tremors, had already dissipated. Animals treated with paraoxon, either alone or with either antidote, showed deficits in avoidance percentages on day 1 following treatment, indicating that they were requiring some shocks following treatment to stimulate movement that they did not require beforehand (Table 1). With the lethal dose of paraoxon antidoted with either form of atropine, this deficit on day 1 was showing some improvement on day 2, although there was not complete recovery. However, with the sublethal dose of paraoxon, the avoidance percentage deficit worsened on day 2 compared to day 1, suggesting that antidoting at least the peripheral muscarinic receptors is useful in accelerating behavioral recovery. With the controls and all treatments other than the two AMB treatments, virtually all non-avoidance trials resulted in escape, indicating that sensory perception and nociception appeared to be unaffected. With both AMB treatments many of the non-avoidance trials did not result in escape. Although the possible sedative action of bromide was suspected as the cause of the movement deficiencies, rats injected with an equimolar dosage of sodium bromide did not demonstrate avoidance deficits, so bromide sedation does not appear to be responsible for the observed effects.

Thus, deficits in learned behavior did occur as a result of antidoted lethal dose paraoxon administration and recovery of pre-treatment performance levels had started to occur despite the presence of reasonably

Table 1. Shock Avoidance and Escape Behavior Following Acute Administration of Paraoxon or Atropine[a].

		Avoidance[b]		Escape[c]			Failure[d]		
Group	N	Day 1	Day 2	Day 0	Day 1	Day 2	Day 0	Day 1	Day 2
Control	22	-10.7	-17.0	91.1	96.2	92.4	4.4	2.1	4.4
AMB	8	-44.8*	-32.6*	98.6	70.0	69.6	0.6	20.6	18.8
AS	8	-8.4	+2.2	90.6	87.0	70.0	2.5	4.3	7.5
AMB/P	7	-48.7*	-23.5*	99.0	85.4	54.7	0.3	9.3	20.7
AS/P	9	-40.9*	-21.5*	100.0	98.0	100.0	0.0	0.6	0.0
P	8	-10.3	-36.7*	98.2	97.7	97.3	0.4	0.6	1.3

[a]Rats were trained during daily tests of 20 trials to avoid (within 5 sec) or escape from (between 5-40 sec) a footshock following a buzzer. Trials were separated by a 45-115 sec random interval. Treatment group abbreviations: AMB = 100 mg/kg atropine methyl bromide; AS = 90 mg/kg atropine sulfate; AMB/P = 100 mg/kg atropine methyl bromide plus 2.0 mg/kg paraoxon; AS/P = 90 mg/kg atropine sulfate plus 2.0 mg/kg paraoxon; P = 1.3 mg/kg paraoxon. Day 0 is the mean of the last 3 pre-treatment tests.

[b]Avoidance data were expressed in terms of per cent change relative to the Day 0 values for percentages of total trials which resulted in avoidance. Asterisks indicate a significant difference from pre-treatment values ($P < 0.05$).

[c]Escape data are expressed in terms of percentage of non-avoidance trials which resulted in escape.

[d]Failure to avoid or escape data are expressed in terms of percentage of total trials in which there was neither avoidance nor escape.

high levels of AChE inhibition. It would appear that antidoting the central muscarinic receptors allowed faster recovery of performance when compared to the non-antidoted sub-lethal dose of paraoxon.

HIGH AFFINITY CHOLINE UPTAKE

Sodium dependent high affinity choline uptake is the rate limiting step in the production of the neurotransmitter ACh (Barker and Mittag, 1975; Kuhar and Murrin, 1978; Simon et al., 1976). It is a specific, saturable and energy dependent system localized in cholinergic nerve terminals serving to bring the ACh precursor choline into the nerve terminal so that it can be a substrate for ChAT (Yamamura and Snyder, 1973). A reduction in choline uptake has been noted following acute or chronic administration of anticholinesterases (Atweh et al., 1975; Yamada et al., 1983). Although this has been interpreted to indicate a depression in presynaptic cholinergic activity because of the reduction in choline turnover resulting from decreased ACh catabolism (Atweh et al., 1975), this reduction can also be viewed as a compensatory mechanism for the excess ACh accumulating in the presence of the anticholinesterase. The result of this decrease in choline uptake would be to reduce the pool of ACh available for release.

High affinity choline uptake was monitored in vitro using ^{3}H-choline uptake into crude synaptosomal (P_2) preparations of rat corpus striatum (Yamamura and Snyder, 1973). A kinetic assessment was performed using 6 ^{3}H-choline concentrations between 0.2 and 2.0 μM, using uptake in the presence of the inhibitor hemicholinium-3 to correct for non-specific (low affinity) uptake. Synaptosomes containing ^{3}H-choline were isolated using reduced pressure on glass fiber filters and the radioactivity was determined by liquid scintillation spectrometry. Michalis-Menten kinetic analysis of the saturation curves was by linear regression of Lineweaver-Burk (double reciprocal) plots, and an analysis of variance was run on the resultant maximal velocity (V_{max}) and dissociation constant (K_m) values. Samples were taken 0.5 and 2 hours and 1, 2 and 4 days after treatment.

No significant differences in K_m occurred between any treatment group compared to controls, indicating that the affinity of the carrier had not been altered by the anticholinesterase or either muscarinic antagonist (Table 2). These results are consistent with the reports of others who tested some cholinergic drugs (Simon et al., 1976; Yamamura and Snyder, 1973). There were also no significant differences between the V_{max} for any treatment and the controls. Effects on choline uptake have been found by others to occur quite rapidly, within 30 min of treatment, and that the effects were transient, disappearing by 24 hr after treatment. However, the drugs tested were considerably more transient than the OP anticholinesterases. In our experiments there was a suggestion of a decrease in choline uptake in the high sublethal dose of paraoxon and the lethal dose of paraoxon antidoted with the non-centrally acting form of atropine (AMB), whereas a lesser decrease from controls occurred in the lethal dose of paraoxon antidoted with the centrally acting form of atropine (AS) at 30 min after treatment. If these decreases do represent real phenomena, despite the lack of statistical significance, then it could be that a 25% reduction of high affinity choline uptake serves a role in compensating for the hypercholinergic activity induced by the anticholinesterase whose effects have not been antagonized in the brain. However, a 25% reduction is not necessary when the effects of paraoxon were antagonized in the brain. No indication of changes was apparent at two hours, so such an effect, if real, was a transient one. Based on results in the literature, a lack of significant differences in the later sampling times (1, 2, and 4 days) is expected. These possible changes at early sampling times should be investigated further.

CHOLINE ACETYLTRANSFERASE

ChAT is a mitochondrial enzyme which catalyzes the formation of ACh from choline and acetyl coenzyme A. ChAT activity was monitored in fresh whole homogenates of cerebral cortex or corpus striatum in which choline and ^{14}C-acetyl CoA were added as substrates (Fonnum, 1975). The labeled product was extracted out of the incubation mixture using liquid cation exchange with tetraphenyl boron in butyronitrile, and the radioactivity was measured using liquid scintillation spectrometry. Data were analyzed using an analysis of variance. Samples were assayed at 0.5 and 2 hr and 1, 2 and 4 days after treatment. There were no significant differences among any of the experimental groups within either tissue (Table 3) (Chambers and Chambers, 1989). Although ChAT activity can be inhibited by ACh (Kaita and Goldberg, 1969), ChAT is not the rate limiting step in ACh formation (high affinity choline uptake is), so it is not surprising that ChAT specific

Table 2. *In vitro* High Affinity Choline Uptake Following Acute Administration of Paraoxon or Atropine.

Group[b]	Time after Treatment 0.5 hr	2 hr	1 d	2 d	4 d
K_m					
Control	100.0 (6)	100.0 (5)	100.0 (10)	100.0 (20)	100.0 (13)
AMB	90.4 (6)	110.9 (4)	93.6 (4)	81.6 (5)	82.0 (5)
AS	94.7 (6)	119.5 (5)	147.8 (3)	102.3 (9)	108.4 (5)
AMB/P	90.6 (6)	118.3 (4)	104.8 (4)	96.4 (6)	92.8 (5)
AS/P	79.8 (5)	98.2 (4)	109.0 (4)	111.4 (11)	149.3 (4)
P	87.2 (6)	105.6 (4)	102.2 (4)	120.6 (4)	126.6 (3)
V_{max}					
Control	100.0 (6)	100.0 (5)	100.0 (10)	100.0 (20)	100.0 (13)
AMB	97.7 (6)	120.9 (4)	109.9 (4)	102.8 (5)	84.7 (5)
AS	87.9 (6)	104.8 (5)	131.0 (3)	102.8 (9)	114.3 (5)
AMB/P	72.7 (6)	93.8 (4)	105.3 (4)	105.4 (6)	84.1 (5)
AS/P	86.3 (5)	102.5 (4)	125.6 (4)	98.7 (11)	121.2 (4)
P	74.9 (6)	104.0 (4)	102.1 (4)	83.4 (4)	91.2 (3)

[a]Data are expressed as per cent relative to control values. Numbers of observations are in the parentheses.

[b]Treatment groups are defined in Table 1.

activity was not affected. Others have also reported a lack of effects on ChAT following acute and chronic OP exposures (Wecker et al., 1977).

ACETYLCHOLINE RELEASE

The neurotransmitter ACh is stored presynaptically in synaptic vesicles. Depolarization of the presynaptic neuron induces an increase in the cytosolic Ca^{++} concentration in the nerve terminal, thus inducing fusion of the synaptic vesicles with the cell membrane and consequently release of ACh into the extracellular space. ACh release was studied in crude synaptosomal (P_2) fractions which had been loaded with ^{3}H-ACh by exposing them to ^{3}H-choline under high affinity uptake conditions (Haga, 1971), similar to what was used to study the high affinity uptake of choline described above. Both spontaneous (using 5 mM K^+) and evoked (using 35 mM K^+) release were studied. The ^{3}H-ACh in the synaptosomal supernatants was assessed using an enzymatic liquid cation exchange process in which initially the ^{3}H-choline present was phosphorylated by choline kinase, followed by the extraction of the ^{3}H-ACh into tetraphenyl boron in butyronitrile (Nemeth and Cooper, 1979). The ^{3}H-ACh was quantitated by liquid scintillation spectrometry. Only a limited series of experiments were conducted on this parameter using only the control, AS, and AS/P groups, since these experiments did not indicate any obvious changes from

Table 3. Choline Acetyltransferase Activity Following Acute Administration of Paraoxon or Atropine.[a]

Group[b]	Time after Treatment				
	0.5 hr	2 hr	1 d	2 d	4 d
Cerebral cortex					
Control	100.0 (3)	100.0 (3)	100.0 (12)	100.0 (13)	100.0 (12)
AMB	105.5 (3)	110.0 (3)	113.5 (3)	133.3 (3)	109.2 (4)
AS	97.3 (3)	90.0 (3)	97.3 (5)	101.0 (7)	107.1 (5)
AMB/P	94.5 (3)	90.0 (2)	102.7 (4)	104.2 (3)	115.3 (4)
AS/P	100.0 (3)	103.3 (3)	102.7 (5)	103.1 (7)	107.1 (5)
P	94.5 (3)	95.0 (3)	95.5 (3)	78.1 (3)	75.5 (3)
Corpus striatum					
Control	100.0 (3)	100.0 (3)	100.0 (12)	100.0 (13)	100.0 (12)
AMB	122.8 (3)	106.8 (3)	101.9 (3)	109.6 (3)	117.4 (4)
AS	106.1 (3)	116.4 (3)	96.1 (5)	111.3 (7)	87.1 (5)
AMB/P	111.9 (3)	102.4 (2)	111.2 (4)	97.1 (3)	118.4 (4)
AS/P	114.6 (3)	105.2 (3)	106.1 (5)	116.3 (7)	89.4 (5)
P	103.7 (3)	103.2 (3)	93.5 (3)	104.9 (3)	81.6 (3)

[a]Data are expressed as per cent relative to control values. Number of observations are in parentheses.

[b]Treatment groups are defined in Table 1.

the controls (data not shown). Again, it is not surprising that this parameter was unchanged since other parameters which are rate-limiting (i.e., choline uptake) or which directly modulate biological activity (i.e., receptors) would be more logical points of control. However, one point worth investigating further would be a study of ACh release as an index of the level of functioning cholinergic autoreceptors.

MUSCARINIC RECEPTORS

Muscarinic receptors are the population of cholinergic receptors having a high affinity for the alkaloid muscarine and occur in both the central and peripheral nervous system. Since they are heavily involved in the autonomic nervous system, they are very important in the control of visceral function. They also occur prominently in the brain. They induce their physiological effect by stimulation of the phosphoinositol pathway. Their population can be readily monitored using the high affinity ligand ^{3}H-QNB. However, this ligand does not discriminate between the post-synaptic receptors and the pre-synaptic autoreceptors which help regulate neurotransmitter release.

A down-regulation of muscarinic receptor densities (B_{max}) has been observed by others following chronic administration of the two potent anticholinesterases DFP and soman, without a change in affinity (K_d), as monitored with ^{3}H-QNB binding (Gardner et al., 1984; Sivam et al., 1983;

Yamada et al., 1983). Little information is available on the effects of a single dose of an insecticidal OP on muscarinic receptors.

Muscarinic cholinergic receptors were assayed by the binding of ^{3}H-QNB to membranes of the corpus striatum (Yamamura and Snyder, 1974; Sivam et al., 1983). Membrane fractions were isolated from individual rats and incubated with six concentrations of ^{3}H-QNB to yield saturation curves. Non-specific binding was assessed in the presence of the muscarinic antagonist atropine sulfate. Membranes bound to ^{3}H-QNB were isolated by filtration under reduced pressure onto glass fiber filters and the radioactivity was quantitated using liquid scintillation spectrometry. B_{max} and K_d for specific binding were calculated by Scatchard plot analysis, and the data were then compared by an analysis of variance. Data were obtained 2 days after treatment in rats which had been run through the behavioral tests.

No significant differences in the affinity (K_d) of ^{3}H-QNB for striatal muscarinic receptors were induced by any treatment (Table 4) (Chambers and Chambers, 1988). This is as expected, consistent with the results others have reported following chronic exposures to other OP anticholinesterases. Although there was no statistical difference between the B_{max} for the sublethal paraoxon treatment and the control, there did appear to be a trend for a decrease, as expected if there was down-regulation of receptor numbers. There also appeared to be a trend toward an increase in receptor densities in all of the atropine-treated groups. Although this is surprising in the AS and AS/P groups because atropine is cleared quite readily, the dosage (90 mg/kg) was high, so the effect may have lasted 2 days. A dose of 50 mg/kg AS has been reported to totally block cholinergic receptors in the brain (Vanderwolf, 1987). Inexplicably, an increase in muscarinic receptors was also suggested in the AMB and AMB/P groups, despite the fact that AMB should have had little ability to enter the brain. If these suggested changes are real, it may be that such changes were more apparent earlier and had substantially dissipated by two days after treatment when these studies were done. These studies were performed only on the rats in the behavioral test, which were scheduled for sacrifice on day 2 alone. Observing a shorter time frame is warranted since it has been reported that a single dose of DFP yielded a reduction in cortical muscarinic receptors at one day (Lerer et al., 1984). Another possibility which has not been checked is that the receptors have become uncoupled from their physiological actions, or that they have been internalized and therefore, are not accessible to the neurotransmitter in an intact situation. If either of these latter possibilities were true, a decrease in muscarinic receptor-mediated action would have occurred without an effect on the overall number of sites capable of binding ^{3}H-QNB.

CONCLUSION

Apparently the cholinergic system regulates its activity in different ways following acute and chronic exposures to OP anticholinesterases. There is ample evidence in the literature that brain muscarinic receptors become down-regulated in response to chronic exposures to OP's. Even though the AChE inhibition that occurred in response to our high level acute poisoning protocol was very persistent, the animals did not seem to react by a long-term down-regulation of muscarinic receptor numbers. The high affinity choline uptake system also showed small but not statistically significant decreases which suggest a short-term and transient down-regulation of transport when the OP-induced hypercholinergic activity is not antagonized centrally. We were expecting to observe more pronounced neurochemical changes which could have effected a partial compensation for the paraoxon-induced hypercholinergic activity. We expected the presence

Table 4. Striatal Muscarinic Receptors Two Days Following Acute Administration of Paraoxon or Atropine, as Determined by ^{3}H-Quinuclidinyl Benzilate Binding.[a]

Group[b]	N	K_d	B_{max}
Control	11	100.0^A	100.0^B
AMB	6	105.3^A	130.4^{AB}
AS	4	107.9^A	127.5^{AB}
AMB/P	5	84.2^A	143.5^A
AS/P	5	126.3^A	137.7^A
P	5	79.0^A	72.5^B

[a]Data are expressed as per cent relative to control values.

[b]Treatment groups are defined in Table 1.

[A,B]Statistical grouping within parameters based on analysis of variance and Student-Newman-Keuls test.

of the persistent inhibition by paraoxon to be detected as though this were a continuous, chronic exposure, but apparently the animal reacts differently to an acute, albeit persistent, exposure to an OP and to a chronic exposure. Perhaps a certain level of AChE inhibition must be maintained over a several day period in order to elicit a significant decrease in [^{3}H]QNB binding sites, particularly if the initial response is receptor internalization and a later response will be receptor destruction. This possibility could be investigated using a muscarinic receptor ligand incapable of penetrating membranes, such as [^{3}H]N-methylscopolamine.

There is undoubtedly a great excess of AChE present in the brain over the minimum required for comparatively normal physiological and behavioral function. Our animals showed severe symptomology for several hours following treatment when the AChE was very greatly inhibited, but they had recovered from the majority of the gross symptoms by the day after treatment when brain AChE activity was only about 30% of control activity. At this point deficits in learned behavior were apparent, but not a severity proportional to the severity of the AChE inhibition. Recovery of the performance to normal levels also appeared to occur faster than recovery of the AChE activity. It would appear that recovery of a relatively small amount of the AChE activity is certainly enough to bring gross behavior back to normal and that even relatively high levels of AChE inhibition, such as 50-70%, lead to relatively small effects on memory retention, and no effects on nociception. Recovery of the behavioral deficits improved when the centrally active muscarinic antagonist was used. Although task performance was not totally recovered to control levels on the second day after treatment, performance levels were approaching those of the controls (especially in the animals antidoted with AS) and would probably have been back to normal within 1 or 2 more days. Data accumulated on a positive reinforcement paradigm using the same experimental conditions as reported here have indicated far more severe performance deficits on the day after treatment with complete recovery by day 3 (unpublished observations).

In conclusion, a severe accidental acute poisoning with an OP anticholinesterase at a lethal or near lethal level leads to a high degree of inhibition of brain AChE which persists for several days, a relatively small and transient effect on the performance of negatively-reinforced learned tasks which may reflect a transient memory loss but no effect on nociception, and little if any effect on other cholinergic neurochemistry. Unlike the response to chronic exposures to OP compounds, an acute OP intoxication, even at this high dose level, has not resulted in the employment of cholinergic compensatory mechanisms to a very great extent, if any at all, to counteract the hypercholinergic activity induced by the accumulation of acetylcholine.

ACKNOWLEDGEMENTS

This research was supported by EPA Grant R-811295. The laboratory assistance of Cheryl Plunkett, Sherrill Wiygul, Ellen Bickley, Marilynn Alldread, Amanda Holland, John Snawder and Jeff Stokes is greatly appreciated.

REFERENCES

Aas, P., Veiteberg, T. A., and Fonnum, F., 1987, Acute and sub-acute inhalation of an organophosphate induce alteration of cholinergic muscarinic receptors, Biochem. Pharmacol., 36:1261.

Atweh, S., Simon, J. R., and Kuhar, M. J., 1975, Utilization of sodium-dependent high affinity choline uptake *in vitro* as a measure of the activity of cholinergic neurons *in vivo*, Life Sci., 17:1535.

Barker, L. A., and Mittag, T. W., 1975, Comparative studies of substrates and inhibitors of choline transport and choline acetyltransferase. J. Pharmacol. Exptl. Therapeut., 192:86.

Beninger, R. J., Jhamandas, K., Boegman, R. J., and El-Defrawy, S. R., 1986a, Effects of scopolamine and unilateral lesions of the basal forebrain on T-maze spatial discrimination and alteration in rats, Pharmacol. Biochem. Behav., 24:1353.

Beninger, R. J., Wirsching, B. A., Jhamandas, K., Boegman, R. J., and El-Defrawy, L.R., 1986b, Effects of altered cholinergic function on working and reference momory in the rat, Can. J. Physiol. Pharmacol., 64:376.

Berends, F., Posthumus, C. H., Van der Sluys, I., and Deierkauf, F. A., 1959, The chemical basis of the "aging process" of DFP-inhibited pseudocholinesterase, Biochem. Biophys. Acta., 34:576.

Bermudez-Rattoni, F., Mujica-Gozalez, M., and Prado-Alcala, R. A., 1986, Is cholinergic activity of the striatum involved in the acquisition of positively-motivated behavior? Pharmacol. Biochem. Behav., 24:715.

Bignami, G., and Giardini, V., 1983, Cholinergic effects: critical treatment-behavior interactions in paraoxon tolerance, *in*: "Applications in Behavior, Pharmacology and Toxicology" (Lecture Workshop), G. Zbinden, V. Cuomo, and G. Racagni, ed., Raven Press, New York.

Bignami, G., Rosic, N., Michalek, H., Milosevic, M., and Gatti, G. L., 1975, Behavioral toxicity of anticholinesterase agents: Methodological, neurochemical and neuropsychological aspects, *in*: "Behavioral Toxicology", B. Weiss and V.G. Laties, ed., Plenum Press, New York.

Brezenoff, H. E., McGee, J., and Hymowitz, N., 1985, Effect of soman on schedule-controlled behavior and brain acetylcholinesterase in rats, Life Sci., 37:2421.
Brush, F. R., 1962, The effects of intertrial interval on avoidance learning in the rat, J. Comp. Physiol. Psychol., 55:888.
Chambers, J. E., Wiygul, S. M., Harkness, J. E., and Chambers, H. W., 1988, Effects of acute paraoxon and atropine exposures on retention of shuttle avoidance behavior in rats, Neurosci. Res. Comm., In press.
Chambers, J. E., and Chambers, H. W., 1988, Effect of the anticholinesterase paraoxon on shuttle avoidance behavior and brain muscarinic receptors, Toxicologist, 8:76.
Chambers, H. W., and Chambers, J. E., 1989, An investigation of acetylcholinesterase inhibition and aging and choline acetyltransferase activity following a high level acute exposure to paraoxon, Pestic. Biochem. Physiol. 33:125
Churchill, L., Pazdernik, T. L., Jackson, J. L., Nelson, S. R., Samson, F.E., and McDonough, J. H., 1984a, Topographical distribution of decrements and recovery in muscarinic receptors from rat brains repeatedly exposed to sublethal doses of soman, J. Neurosci., 4:2069.
Churchill, L., Pazdernik, T. L. Samson, F., and Nelson, S. R., 1984b, Topographical distribution of down-regulated muscarinic receptors in rat brains after repeated exposure to diisopropylphosphonofluoridate, Neurosci., 11:463.
Davis, W. M., and Hatoum, H. T., 1987, Comparison of stimulants and hallucinogens on shuttle avoidance in rats, Gen. Pharmacol., 18:123.
Desi, I., 1983, Neurotoxicological investigation of pesticides in animal experiments, Neurobeh. Toxicol. Teratol., 5:503.
Ellman, G. L., Courtney, K. D., Andres, V., and Featherstone, R. M., 1961, A new and rapid colorimetric determination of acetylcholinesterase activity, Biochem. Pharmacol., 7:88.
Fonnum, F., 1975, A rapid radiochemical method for the determination of choline acetyltransferase, J. Neurochem., 24:407.
Gaines, T. B., and Linder, R. E., 1986, Acute toxicity of pesticides in adult and weanling rats, Fundam. Appl. Toxicol., 7:299.
Gardner, R., Ray, R., Frandkenheim, J., Wallace, K., Loss, M., and Robichaud, R., 1984, A possible mechanism for diisopropylfluorophosphate-induced memory loss in rats, Pharmacol. Biochem. Behav., 21:43.
Gazit, H., Silman, I., and Dudai, Y., 1979, Administration of an organophosphate causes a decrease in muscarinic receptor level in rat brain, Brain Res., 174:351.
Geller, I., Hartmann, R. J., Leal, B. Z., Haines, R. J., and Gause, E. M., 1984, Effects of the irreversible acetylcholinesterase inhibitor soman on match-to-sample behavior of the juvenile baboon, Proc. West. Pharmacol. Soc., 27:217.
Gonzalez, L. P., and Ellinwood, Jr., E. H., 1984, Cholinergic modulation of stimulant-induced behavior, Pharmacol. Biochem. Behav., 20:397.
Haga, T., 1971, Synthesis and release of [^{14}C]acetylcholine in synaptosomes, J. Neurochem., 18:781.
Haggerty, G. C., Kurtz, P. J., and Armstrong, R. D., 1986, Duration and intensity of behavioral change after sublethal exposure to soman in rats, Neurobeh. Toxicol. Teratol., 8:695.
Hayes, W. J., Jr., 1969, Pesticides and human toxicity, Ann. New York Acad. Sci., 160:40.
Kaita, A. A., and Goldberg, A. M., 1969, Control of acetylcholine synthesis--the inhibition of choline acetyltransferase by acetylcholine, J. Neurochem., 16:1185.
Kuhar, M. J., and Murrin, L. C., 1978, Sodium-dependent, high affinity choline uptake, J. Neurochem., 30:15.
Lehotzky, K., 1982, Effect of pesticides on central and peripheral nervous system function in rats, Neurobeh. Toxicol. Teratol., 4:665.

Lerer, B., Altman, H., and Stanley, M., 1984, Enhancement of memory by a cholinesterase inhibitor associated with muscarinic receptor down-regulation, Pharmacol. Biochem. Behav., 21:467.
Lim, D. K., Hoskins, B., and Ho, I. K., 1986, Correlation of muscarinic receptor density and acetylcholinesterase in repeated DFP-treated rats after termination of DFP administration, European J. Pharmacol., 123:223.
McMillan, D. E., 1982, Effects of chronic administration of pesticides on schedule-controlled responding by rats and pigeons, in: "Effects of Chronic Exposures to Pesticides on Animal Systems", J.E. Chambers and J.D. Yarbrough, ed., Raven Press, New York.
Nemeth, E. F., and Cooper, J. R., 1979, Effects of somatostatin on acetylcholine release from rat hippocampal synaptosomes, Brain Res., 165:166.
Raffaele, K., Hughey, D., Wenk, G., Olton, D., Modrow, H., and McDonough, J., 1987, Long-term behavioral changes in rats following organophosphate exposure, Pharmacol. Biochem. Behav., 27:407.
Rauch, S. L., and Raskin, L. A., 1984, Cholinergic mediation of spatial memory in the preweanling rat: application of the radial arm maze paradigm, Behav. Neurosci., 98:35.
Sanberg, P. R., Lehmann, J., and Fibiger, H. C., 1978, Impaired learning and memory after kainic acid lesions of the striatum: a behavioral model of Huntington's disease, Brain Res., 149:546.
Sandberg, K., Sanberg, P., Hanin, I., Fisher, A., and Coyle, J. T., 1984, Cholinergic lesion of the striatum impairs acquisition and retention of a passive avoidance response, Behav. Neurosci., 98:162.
Schiller, G. D., 1979, Reduced binding of (^{3}H)-quinuclidinyl benzilate associated with chronically low acetylcholinesterase activity, Life Sci., 24:1159.
Simon, J. R., Atweh, S., and Kuhar, M. J., 1976, Sodium-dependent high affinity choline uptake: a regulatory step in the synthesis of acetylcholine, J. Neurochem., 26:909.
Sivam, S. P., Norris, J. C., Lim, D. K., Hoskins, B., and Ho, I. K., 1983, The effect of acute and chronic cholinesterase inhibition with diisopropylfluorophosphate on muscarinic dopamine and GABA receptors of the rat striatum, J. Neurochem., 40:1414.
Smolen, T. N., Smolen, A., and Collins, A. C., 1986, Dissociation of decreased numbers of muscarinic receptors from tolerance to DFP, Pharmacol. Biochem. Behav., 25:1293.
Uchida, S., Takeyasu, K., Matsuda, T., and Yoshida, H., 1979, Changes in muscarinic acetylcholine receptors of mice by chronic administrations of diisopropylfluorophosphate and papaverine, Life Sci., 24:1805.
Upchurch, M., and Wehner, J. M., 1987, Effects of chronic diisopropylfluorophosphate treatment on spatial learning in mice, Pharmacol. Biochem. Behav., 27:143.
Walsh, T. J., Tilson, H. A., DeHaven, D. L., Mailman, R. B., Fisher, A., and Hanin, I., 1984, AF64A, a cholinergic neurotoxin, selectively depletes acetylcholine in hippocampus and cortex, and produces long-term passive avoidance and radial-arm maze deficits in rats, Brain Res., 321:91.
Wecker, L., Mobley, P. L., and Dettbarn, W.-D., 1977, Central cholinergic mechanisms underlying adaptation to reduce cholinesterase activity, Biochem. Pharmacol., 26:633.
Wenk, G., Hughey, D., Boundy, V., Kim, A., Walker, L., and Olton, D., 1987, Neurotransmitters and memory: role of cholinergic, serotinergic and noradrenergic systems, Behav. Neurosci., 101:325.
Woolley, D. E., Chernobieff, J. R., and Reiter, L. W., 1979, Effects of parathion on the mammalian nervous system, in: "Neurotoxicity of Insecticides and Pheromones", T. Narahashi, ed., Plenum Press, New York.
Yamada, S., Isogai, M., Okudaira, H., and Hayashi, E., 1983, Correlation

between cholinesterase inhibition and reduction in muscarinic receptors and choline uptake by repeated diisopropylfluorophosphate administration: antagonism by physostigmine and atropine, J. Pharmacol. Exptl. Therapeut., 226:519.
Yamamura, H. I., and Snyder, S. H., 1973, High affinity transport of choline into synaptosomes of rat brain, J. Neurochem., 21:1355.
Yamamura, H. I., and Snyder, S. H., 1974, Muscarinic cholinergic binding in rat brain, Proc. Nat. Acad. Sci., 71:1725.

MONOOXYGENATIONS: INTERACTIONS AND EXPRESSION OF TOXICITY

Patricia E. Levi and Ernest Hodgson

Toxicology Program
North Carolina State University
Raleigh, N.C.

INTRODUCTION

The toxic action of an insecticide is dependent on its reactivity to the biochemical target (e.g., reactivity of paraoxon toward acetylcholinesterase). Toxicity in vivo, however, is considerably modulated by rates of absorption, metabolism, and elimination of the insecticide; thus, metabolic fate plays a pivotal role in the toxicity and persistence of insecticides. Species variance in toxicity is often due to differential rates or routes of biotransformation of the parent compound or one of the metabolites.

Even though insecticides are subject to a wide array of phase I and phase II xenobiotic-metabolizing enzymes, the role of the monooxygenases is of primary and critical importance. In addition to detoxication reactions, the monooxygenases also produce highly reactive intermediates that play an important part in activation reactions and hence, in both acute and chronic toxicity. Since the monooxygenases are enzymes involved in the metabolism of a wide variety of xenobiotics, many of which can act as inducers or inhibitors as well as substrates, the monooxygenases are a focal point for interactions between different compounds.

While the cytochrome P-450 monooxygenases (P-450) are probably the most important enzymes involved in the initial metabolism of insecticides and other xenobiotics, the flavin-containing monooxygenases (FMO) also make a significant contribution to xenobiotic metabolism. As can be seen from Table 1, there is considerable overlap between P-450 and FMO in the types of reactions catalyzed and the substrates for these reactions. Thus, it is of considerable importance in understanding insecticide metabolism to define the role of the FMO enzymes relative to that of the P-450 system. Table 2 summarizes some of the similarities and differences between the FMO and P-450 monooxygenases. The flavin-containing monooxygenase, like cytochrome P-450, is located in the endoplasmic reticulum and is involved in the monooxygenation of a wide variety of xenobiotics. Although the FMO was originally characterized as an amine oxidase, it is now recognized that, like P-450, this enzyme can catalyze the oxygenation of a wide variety of xenobiotic nitrogen-, sulfur-, and phosphorus-containing xenobiotics (Hodgson and Levi, 1988, Ziegler, 1980, Ziegler, 1984, Ziegler, 1988) but unlike P-450 cannot catalyze carbon hydroxylations.

Table 1. Summary of major reactions catalyzed by P-450 and FMO

Reactions	Examples
	Cytochrome P-450 (P-450)
Epoxidation	Aldrin, benzo(a)pyrene, aflatoxin
Hydroxylation	Nicotine, bromobenzene
N-dealkylation	Ethylmorphine, atrazine, dimethylaniline
O-dealkylation	p-Nitroanisole, chlorfenvinphos
S-dealkylation	Methylmercaptan
S-oxidation	Thiobenzamide, phorate, chlorpromazine
N-oxidation	2-Acetylaminofluorene
P-oxidation	Diethylphenylphosphine
Desulfuration	Parathion, fonofos, carbon disulfide
Dehalogenation	CCl_4, $CHCl_3$
	Flavin-containing monooxygenase (FMO)
N-oxygenation	Nicotine, dimethylaniline, imipramine
S-oxygenation	Thiobenzamide, phorate, thiourea
P-oxygenation	Diethylphenylphosphine
Desulfuration	Fonofos

Table 2. Comparison of P-450 and FMO

Feature	FMO	P-450
Cofactors	NADPH, O_2	NADPH, O_2, reductase
Location	Microsomes	Microsomes
Inducers	None	PB, PAHs, EtOH, PCN
Inhibitors	None	CO, SKF-525A, PBO
Isozymes	Few	Many
Substrates	N,S,P compounds	N,S,P,C compounds
Reactions	Oxygenation	Oxygenation, epoxidation, reduction, dealkylation

The FMO was first purified to homogeneity from pig liver microsomes (Ziegler and Poulsen, 1978). In our laboratory we have purified the FMO from both pig liver and mouse liver (Sabourin et al., 1984, Sabourin and Hodgson, 1984). Subsequently, our group (Tynes et al., 1985) as well as Williams et al. (1984) purified from rabbit lung an FMO that was shown to be catalytically and immunologically distinct from the liver enzyme. The mouse and rabbit lung FMOs have a unique ability for N-oxidation of the primary aliphatic amine, n-octylamine, a chemical commonly included in microsomal incubations to inhibit P-450. In the mouse lung, this compound not only serves as a substrate, but also is a positive effector of metabolism. The mouse and rabbit lung enzymes have a higher pH optimum, near 9.8, compared to the liver form, which has an optimum near 8.8. In addition, antibodies to the liver and lung proteins have shown that these enzymes are immunochemically dissimilar (Tynes and Hodgson, 1985). It has now become evident that there are several FMO enzymes with overlapping substrate specificities, and it is likely that the relative proportions of these isozymes vary in different tissues within and between species (Tynes and Philpot, 1987).

Since a wide variety of pesticides and other xenobiotics are metabolized by both P-450 and FMO systems, it is important to delineate the relative contributions of these two enzyme systems in microsomal oxidations as well as to examine isozyme specific activities of the individual FMO and P-450 isozymes. Changes in the activities of these enzymes either through environmental, physiological, or genetic factors will ultimately affect the efficacy and/or toxicity of pesticides and other xenobiotics.

RELATIVE CONTRIBUTIONS OF FMO AND P-450 IN MICROSOMAL OXIDATIONS

Often the same substrate is metabolized by both P-450 and FMO; this situation is especially prevalent with many N- and S-containing pesticides. To study the relative contributions of these two enzymes with common substrates, we developed methods to measure each separately in the same microsomal preparation. The most useful of these techniques is the inhibition of P-450 activity by using an antibody to NADPH cytochrome P-450 reductase, thus permitting measurement of FMO activity alone. A second procedure is heat treatment of microsomal preparations (50° for 1 min), which inactivates the FMO, thus allowing determination of P-450 activity, which is unchanged by heat treatment (Tynes and Hodgson, 1983). Thermal inactivation, however, is ineffective with lung microsomes, since the lung FMO is more heat stable than the liver FMO; this necessitates the use of anti-reductase with lung microsomal preparations to define the relative roles of P-450 and FMO.

The contribution of the two enzyme systems to the metabolism of the insecticide phorate in different tissues has been examined using these procedures (Table 3; see Figure 1 for structure of phorate). While the oxidation of phorate is relatively similar between males and females,

Table 3. Relative contribution of P-450 and FMO to microsomal oxidation of phorate in mouse

		Product Formation[a]			
Tissue	Sex	Control	+AR[b]	% FMO	% P-450
Liver	M	12.7	2.8	21.7	78.3
Liver	F	14.4	3.7	24.0	76.1
Lung	M	3.3	1.9	59.1	41.3
Lung	F	5.7	3.1	54.0	46.0
Kidney	M	1.6	1.2	72.0	28.1
Kidney	F	69.7	10.1	14.3	85.5

[a]nmols phorate sulfoxide/mg protein/min.
[b]Antibody to P-450 reductase.
[c]Phenobarbital-treated mice.

SOURCE: Kinsler et al., 1988, 1989.

there is a striking difference in the proportion of sulfoxidase activity resulting from FMO in the different tissues. In the liver of untreated animals, P-450 is more important in the sulfoxidation of phorate (78 percent by P-450, 22 percent by FMO in males). By contrast, in the kidney and lung, although the overall activity is low compared to that in the liver, the relative contribution by FMO is significantly higher. This is especially dramatic in kidney microsomes from female mice, where about 90 percent of the activity is associated with FMO and only 10 percent with P-450. It is interesting to note that in the absence of P-450 activity, the FMO rate per mg of microsomal protein is similar in the liver and the lung, suggesting that the lung form is more efficient than the liver form in catalyzing phorate sulfoxidation.

This balance of enzyme activity, especially in the liver, is easily disturbed by compounds that alter the concentration of P-450 isozymes. Of special interest is the change in the balance of microsomal enzyme activity after in vivo exposure of animals to either inducers or inhibitors of monooxygenase activity (Kinsler et al., 1989). Thus, pretreatment of mice with phenobarbital increased not only the total rate of phorate oxidation, but the proportion metabolized by P-450.

An interesting situation is observed in mice pretreated with the insecticide synergist piperonyl butoxide (PBO). Piperonyl butoxide functions as a synergist by inhibiting P-450 activity in vivo; this inhibition, however, is followed by induction of P-450 activity. As shown in Table 4, an initial inhibition of P-450 activity at 2 hr results in a decrease in both the P-450 rate and the percent contribution by P-450. This phase is followed by induction of P-450; by 4 hr the level of activity is similar to control, and by 36 hr the rate is elevated above control, with the contribution by P-450 being approximately 84 percent. Such alterations in the relative contributions of the two enzyme systems may assume toxicological importance when the products from the two enzymes differ, and particularly when one metabolite is more toxic than the others.

COMPLEX METABOLIC PATHWAYS

Purified enzymes/isozymes are essential for elucidating the reactions and products involved in complex metabolic pathways, especially when more

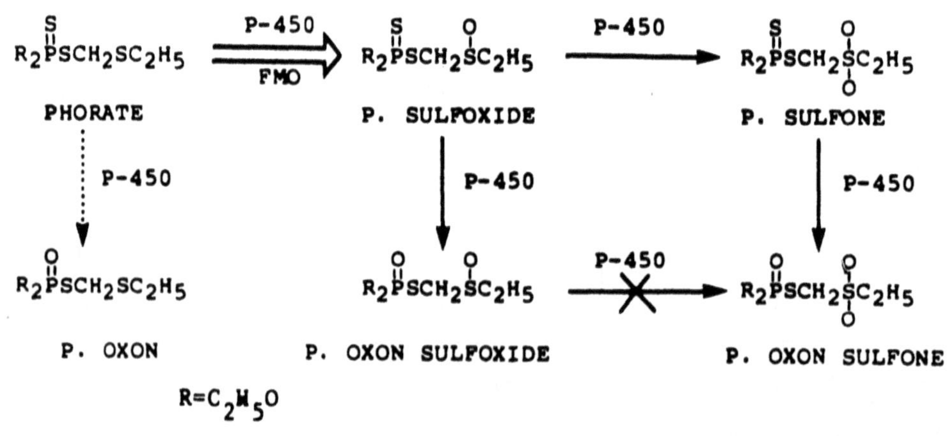

Fig. 1. Oxidation of phorate by cytochrome P-450 and FMO.

Table 4. Effect of in vivo treatment with piperonyl butoxide on phorate oxidation by mouse hepatic microsomes

	Product formation[a]			
Hours	Total rate	+ Heat	% P-450	% FMO
0	13.8	10.5	76.0	24.0
2	11.1	6.5	58.1	47.9
4	12.1	9.3	77.0	23.0
8	13.3	10.9	82.2	17.8
12	15.4	13.8	89.2	10.8
36	19.4	16.3	83.9	16.1

[a]nmols phorate sulfoxide/min/mg protein.

SOURCE: Kinsler et al. (1989).

than one enzyme/isozyme is involved, e.g., the oxidation of phorate by FMO and P-450 isozymes (Levi and Hodgson, 1988) which is illustrated in Figure 1. Both P-450 and FMO readily catalyze the initial sulfoxidation of the thioether moiety of phorate to form the sulfoxide. Subsequent oxidation reactions, however, such as formation of the sulfone and oxidative desulfuration to the corresponding oxons are catalyzed entirely by P-450. Although both FMO and P-450 form phorate sulfoxide, the products are stereochemically different, with the FMO forming (-) phorate sulfoxide and two of the P-450 isozymes yielding (+) phorate sulfoxide. The other three P-450 isozymes examined gave racemic mixtures. Thus, in vivo, the net optical activity of sulfoxidation would be a function of such factors as the presence of activators, inhibitors, and inducers, as well as of sex and organ.

Although two of the P-450s yielded (+) phorate sulfoxide as the first oxidation product, either (+) or (-) phorate sulfoxide served as substrate for additional oxidation by P-450; the preferred substrate, however, is (+) phorate sulfoxide. The products formed are either the oxon sulfoxide (an activation reaction) or phorate sulfone (a detoxication pathway), with the percent oxon sulfoxide being greater with (+) phorate sulfoxide as the substrate than with (-) phorate sulfoxide (Table 5). P-450 PB, which rapidly forms the (+) phorate sulfoxide, also produces the highest percentage of oxon sulfoxide of any of the P-450s. Thus, it is interesting to speculate that environmental or physiological factors that increase the level of this enzyme in vivo could potentially increase the production of the more toxic metabolite.

P-450 ISOZYME DIFFERENCES IN ACTIVATION/DETOXICATION REACTIONS

A number of phosphorothionate triesters such as parathion and fenitrothion are metabolically activated in vivo by P-450 to the corresponding oxons, which are potent acetylcholinesterase inhibitors. It has been proposed, using parathion as the experimental model (Neal and Halpert, 1982, Neal, 1985), that a cyclic phosphorus-sulfur-oxygen intermediate is formed initially that can then break down via two different pathways: oxidative

Table 5. Oxidation of (+) and (-) phorate sulfoxide by purified cytochrome P-450 isozymes

	(+) Phorate sulfoxide			(-) Phorate		
Enzyme	Sulfone[a]	Oxon[a]	% oxon	Sulfone[a]	Oxon[a]	% oxon
P-450 PB	4.8	5.3	52.5	3.9	2.4	38.1
P-450 A1	1.4	0.5	26.3	0.6	0.1	14.3
P-450 B1	1.2	0.4	25.0	0.6	0.2	25.0
P-450 B2	2.3	2.0	46.5	1.3	0.9	40.9
P-450 B3	0.9	0.6	40.0	0.6	0.2	25.0

[a]nmol product formed/30 min/0.4 nmol P-450.

SOURCE: Levi and Hodgson (1988).

desulfuration to yield the oxon (an activation reaction) or oxidative dearylation to form the nitrophenol (a detoxication reaction).

Recently we have shown, using purified enzymes and fenitrothion as the substrate, that the ratio of activation/ detoxication products varies considerably with different P-450 isozymes (Figure 2) (Levi et al., 1988). For example, the P-450 induced by phenobarbital (P-450 PB) yielded the highest percentage of oxon (84%), whereas one of the uninduced isozymes, P-450 A2, produced primarily the nitrophenol (63%). The other isozymes formed oxon:cresol ratios in between these limits. These findings are of interest toxicologically, since the potential effect of PB-like inducers

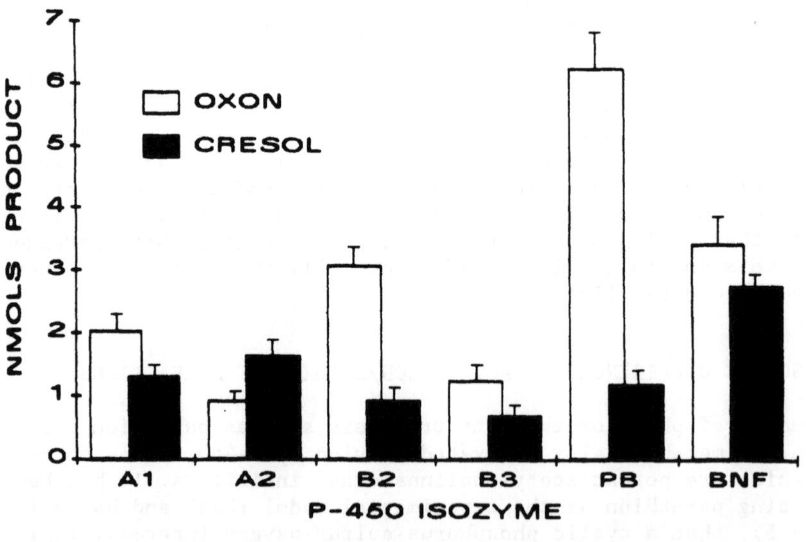

Fig. 2. Oxidation of fenitrothion by purified P-450 isozymes

would be to increase the percent of activation products relative to detoxication products.

Differences in the ratios of activation/detoxication products in the metabolism of fenitrothion by the P-450 system have been observed in organophosphorous resistant and insecticide susceptible houseflies (Ugaki, et al., 1985a). The metabolism of fenitrothion was investigated in highly resistant (Akita-f) and susceptible (SRS) housefly (Musca domestica L.) strains. In vitro studies showed that the P-450 system metabolized fenitrothion into fenitroxon and 3-methyl-4-nitrophenol as major metabolites, with the resistant strain producing 2.2-fold the amount of the phenol as the susceptible strain. In addition, the overall metabolism was greater in the resistant strain. It should be noted, however, that although differences in activation/detoxication of fenitrothion by the P-450 systems occurs in the two strains, this is only one of the mechanisms contributing to fenitrothion resistance in the Akita-f strain; an increase in glutathione S-transferase activity and a decrease in acetylcholinesterase sensitivity are considered to be major mechanisms of organophosphorous insecticide resistance (Ugaki et al., 1985b).

INDUCTION OF MONOOXYGENASE ACTIVITY AND RESISTANCE TO PESTICIDES

Induction of detoxication enzyme systems as a result of feeding on particular host plants is known to occur in several insect species. This induction may be caused by naturally occurring substances in the host plants; often this effect on microsomal monooxygenases is bimodal, increasing some activities but inhibiting others. A species of wild tomato, Lycopersicum hirsutum F. glabratrum is resistant to a number of insect pests because of the presence of 2-tridecanone in the plant. The effect of the wild tomato on the susceptibility to and metabolism of diazinon in the larvae of the tobacco budworm, Heliothis virescens F., has been studied recently in our laboratory (Riskallah et al., 1986a, Riskallah et al., 1986b). The larvae were over four fold more tolerant to diazinon toxicity when fed on leaves of wild tomato than when fed on artificial diet (Table 6). This tolerance developed within 24 hr of feeding on the host plants. Interactions between diazinon or diazoxon and piperonyl butoxide, a P-450 inhibitor, or TOTP (tri-o-tolyl phosphate), an esterase inhibitor, suggest that both compounds are metabolized in H. virescens larvae, primarily by oxidative enzymes that are induced in the larvae fed on this species of tomato.

Comparative in vitro studies with microsomal and soluble fractions provided additional evidence of enhanced degradation in larvae fed tomato leaves (Table 7). With both diazinon and diazoxon, the largest amount of degradation occurred in the microsomal fraction. Piperonyl butoxide and SKF-525A, both inhibitors of monooxygenase activity, inhibited microsomal degradation of diazinon and diazoxon greater than 80 percent with control larvae and larvae fed tomato leaves, thus confirming that both diazinon and its oxon are degraded in H. virescens larvae, mainly through microsomal monooxygenases. Table 8 shows that the P-450 content is increased up to twofold by feeding the larvae on tomato leaves. The same pattern is apparent in 2-tridecanone-treated larvae, with the increase ranging between 1.5- and 2.1-fold, depending on the tridecanone concentration. In addition to an increase in P-450 content, the induction of P-450 was associated with a shift in the λ max of the CO P-450 complex, suggesting a qualitative as well as a quantitative change in the composition of the P-450 isozymes. Current studies are in progress to characterize the P-450 isozymes involved.

Table 6. Effect of diet on toxicity of diazinon and diazoxon in *Heliothis virescens*

Diet[a]	LD50[b]		
	Pesticide only	Pesticide + PBO	Pesticide + TOTP
Diazinon			
Control diet	139	246	111
Tomato leaves	576	281	438
Diazoxon			
Control diet	61	49	52
Tomato leaves	256	128	212

[a]3rd instar larvae fed diet 24 hr prior to assay.
[b]μg/g body wt.

SOURCE: Riskallah et al. (1986a).

Table 7. In vitro degradation of diazinon and diazoxon by *Heliothis virescens*

Diet[a]	Microsomes[b]	Cytosol[b]
Diazinon		
Control diet	95.4 ± 16.2	3.9 ± 0.6
Tomato leaves	438.0 ± 25.2	14.2 ± 2.0
Diazoxon		
Control diet	30.7 ± 1.6	3.8 ± 1.1
Tomato leaves	141.1 ± 15.4	12.8 ± 3.8

[a]5th instar fed diet 24 hr prior to assay.
[b]Rate: 10^{-2} μg/hr/mg protein.

SOURCE: Riskallah et al. (1986a).

These results suggest that some tomato species that may prove to be resistant to various insect pests could also induce tolerance to insecticides in the same or other pest species. The implications of this study are also important for the development of insect resistance in other crops. The potential for adverse effects associated with a specific type of resistance should be considered early in the breeding effort in order to develop crop cultivars resistant to insect pests without adverse interactions between plant resistance and chemical control.

Table 8. Degradation of diazinon/diazoxon and cytochrome P-450 content in microsomes of <u>Heliothis virescens</u>

	Degradation[b]		Cytochrome P-450	
Diet/Treatment[a]	Diazinon	Diazoxon	nmol/mg (%)	λ max
Control diet	18.7	7.8	0.034 (100)	450.5
Tomato leaves	46.9	23.5	0.068 (200)	449.5
2-Tridecanone[c]				
400	40.5	19.4	0.054 (159)	449.5
800	61.5	30.8	0.073 (215)	448.5

[a]5th instar fed diet 24 hr prior to assay.
[b]$\mu g/hr/g$ larvae
[c]$\mu g/cm^2$

SOURCE: Riskallah et al. (1986a).

INHIBITION AND INDUCTION OF MONOOXYGENASE ACTIVITY BY METHYLENEDIOXYPHENYL COMPOUNDS

Sometimes a compound may be a substrate, an inhibitor, and an inducer of the same enzyme and associated activities. The interactions of methylenedioxyphenyl compounds (MDP) illustrate this type of relationship. As a chemical class, MDP compounds are widespread in the environment, and include carcinogens (e.g., isosafrole), food constituents (e.g., myristicin), and insecticide synergists (e.g., piperonyl butoxide).

The activity of MDP compounds as insecticide synergists is a result of their inability to inhibit cytochrome P-450 and the associated reactions involved in the metabolism of insecticides. MDP compounds owe this inhibitory effect to their ability to form a metabolite inhibitory complex at the active site of the P-450 molecule. The formation of this stable MDP metabolite cytochrome P-450 complex is responsible for the appearance of a Type III double Soret optical difference spectrum as well as inhibition of monooxygenase activity. Methylenedioxyphenyl compounds have a biphasic effect on P-450 levels and related monooxygenase activities, namely inhibition followed by induction.

While inhibition is known to be due to the formation of the P-450 metabolite inhibitory complex, neither the mechanism for MDP induction of P-450 nor the pattern of isozyme induction are well known. Recently we have focused on these events by examining the structure activity relationship between methylenedioxyphenyl compounds with different substituents on the phenyl ring and the ability of these compounds to induce specific P-450 isozymes and associated monooxygenase activities (Lewandowski et al., 1989).

To determine if the induction patterns are modulated by interactions with the Ah receptor, two strains of mice were used, C57/BL6 (Ah positive) and DBA (Ah negative). Initial studies with safrole, isosafrole, and dihydrosafrole showed no difference in induction of monooxygenase activities between the two strains of mice, indicating that P-450 induction was not mediated by the Ah receptor (Lewandowski et al., 1989). Thus, for the

Table 9. Effect of in vivo administration of MDP compounds on in vitro microsomal monooxygenase activities

	Compound			
Treatment	Ethyl-morphine[a]	Acet-anilide[a]	Benzo(a)-pyrene[b]	Ethoxy-resorufin[b]
Control	3.07	2.88	0.99	0.10
3-MC	3.18	12.79*	7.46*	0.88*
PB	7.61*	4.88*	2.78*	0.30*
Safrole	5.10*	6.13*	1.02	0.29*
Isosafrole	5.22*	7.40*	0.95	0.48*
Dihydrosafrole	3.10	2.99	0.50*	0.22*
n-Butyl-BD	5.09*	3.44	1.09	0.21*
t-Butyl-BD	6.23*	4.35*	0.96	0.21*
Methyl-BD	3.00	3.27	0.70	0.11
Nitro-BD	3.21	3.15	0.74	0.12
Bromo-BD	3.89	3.58	0.79	0.10

*Significantly different ($p < 0.05$) from control values.
[a]nmol product/mg protein/min.
[b]nmol substrate consumed/mg protein/min.

SOURCE: Lewandowski et al. (1989).

remainder of the studies, only C57/BL6 mice were used. Table 9 summarizes the inductive effects of eight MDP compounds on P-450 monooxygenase activities. Microsomes from phenobarbital (PB) and 3-methylcholanthrene (3-MC)-induced mice were included for comparison. Two of the propyl-substituted compounds (safrole and isosafrole but not dihydrosafrole) and both butyl-substituted compounds were effective inducers of P-450 enzyme activities, while methyl-, bromo- and nitro-substituted compounds were very poor inducers. As can be seen from examining the induction of enzyme activities, there was a significant increase in ethylmorphine N-demethylase activity, suggesting PB-type induction. On the other hand, there was also an increase in ethoxyresorufin-O-demethylase and acetanilide hydroxylase activities, both of which are associated with 3-MC-type induction; however, no increase in benzo(a)pyrene hydroxylase activity was detected, as would be expected with classical 3-MC-type induction. Evaluating MDP induction on the basis of enzyme activities is complicated by the fact that both induction and inhibition (resulting from formation of the MDP metabolite P-450 complex) are occurring simultaneously. Since a significant portion of the induced isozyme is complexed, and there is no reliable method of removing the complex in vitro, the level of enzyme activity represents only the uncomplexed isozyme, and this level varies both with the P-450 isozyme and the MDP compound.

The use of specific antibodies to individual P-450 isozymes provides an additional tool for assessing the induction pattern of individual isozymes. Western blots and immunoquantitation using antibodies to the P-450s induced by PB and 3-MC showed that, in fact, proteins that reacted with both antibodies were induced by the MDP compounds (Table 10). In addition, immunoquantitation showed no significant difference in the level of the induced isozyme, which reacted with the 3-MC antibody in either Ah

Table 10. Quantitation of MDP induction of P-450 isozymes from Western blots

Treatment	P-450 PB[a]	P-450 3-MC[a]
Control	ND[b]	ND
3-MC	ND	4.14
PB	1.28	ND
Safrole	0.38	0.85
Isosafrole	0.61	1.88
Dihydrosafrole	0.11	0.88
n-Butyl-BD	0.35	2.52
t-Butyl-BD	0.48	0.85
Methyl-BD	ND	ND
Nitro-BD	ND	ND
Bromo-BD	ND	ND

[a]pmol P-450 /5μg microsomal protein.
[b]No P-450 detected.

SOURCE: Lewandowski et al. (1989).

positive (C57) or Ah negative (DBA) mice. Additional studies with purified isozymes will be essential, however, to elucidate the ability of the MDP compounds to complex with and inhibit specific P-450 isozymes so as to evaluate more accurately the inhibition and induction of monooxygenase activities.

In conclusion, the studies discussed in this paper illustrate some of the many diverse conditions that can affect monooxygenase activity and the expression of toxicity. Among the more important factors influencing monooxygenase activity are the relative activity of different isozymes/ enzymes toward the same substrate, the organ specificity of the different isozymes/enzymes, the ratio of activation/detoxication reactions, and the induction/inhibition of monooxygenase activity.

REFERENCES

Hodgson, E. and Levi, P. E., 1988, The flavin-containing monooxygenase as a sulfur oxidase, in "Metabolism of Xenobiotics," J.W. Gorrod, H. Oelschlager and J. Caldwell, eds., Taylor and Francis, London.

Kinsler, S., Levi, P. E. and Hodgson, E., 1988, Hepatic and extrahepatic microsomal oxidation of phorate by the cytochrome P-450 and FAD-containing monooxygenase systems in the mouse, Pest. Biochem. Physiol., 31:54-60.

Kinsler, S., Levi, P. E. and Hodgson, E., 1989, Effects of pretreatment with xenobiotics on the relative contributions of the cytochrome P-450 monooxygenase and flavin-containing monooxygenase systems in the microsomal oxidation of phorate in the mouse, in progress.

Levi, P. E. and Hodgson, E., 1988, Stereospecificity in the oxidation of phorate and phorate sulphoxide by purified FAD-containing monooxygenase and cytochrome P-450 isozymes, Xenobiotica, 18:29-39

Levi, P. E., Hollingworth, R. M. and Hodgson, E., 1988, Differences in oxidative dearylation and desulfuration of fenitrothion by cytochrome

P-450 isozymes and in the subsequent inhibition of monooxygenase activity, Pest. Biochem. Physiol., 32:224-231.
Lewandowski, M., Chui, Y. C., Levi, P. E. and Hodgson, 1989, Induction of hepatic cytochrome P-450 monooxygenase activities by methylenedioxy-pehnyl compounds in mice, in progress.
Riskallah, M. R., Dauterman, W. C. and Hodgson, E., 1986a, Host plant induction of microsomal monooxygenase activity in relation to diazinon metabolism and toxicity in larvae of the tobacco budworm Heliothis virescens (F.), Pest. Biochem. Physiol. 25:233-247.
Riskallah, M. R., Dauterman, W. C. and Hodgson, E., 1986b, Nutritional effects on the induction of cytochrome P-450 and glutathione transferase in larvae of the tobacco budworm, Heliothis virescens (F.), Insect Biochem., 16:491-499.
Sabourin, P. J. and Hodgson, E., 1984, Characterization of the purified microsomal FAD-containing monooxygenase from mouse and pig liver, Chem.-Biol. Interact., 51:125-139.
Sabourin, P. J., Smyser, B. P. and Hodgson, E., 1984, Purification of the flavin-containing monooxygenase from mouse and pig liver microsomes, Int. J. Biochem., 16:713-720.
Tynes, R. E. and Hodgson, E., 1985, Catalytic activity and substrate specificity of the flavin-containing monooxygenase in microsomal systems: Characterization of the hepatic, pulmonary, and renal enzymes of the mouse, rabbit, and rat, Arch. Biochem. Biophys., 240:77-93.
Tynes, R. E. and Philpot, R. M., 1987, Tissue- and species-dependent expression of multiple forms of mammalian microsomal flavin-containing monooxygenase, Mol. Pharmacol., 31:569-74.
Tynes, R. E., Sabourin, P. J. and Hodgson, E., 1985, Identification of distinct hepatic and pulmonary forms of microsomal flavin-containing monooxygenase in the mouse and rabbit, Biochem. Biophys. Res. Commun., 126:1069-75.
Ugaki, M., Shono, T. and Fukami, J., 1985a, Metabolism of fenitrothion by organophosphorous-resistant and -susceptible house flies, Musca domestica L., Pest. Biochem. Physiol., 23:33-40.
Ugaki, M., Shono, T. Tsukamoto, M. and Fukami, J., 1985b, Linkage group analysis of glutathione S-transferase and acetylcholinesterase in an organophosphorus resistant strain of Musca domestica L. (Diptera: Muscidae), Appl. Ent. Zool., 20:73-81.
Williams, D. E., Ziegler, D. M., Nordin, D. J, Hale, S. E. and Master, B. S. S., 1984, Rabbit lung flavin-containing monooxygenase is immunochemically and catalytically distinct from the liver enzyme, Biochem. Biophys. Res. Commun., 125: 116-122.
Ziegler, D. M., 1980, Microsomal flavin-containing monooxygenation of nucleophilic nitrogen and sulfur compounds, in "Enzymatic Basis of Detoxification," vol. 1, W.B. Jacoby, ed., Academic Press, New York.
Ziegler, D.M. and Poulsen, L.L., 1978, Hepatic microsomal mixed-function amine oxidase, Methods in Enzymol., 52:142-155.

NON-CATALYTIC DETOXICATION OF ACETYLCHOLINESTERASE INHIBITORS BY LIVER AND PLASMA PROTEINS

Howard W. Chambers and Janice E. Chambers

Departments of Entomology and Biological Sciences
Mississippi State University
Mississippi State, MS 39762

Metabolic detoxication of xenobiotics has long been a subject of major concern to pharmacologists and toxicologists. Not surprisingly, enzymes involved in catalytic degradation have received the greatest attention. Some years ago, however, it was proposed that interaction of toxicants with non-target sites could serve to reduce the potential toxicity of the chemical. This process would be particularly effective with irreversible enzyme inhibitors where inhibition of non-target enzymes chemically destroys the inhibitor molecule. Assuming that inhibition of the non-target enzymes exerts little or no toxic effect, the stoichiometric elimination of the inhibitor would represent a non-catalytic detoxication; that is, each non-target enzyme molecule would be responsible for detoxication of a single molecule of inhibitor.

Non-catalytic detoxication is exemplified by reactions involving the organophosphorus (OP) anticholinesterases. It is well established that the acute lethality of most OP's can be attributed to the inhibition of acetylcholinesterase (AChE), the enzyme responsible for termination of synaptic impulse transmission in cholinergic nerve synapses and neuromuscular junctions. OP's, however, are not selective for AChE, and are known to inhibit a variety of serine-containing carboxylesterases. The inhibition of non-neural esterases is therefore a potential mechanism of non-catalytic detoxication. Strong evidence for the reality and significance of this mechanism was first reported by Lauwerys and Murphy (1969a, b). As will be discussed in greater detail later in this manuscript, the authors concluded that the "paraoxon binding" responsible for decreasing the potential toxicity of paraoxon was in fact the inhibition of liver and plasma esterases, presumably butyrylcholinesterase and aliesterase (AliE). AliE is herein defined as that portion of carboxylesterase activity that is inhibited by paraoxon but not by esterine, i.e., B-esterases excluding the cholinesterases. Thus, the AChE of erythrocytes, butyrylcholinesterase of plasma and AliE of liver, plasma and other tissues represent a large pool of non-target sites available for removal of available OP.

A second phenomenon deserving consideration is that of reversible binding of OP's to serum albumin. Although this is not detoxication in the strict sence, a decrease in the effective concentration of OP available to penetrate into target tissues would occur. While the reversibility of the interaction would limit the efficacy of the process, this binding may well temporarily sequester sufficient OP to reduce toxicant concentrations while detoxication and elimination of other OP molecules is occurring, and thus could alter the overall level of toxicity.

The persistent question concerning the importance of non-catalytic detoxication is its relatively low efficiency compared to catalytic processes. The A-esterases, glutathione-S-alkyl transferases and cytochrome P-450 monooxygenases are all capable of metabolizing multiple OP molecules per molecule of enzyme. How likely is it, then, that non-target esterases which detoxify OP's at a mere one-to-one ratio could contribute significantly to the total picture?

Factors which must be considered include availability, affinity and selectivity of the enzymes and enzyme reactions in question. Potential value and limitations of the three major catalytic processes are compared to AliE's below.

A-esterases are of considerable theoretical value in the detoxication of OP's because of their ability to hydrolyze the active AChE inhibitors. Many OP's, however, have rather low affinity for A-esterases and, while they are good substrates at high concentrations, little or no hydrolysis occurs at the low concentrations required for AChE inhibition. Additionally, some OP's are not substrates for A-esterases at any concentration and may even be inhibitors. AliE's, on the other hand, usually interact with OP's at concentrations at or below those required for AChE inhibition. This high affinity should give AliE's the ability to sequester OP's throughout the concentration range likely to be encountered in *in vivo* poisonings.

Similarly, the glutathione-S-alkyl transferases are highly variable in their responses to different OP's. While they are often highly effective with dimethyl phosphates, they are often only marginally effective against diethyl OP's and virtually ineffective against higher homologs. AliE's show no such selectivity.

The cytochrome P-450 monoxygenases, or mixed function oxidases (mfo's), are undoubtedly the most important phase I detoxifying enzymes in the overall scheme of xenobiotic metabolism. This is probably not the case for OP's, however. Many OP insecticides are applied as pro-insecticides, the phosphorothionates. In these cases, mfo's are responsible for the desulfuration (oxon formation) which is an essential activation reaction. Further, sulfoxidations and certain N-dealkylations catalyzed by mfo's also lead to increased, rather than decreased, anti-AChE activity of the chemical. Again, AliE's are devoid of such limitations since they always destroy the OP molecule.

The remaining question, then, is whether there is sufficient AliE to decrease OP concentrations to an appreciable extent. Classic studies by Lauwerys and Murphy (1969a,b), as well as other work in both mammals and insects, strongly suggest that indeed AliE's are important in decreasing available OP's following exposure.

Lauwerys and Murphy (1969a, b) investigated the comparative roles of catalytic hydrolysis of paraoxon by paraoxonase (an A-esterase) and non-catalytic detoxication by what the authors called paraoxon binding in the overall detoxication of paraoxon in rats. As may be seen in Table 1, under the conditions tested, paraoxonase detoxifies more paraoxon than does paraoxon binding. It should be emphasized, however, that paraoxon was present at 400 μM in the paraoxonase assay but at only 0.1 μM in the paraoxon binding assay. Based on the kinetics of paraoxonase, the authors calculated that hydrolysis of paraoxon by this enzyme would be almost negligible at 0.1 μM, and that less than 1% of the paraoxon loss in the paraoxon binding assay could be attributed to paraoxonase activity. This conclusion was substantiated by the finding that while tri-o-tolyl phosphate (TOTP) pretreatment increased both the in vivo anti-AChE potency and acute toxicity of paraoxon (Table 2), it decreased paraoxon binding but not paraoxonase activity in liver and plasma (Table 3). Though it was not specifically investigated, the authors suggested that observed paraoxon binding was possibly phosphorylation of non-vital esterases such as AliE and butyrylcholinesterase.

Table 1. Paraoxonase activity and paraoxon binding in rat liver and plasma[a]

	Liver	Plasma
Paraoxonase activity[b]	232	94
Paraoxon binding	15.7	3.4

[a]Data from Lauwerys and Murphy (1969b)
[b]nmo p-nitrophenol released/min/g tissue using 400 μM paraoxon as substrate.
[c]decrease in available paraoxon (nmol/g tissue) using 0.1 μM paraoxon initial concentration

Table 2. Effect of TOTP treatments on AChE inhibition and mortality following paraoxon in vivo[a]

	AChE inhibition[b]		Mortality[c]	
Paraoxon treatment	Control	TOTP[d]	Control	TOTP[d]
0.25 mg/kg, ip	5%	60%		
0.375 " , "	10%	75%		
0.75 " , "			0/5	5/5
0.333 " , sc			1/6	6/6

[a]See footnote a, Table 1
[b]Inhibition relative to vehicle controls 10 min after treatments.
[c]No. dead/no. treated 24 hrs after treatments (vehicle/vehicle and TOTP/vehicle controls yielded no mortality)
[d]TOTP administered at 125 mg/kg ip, 16 hrs prior to paraoxon, controls received vehicle only.

Table 3. Effect of TOTP treatments on paraoxonase and paraoxon binding in rat liver and plasma[a]

	% of control activity after TOTP			
	125 mg/kg ip[b]		500 mg/kg oral[b]	
	Liver	Plasma	Liver	Plasma
Paraoxonase activity	130%	90%	120%	118%
Paraoxon binding	0%	0%	7%	17%

[a]See footnotes a, b & c, Table 1
[b]Administered 16 hrs prior to sacrifice

Further evidence of the protective role of non-target esterases has been obtained in insects. In the aphid, Myzus persicae, OP resistance is associated with high levels of an esterase designed E4. Even though E4 is capable of measurable rates of hydrolysis of dimethyl phosphates, diethyl phosphates inhibit the esterase with only extremely slow recovery. Yet, the insects exhibit resistance to both types of OP's. Resistance to diethyl phosphates was attributed to the increased capacity, as much as a 60-fold increase in one strain, of E4 to sequester the OP by acting as an alternate target (Devonshire and Moores, 1982).

In studies in guinea pigs involving TOTP pretreatments, Fonnum and Sterri (1981) similarly concluded that AliE and butyrylcholinesterase significantly reduce the potential toxicities of soman and sarin by acting as alternate sites for phosphorylation. They further proposed that the extreme toxicity of another nerve agent, VX, may be in part due to the fact that it contains a quaternary ammonium group and thus has a low affinity for AliE.

Recent studies in our laboratory have used in vitro methods to investigate non-catalytic detoxication of OP's in rat liver homogenates and in plasma. The assay methods were similar to those used by Lauwerys and Murphy in which inhibition of exogenous AChE was measured to determine the amount of residual free inhibitor. A 1,000-*g* pellet of bovine brain homogenate was used as the AChE source. This preparation was chosen because (1) brains were readily available from local meat processors, (2) the membrane-bound enzyme is quite stable, and (3) the AChE could be isolated from other components of the reaction mixture by centrifugation.

Test preparations used were rat liver (1,000-*g* supernate of a 12.5 mg/ml homogenate) and rat plasma (20 μl/ml). Average protein contents were 2.0-2.4 mg/ml for liver and 1.6-2.0 mg/ml for plasma. OP's (10 μM) in ethanol were added at 10 μl/ml to test preparations to yield a final concentration of 0.1 μM. Since preliminary studies indicated that the extent of detoxication was essentially the same for incubation times of 5-60 minutes, 15 minutes was selected as a standard convenient incubation time for the preparations prior to addition of the AChE. Following a second 15-minute incubation period to allow residual inhibitor to react with AChE, the brain particles were sedimented by centrifugation. AChE activity in the resuspended pellets was measured by a modification of the procedure of Ellman et al. (1961) as described by Chambers et al. (1988). Per cent inhibition was calculated against untreated control samples. Residual inhibitor concentration was estimated by comparing the observed per cent inhibition to regression lines for inhibitors to the AChE following incubation in buffer only. The residual inhibitor concentrations were then used to calculate the amount of inhibitor detoxified and expressed (in subsequent tables) as nanomoles OP detoxified per gram tissue, and as percentage of total the OP detoxified.

Inhibition of rat liver AliE was determined by incubation of rat liver homogenate with OP and measurement of residual activity. Substrate used was 4-nitrophenyl valerate and enzyme activity was determined by measuring the release of 4-nitrophenol at 400 nm.

Ten OP's were used in detoxication studies, 5 dialkyl phosphates (paraoxon series) and 5 alkyl phenylphosphonates (EPN-oxon series); structures are shown in Figure 1. All compounds were synthesized in our laboratory by published methods (Eto, 1974) and were judged as greater than 95 per cent pure by thin layer chromatography.

All OP's tested were effective inhibitors of both bovine brain AChE and rat liver AliE (Table 4) with I_{50} values of less than 1.0 μM. In all cases, AliE was more sensitive to inhibition than AChE, with differences ranging from 1.4- to 154-fold. Methyl paraoxon was the least potent in inhibiting both enzymes, but its effect on AliE was most striking, being more than 100-fold less potent than any other compound. The methyl homolog of the EPN-oxon series (MPNxn), on the other hand, was the most effective AliE inhibitor tested.

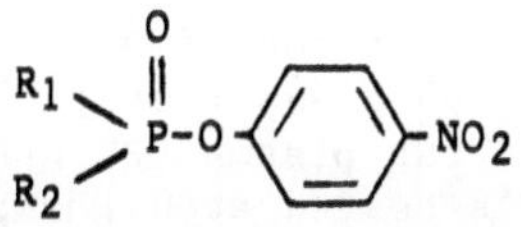

Paraoxon series		EPN-oxon series	
Compound	$R_1 = R_2 =$	Compound	R_1 = Ph-; R_2 =
MePxn	CH_3-O-	MPNxn	CH_3-O-
EtPxn	CH_3CH_2-O-	EPNxn	CH_3CH_2-O-
PrPxn	$CH_3CH_2CH_2$-O-	PPNxn	$CH_3CH_2CH_2$-O-
BuPxn	$CH_3(CH_2)_2CH_2$-O-	BPNxn	$CH_2(CH_2)_2CH_2$-O-
AmPxn	$CH_3(CH_2)_3CH_2$-O-	APNxn	$CH_3(CH_2)_3CH_2$-O-

Fig. 1. Structures of OP's used in study of non-catalytic detoxication by liver and plasma proteins.

Table 4. Inhibitory potency of experimental OP's against bovine brain AChE and rat liver AliE

Paraoxon series			EPN-oxon series		
	I_{50} (nM) vs:			I_{50} (nM) vs:	
OP	AChE	AliE	OP	AChE	AliE
MePxn	408.3	290.0	MPNxn	70.0	0.68
EtPxn	74.2	1.29	EPNxn	171.0	1.11
PrPxn	35.0	1.16	PPNxn	106.3	1.11
BuPxn	25.8	2.28	BPNxn	54.6	0.73
AmPxn	35.2	1.58	APNxn	52.9	1.46

Both liver homogenate and plasma were capable of reducing the concentration of available OP (Table 5). With liver, MePxn again stands out. While liver homogenates detoxifies an average of 6.8 nmol/g (85% of total) of the other 9 OP's, only 3.4 nmol/g (43%) of MePxn is detoxified. It seems likely that the low potency of MePxn against AliE is the result of low affinity and therefore leads to the lesser degree of detoxication.

This explanation becomes less plausible, however, when results with plasma are considered. While the decreased overall detoxication in plasma is expected because of lower AliE activity in plasma compared to liver, detoxication of MePxn in plasma is only slightly less than that of paraoxon. Additional data showed that the difference in sensitivity of plasma AliE to the two OP's is almost as great as that in liver. The possibility that plasma butyrylcholinesterase contributes to the unexpected high detoxication of MePxn has not been investigated.

Additional studies not yet completed also lend support to the hypothesis that AliE's significantly detoxify OP's by non-catalytic processes. The selective AliE inhibitor 4-nitrophenyl diphenylphosphinate decreases detoxication of EtPxn by rat liver in vitro in a dose-dependent fashion which correlates with the per cent inhibition of AliE, but does

Table 5. Detoxication of OP's by rat liver (1,000 g supernate; 12.5 mg/ml) and rat plasma (20 μl/ml). All OP's tested at 0.1 μM; see text for methods.

	% reduction of available OP		nmo OP detoxified per gram tissue	
OP	Liver	Plasma	Liver	Plasma
MePxn	43	31	3.4	1.6
EtPxn	80	39	6.4	1.9
PrPxn	82	44	6.6	2.2
BuPxn	93	51	7.4	2.5
AmPxn	91	50	7.3	2.5
MPNxn	75	27	6.0	1.4
EPNxn	80	23	6.4	1.2
PPNxn	87	--	9.9	---
BPNxn	91	--	7.3	---
APNxn	86	--	6.9	---

not inhibit Ca^{++}-activated hydrolysis of paraoxon by A-esterase. The phosphinate also synergizes paraoxon toxicity in vivo at doses which inhibit neither AChE nor A-esterase. The mfo inducing agent β-naphthoflavone decreases liver AliE by an undefined mechanism; it is not an AliE inhibitor in vitro. Livers from β-naphthoflavone-treated rats show decreased in vitro detoxication of EtPxn which corresponds well with the decrease in AliE activity.

In summary, it is clear that AliE's contribute significantly to the detoxication of OP's by acting as alternate phosphorylation sites. The extent of detoxication depends largely upon the affinity of the Op for the esterase, a factor which may exist especially with highly polar compounds and some dimethyl phosphates. More research is needed, however, to more clearly define the effect of structure on affinity toward AliE and on detoxication, and to investigate other potential alternate target proteins which may modulate OP toxicity.

REFERENCES

Chambers, J. E., Wiygul, S. M., Harkness, J. E. and Chambers, H. W., 1988, Effects of acute paraoxon and atropine exposures on retention of shuttle avoidance behavior in rats, Neurosci. Res. Comm., 3:85.

Devonshire, A. L. and Moores, G. D., 1982, A carboxylesterase with broad substrate specificity causes organophophorus, carbamate and pyrethroid resistance in peach-potato aphids (Myzus persicae), Pestic. Biochem. Physiol., 18:235.

Ellman, G. L., Courtney, K. D., Andres, V. and Featherstone, R. M., 1961, A new and rapid colorimetric determination of acetylcholinesterase activity, Biochem. Pharmacol., 7:88.

Eto, M., 1974, "Organophosphorus Pesticides: Organic and Biological Chemistry", CRC Press, Cleveland, OH.

Fonnum, F. and Sterri, S. H., 1981, Factors modifying the toxicity of organophosphorus compounds including soman and sarin, Fund. Appl. Toxicol., 1:143.

Lauwerys, R. R. and Murphy, S. D., 1969a, Comparison of assay methods for studying *O,O*-diethyl *O-p*-nitrophenyl phosphate (paraoxon) detoxication *in vitro*, *Biochem. Pharmacol*, 18:789.
Lauwerys, R. R. and Murphy, S. D., 1969b, Interaction between paraoxon and tri-o-tolyl phosphate in rats, Toxicol. *Appl. Pharmacol*., 14:348.

USE OF RESISTANCE PHENOMENA AS A RESEARCH TOOL FOR STUDYING THE MECHANISMS OF ACTION OF INSECTICIDES

Fumio Matsumura

Department of Environmental Toxicology,
Toxic Substances Research and
Teaching Program, LEHR
University of California, Davis, CA 95616

INTRODUCTION

It has been well recognized that any toxic manifestation in organisms is a result of complex interactions of toxic agents with various biological systems. Many chemical transporting, distribution processes (e.g. uptake, penetration through many layers of cells and tissues, partitioning, transport through blood and body fluid, etc.), metabolic actions (primary and secondary metabolism and sequestration), target interactions (enzymes, receptors, ion channels, etc.) and excretory processes are the major contributing factors to toxic manifestations. Furthermore, there are other factors such as physiological state, age, hormonal, nutritional and environmental influences which are known to often play significant roles. Knowing that all these factors must be considered, in addition to individual differences in susceptibility within a given population, the task for toxicologists to identify a few key toxicological interactions among all these processes has always been difficult.

The basic approaches toxicologists have found useful through trial and error are (a) studies on structure-activity relationships among isomers, analogs and congeners (only active analogs <u>in vivo</u> should be active <u>in vitro</u>), (b) dose-response relationships, (c) pharmacokinetics (e.g. detection of a sufficient quantity of the proposed toxic materials in the target tissues), (d) on-site delivery of the proposed toxicant, (e) selective toxicity studies (comparison of two or more chemicals in one species, or one chemical in several species, sometimes involving different sexes and ages), (f) use of specific inhibitors and in some cases synergists of which the mechanisms of action are well known (e.g. tetrodotoxin for sodium activation mechanism, piperonyl butoxide for cytochrome P450's), (g) use of primitive organisms which possess either less complicated or specific advantages, and finally (h) changes in physiological conditions (e.g. temperature, ionic and nutritional environment) as well as classical observation of symptoms and measurement of tissue specific performances.

I would like to describe in this article another approach which has been instrumental in uncovering basic causes for the toxicity of many chemicals: i.e., the use of resistance (or in vivo susceptibility differences among subpopulations within a species) in toxicological studies. This approach has not been used often due to the scarcity of resistant strains among mammalian species in particular. One notable exception is a study on 2,3,7,8-tetrachlorodibenzo-p-dioxin, where responsive mice were compared to non-responsive strains of mice (Poland and Knutson, 1982). Nevertheless, the logic of this approach and the past accomplishments are very impressive (see Georghiou and Saito, 1982). The increasing availability of resistant or supersensitive strains, particularly those with "target insensitivity" factors, would merit the re-examination of the use of resistance in toxicological studies.

RESULTS

We have recently studied the mechanisms of action of two classes of insecticides. Resistance phenomena as study tools, were very useful in elucidating vital interaction sites. Therefore, I would like to use these examples to discuss the pros and cons of the approach.

Studies on the Mode of Action of Cyclodiene-type Insecticides

Until quite recently, the mechanisms of action of cyclodienes were totally unknown despite their extensive use and the environmental problems caused by these chemicals (Matsumura and Tanaka, 1984). Meanwhile, it has been known for quite some time that various pest insects are capable of developing specific resistance to this group of chemicals which includes gamma-BHC. Studies on cross-resistance patterns have shown that it extends to gamma-BHC, but not to DDT (another chlorinated hydrocarbon insecticide) or any other insecticides. The results of genetic analyses indicate that the resistance in all cases is caused by a single gene mutation (Brown, 1958).

As for their mechanisms of resistance, as early as 1958, Yamasaki and Narahashi have shown electrophysiologically that the central nervous system of the resistant houseflies, was less susceptible to dieldrin than its normal (susceptible) counterpart, when the same concentration was given in situ. All the evidence thus far indicates that a specific change occurring in the nervous system of resistant insects is responsible for these high levels of resistance. The term "target insensitivity" has been coined for such insecticide resistance phenomena.

In view of the observations that the increased transmitter release may be intimately involved in the process of cyclodiene poisoning (see review by Matsumura and Tanaka,

1984), at an early stage of our investigation we tested a number of agents which are known to affect various processes of transmitter release for their toxic actions against cyclodiene-resistant and susceptible strains of the German cockroach. These cockroaches have been extensively studied for the cause of cyclodiene resistance since 1965 (Matsumura, 1971). Two resistant strains (LPP and FRPP) have been genetically purified for dieldrin resistance through eight generations of backcrossing to a standard susceptible strain (CSMA strain).

Among various agents tested, one naturally occurring neuroexcitant, picrotoxinin, was the only compound to which the resistant cockroaches showed a clear sign of cross-resistance (Fig. 1). Many questions were raised due to the resistant strains ability to show no altered susceptibility to well known transmitter releasers such as beta-bungarotoxin or calcium modulators such as theophylline. Therefore, we decided to look further for the cause of this cross-resistance phenomenon.

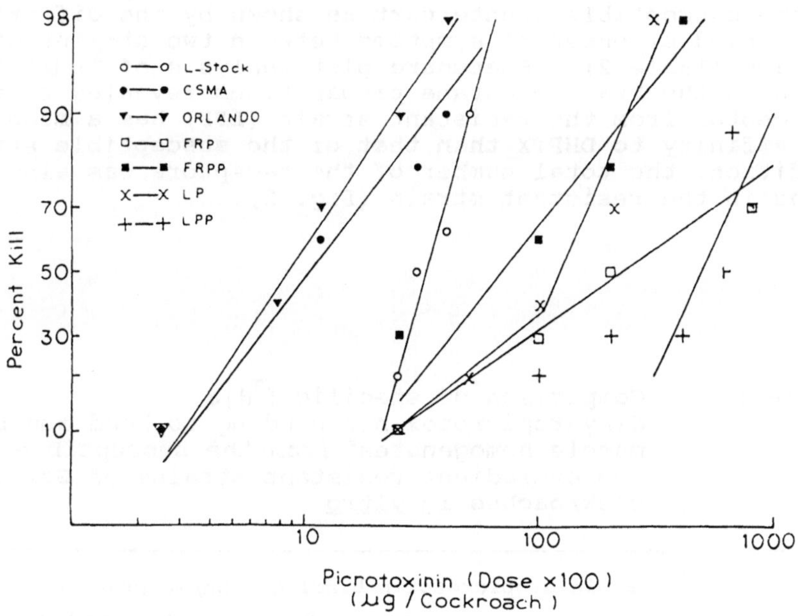

Fig. 1. Dose-mortality relationship of picrotoxinin-treated German cockroach strains. Male cockroaches were treated with piperonyl butoxide (10 µg/cockroach/µl acetone) applied topically 1 hr prior to the administration of a dose of picrotoxinin which was injected in 0.2 µl ethanol. Strain FRP, FRPP, LP and LPP are resistant and others are susceptible.

The mechanism of action of picrotoxinin (PTX) has been studied in detail by a number of pharmacologists (see review by Roberts and Hammerschlag, 1976). In short, it owes its excitatory property to its ability to block the GABA-dependent chloride channel and thereby antagonize the action of the natural chloride channel activator, GABA. Since another excitant, bicuculline, is known to also antagonize GABA's action, we tested the susceptibility of the resistant and the susceptible German cockroaches to this substance. The result was surprising in that none of the resistant strains showed any sign of cross-resistance to bicuculline (Matsumura and Tanaka, 1984). Bicuculline is known to antagonize GABA's action by directly competing with GABA binding, while PTX does not directly bind with the GABA recognition site. Nor does GABA competitively bind with PTX binding site within the GABA receptor. Therefore, the most likely possibility for the cause of PTX cross-resistance is that the PTX receptor of the resistant cockroaches is different from that of the susceptible individuals.

To prove this, a binding test was conducted with ^{3}H-labeled dihydropicrotoxinin (DHPTX) which is slightly less than PTX in binding to the PTX receptor. The results (Table 1) clearly indicate that the nerve components from the resistant cockroaches have less binding capacity to DHPTX than do the susceptible counterpart. At the same time it was possible to show that the isolated abdominal nerve cord of the resistant German cockroach is less susceptible to PTX than the susceptible counterpart as shown by the difference in the speed of onset of symptoms between two strains of the cockroach (Table 2). Scatchard plot analysis of ^{3}H-DHPTX binding to the brain membrane preparations revealed that the PTX receptor from the resistant strain (LPP) has a much lower affinity to DHPTX than that of the susceptible strain. In addition, the total number of the receptors was also reduced in the resistant strain (Fig. 2).

Table 1. Comparison of specific [^{3}H]a-dihydropicrotoxinin binding to head and body muscle homogenates[a] from the susceptible and two cyclodiene-resistant strains of German cockroaches *in vitro*

	[^{3}H] a-dihydropicrotoxin binding (dpm/mg protein)			
	Head		Muscles	
Strain	Total	Specific	Total	Specific
CSMA	46157 ± 2898	6013 ± 1670	41891 ± 256	1769 ± 64
FRPP	47328 ± 1299	4887 ± 1006	37932 ± 173	3178 ± 355
LPP	44150 ± 933	1382 ± 419	44008 ± 136	2768 ± 1340

dData are expressed as means ± SE of two or three experiments, each experiment involving three determinations.

Table 2. Minutes to onset of poisoning symptoms in the abdominal nerve cord of susceptible (CSMA) and cyclodiene-resistant (LPP) German cockroaches

	Strains	
Experiment No.	**CSMA**	**LPP**
Dieldrin (10^{-5}M)		
1	41	65
2	57	87
3	71	77
4	57	>140[a]
5	43	> 90[a]
$\bar{x} \pm$ SE	53.8 ± 5.5	>91.8 ± 12.8[b]
Picrotoxinin (10^{-5}M)		
1	41	>125[a]
2	46	>135[a]
3	91	>135[a]
4	69	> 86[a]
$\bar{x} \pm$ SE	61.8 ± 11.5	>120.3 ± 11.7[b]

[a]Maximum time period observed at which time the experiment was terminated.
[b]Significant difference at P <0.05 as judged by sign test.

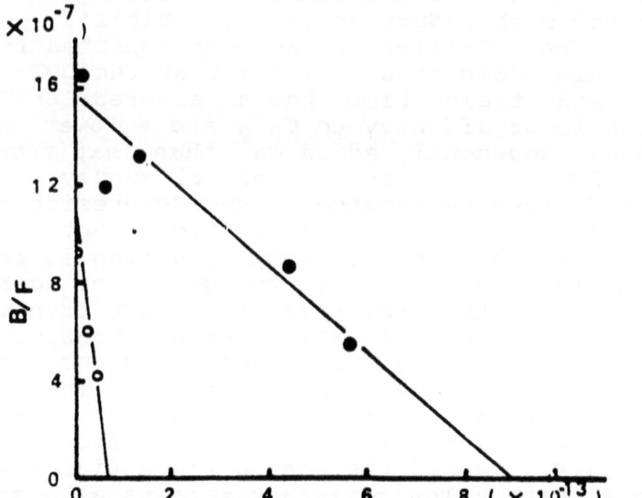

Fig. 2. Scatchard plot analysis of [^{3}H]-DHPTX binding to the brain membrane preparations from two German cockroach strains: Dieldrin-susceptible CSMA (0) and -resistant LPP (0) strains. Bmax (receptor number (mole/mg protein) was: CSMA, 9.0 x 10^{-13}; LPP, 7.1 X 10^{-14}; Kd (dissociation constant (M) was: CSMA, 5.84 x 10^{-7}; LPP, 6.45 x 10^{-8}. Data are expressed as means of two independent experiments, each experiment involving three determinations. Reproduced with permission from Tanaka and Matsumura (1986) (Copyright, 1986 Plenum).

The details of further experimentation to conclude that the PTX receptor is the specific "target" of this group of insecticides including gamma-BHC have been published (Ghiasuddin and Matsumura, 1982; Matsumura and Ghiasuddin, 1983; Matsumura et al., 1986). Therefore, the emphasis has been given to the fact that the resistance phenomenon in this case played an indispensable role in identifying the major site of action of this group of insecticides. Later we have confirmed the same phenomenon in houseflies and yellow-fever mosquitoes, Aedes aegypti, indicating the universal nature of the mechanism of developing cyclodiene resistance among insect species (Tanaka and Matsumura, 1986).

Studies on kdr-type Resistance

The presence of resistance with target insensitivity against DDT has been documented in the German cockroach (Matsumura, 1971). Scott and Matsumura (1981, 1983) have shown that the resistance is due to "target insensitivity", and that the factor confers a high degree of resistance to pyrethroids, particularly when the early "knockdown" symptom was adopted as a criterion for toxicity. These workers could not find any metabolic or penetration differences between the resistant and the susceptible strains against [^{14}C]permethrin administered topically *in vivo*. Based on these observations, they designated this type of resistance in the German cockroach as *kdr*-type resistance.

As for the biochemical mechanism of *kdr*-type resistance, there has been no concrete evidence indicating the underlying cause for such reduction in susceptibility among resistant insects. Earlier it has been reported from this laboratory (Ghiasuddin et al., 1981) that the DDT-resistant German cockroach strain VPIDLS has an altered Ca-ATPase which shows a lower affinity to Ca^{2+}, and a lower maximum stimulation by exogenously added Ca^{2+} than that from the susceptible CSMA strain. In the initial survey on *in vivo* toxicities and cross-resistance of two DDT-resistant strains to various agents, an effort was made to select chemicals for which the mechanism of toxicity or action is relatively well defined (Table 3). Namely, two groups of agents were selected. They are the ones known to affect transmembrane cation transport or calcium metabolism and transport. Among the former group of agents, the levels of cross-resistance were high for grayanotoxin I and veratrine. These two agents are known to affect the sodium channel (Narahashi, 1981). Among calcium modulators, the specific calcium ionophore, A23187, caused the best differential toxicity among these strains. The resistant strains also showed a good cross-resistance to EGTA. On the other hand, their cross-resistance to three calmodulin inhibitors; calmidazolium, trifluoperazine and chlorpromazine, was marginal in VT strain and only modest in VPIDLS. Gramicidin D is known to increase membrane permeabilities to many ions. The degree of cross-resistance to this agent was only marginal. In the next experiment, the original finding of Ghiasuddin et al. (1981) was confirmed (Fig. 3). Fig. 3 shows there is a difference in Ca-ATPase between the DDT-resistant VPIDLS strain and the susceptible CSMA strain in its response to exogenously added Ca^{2+}. The stimulatory effect of Ca^{2+} was much less in the former preparation than in the latter, particularly at low Ca^{2+} concentrations (10^{-7} to 10^{-6}M). At a very high Ca^{2+} concentration (10^{-3}M) the interstrain difference was abolished.

Table 3. Susceptibility levels (24 hr LD_{50}) of kdr-resistant and susceptible strains of German cockroach against various neuroactive agents[a]

	Cockroach strains				
	CSMA	VT	Ratio[b] (resistance)	VPIDLS	Ratio[b] (resistance)
Insecticides					
DDT	60	121	(2.0)	320	(5.3)
Diazinion	0.25	0.30	(1.2)	___	___
Nicotine	37	36	(0.97)	___	___
Carbaryl	18[c]	14[c]	(0.78)	21[c]	(1.2)
Agents affecting Na^+, K^+ channel					
Grayanotoxin I	18	48	(2.7)	80	(4.4)
Valinomycin	0.44	0.50	(1.1)	1.8	(4.0)
Aconitine	6.6	10	(1.5)	16	(2.4)
Veratrine	3.4	7.1	(2.1)	11	(3.2)
Calcium modulator[d]					
Calmidazolium	44	39	(0.88)	108	(2.5)
TPZ[e]	64	96	(1.5)	179	(2.8)
Chlorpromazine	24	17	(0.71)	92	(3.9)
A23187[f]	1.3	6.4	(4.8)	13	(9.4)
Gramicidin D	3.1	5.4	(1.8)	8.6	(2.8)
EGTA[e]	91	206	(2.3)	291	(3.2)
Lanthanum	65	66	(1.1)	73	(1.1)

[a]Data are expressed in terms of dose needed to kill 50% of population. All cockroaches were treated with 30 ug/cockroach of piperonyl butoxide (topical) prior to the test.
[b]Resistance ratio between LD_{50} of resistant (VT or VPIDLS) divided by LD_{50} of susceptible (CSMA) strain. A value less than 1.0 indicates that the susceptible strain is more tolerant.
[c]LD_{50} in hr with piperonyl butoxide.
[d]Other compounds to which VT did not show any significant cross-resistance were: D600, nitrindipine, verapamil and PCMPS.
[e]TPZ (=trifluoroperazine); EGTA (ethylene glycol-bis (p-aminoethyl) ether N,N,N,N-tetraacetic acid).
[f]Calcium ionophore.

In the next series of experiments, we examined the effect of DDT on the same Ca-ATPase with and without 10^{-6}M Ca^{2+} and calmodulin, a universal calcium carrier and modulator. The results shown in Table 4 indicate that Ca-ATPase in the cockroach nerve preparation is activated with calmodulin. The degree of stimulation was more pronounced in the suscep-

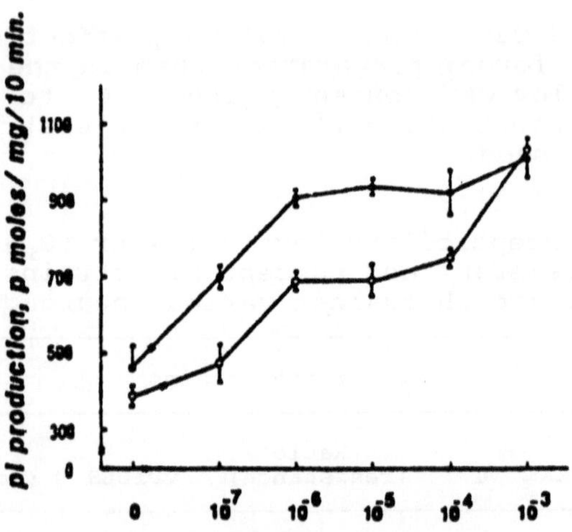

Fig. 3. Effect of changes in Ca^{2+} concentration in the presence of calmodulin (344 nM) on the Na^{+}-Ca^{2+} protein kinase phosphatase (=Ca-ATPase) activity of the synaptosome preparation from DDT-susceptible (CSMA) and DDT resistant strains (VPIDLS) of German cockroach. Protein concentration was 25 μg per assay tube. The assay buffer contained 60 mM NaCl, 60 mM KCl, 0.44 mM $CaCl_2$, 0.5 mM EGTA, 0.1 mM ouabain, 2 mM KCN and 30 mM Tris-HCl (pH 7.4). (0) CSMA, (0) VPIDLS. Experimental variability reported as SEs (Rashatwar and Matsumura, 1985. Copyright Pergamon Press).

Table 4. DDT inhibition and calmodulin stimulation of Na^{+}-Ca^{2+} protein kinase phosphatase (=Ca ATPase) activity of the synaptosome preparations from DDT-susceptible (CSMA) and resistant strain (VPIDLS) of the German cockroach

Treatment[a]	(Pi) production (pmole/mg/10min)[b] CSMA	VPIDLS
Basal (EGTA)	306 ± 22	132 ± 36
Basal + DDT	114 ± 18	78 ± 27
Calcium	432 ± 43	245 ± 27
Calcium + calmodulin	534 ± 25	298 ± 30
Calcium + calmodulin + DDT	245 ± 16	124 ± 29

[a]Concentrations were: DDT 10^{-5}M, calmodulin 344 nM and calcium 10^{-6}M (given as a Ca^{2+}-EGTA complex).
[b]Data expressed as mean ± SE of 3-6 determinations.

tible preparation than in the resistant counterpart. DDT inhibited both basal and the calcium-calmodulin activated enzyme system. Since even the basal preparation with EGTA is expected to contain some amount of endogenous calcium and calmodulin, such a result is not surprising.

Since the proposed target of DDT and type I pyrethroids is the sodium channel, we examined Na^+ uptake by the synaptosomal preparation. The criteria chosen for measurement of sodium channel activities are stimulation of Na^+ uptake by agents known to open the channel (e.g. grayanotoxin I), and inhibition by tetrodotoxin at a low (uM range) concentration. DDT, deltamethrin and grayanotoxin were found to stimulate Na^+ uptake in a dose-dependent manner as expected. Under the experimental conditions, tetrodotoxin at 1 μM was inhibitory, reducing $^{22}Na^+$ uptake below the level of control even in the presence of DDT. It must be noted that in this experiment DDT stimulated only preparations from the CSMA strain, and not those from the resistant strains. If this system truly represents Na^+ channel activity, the effect of DDT should be magnified at low, and suppressed at high Ca^{2+} concentrations. The results of the test on Ca^{2+} effect clearly support such a stimulatory action by DDT which was much higher at low Ca^{2+} concentrations. When the same experiment was repeated with the VPIDLS strain, it has become apparent that the effect of Ca^{2+} is much less pronounced in this resistant strain (Fig. 4). Even at 10^{-8}M Ca^{2+} the extent of DDT stimulation on $^{22}Na^+$ was barely noticeable in the VPIDLS strain.

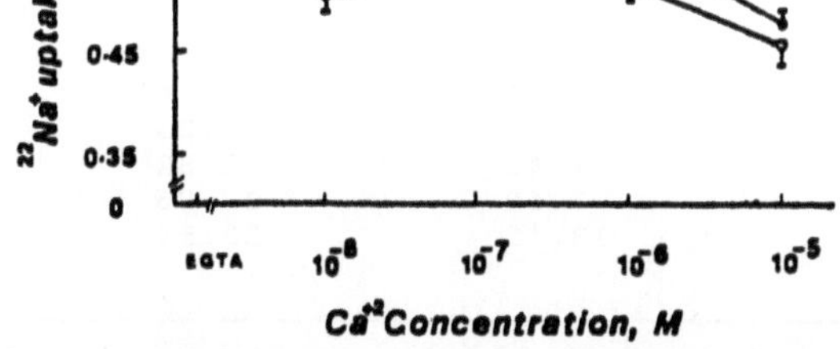

Fig. 4. Effect of changes in Ca^{2+} concentration and DDT 10^{-5}M on the level of $^{22}Na^+$ uptake activity of the synaptosome preparations from DDT-susceptible (CSMA) and resistant strain (VPIDLS) of German cockroaches at room temperature under standard assay condition, for detail see Fig. 3. (O) CSMA, (O) VPIDLS (Rashatwar et al., 1987, Copyright Pergamon Press).

Essentially, an identical result was obtained when kdr houseflies were used in comparison with susceptible strains of houseflies. The results of *in vivo* susceptibility tests indicated that kdr flies are cross-resistant to grayanotoxin I, A23187, ruthenium red and lanthanum. Such a pattern of cross-resistance is very similar to the one obtained from the resistant German cockroaches (Table 5). The most significant phenomenon is their cross-resistance to A23187 which is a well established calcium ionophore, indicating a direct or indirect involvement of a calcium responding system or systems in the development of kdr type resistance. We have also tested Na^+-Ca^{2+} protein kinase activity in the synaptosomal preparations from four strains in terms of Ca^{2+}-induced phosphorylation activity in the presence of Na^+ (Fig. 5). The results indicate that the degree of stimulation by exogenously added Ca^{2+} was much higher in the two DDT susceptible strains than in the resistant strains. Furthermore, the tendency for Ca^{2+} insensitivity is more pronounced in Skdr strain than kdr strain.

Table 5. Susceptibility levels of resistant and susceptible strains of housefly against various neuroactive agents[a]

	Housefly strain		
	cld	Skdr	Relative[b] resistance
Insecticides			
DDT	0.17	835.6[d]	4915
Carbaryl	0.54	0.44	0.81
Malathion	0.50	0.65	1.30
Toxic agents for Na^+ or K^+ channel			
Grayanotoxin I	33.51	230.50[d]	6.88
Valinomycin	0.57	0.89	1.56
Ca^{2+} modulator			
A23187[c]	0.76	12.14[d]	15.91
Gramicidine D	5.52	9.23	1.67
Ruthenium red	2.42	11.82	4.85
Lanthanum	7.69	32.06	4.17
EDTA[c]	13.15	22.54	1.71

[a]Data are expressed in terms of dose needed to kill 50% of population (LD_{50}).
[b]Relative resistance between resistant (Skdr) against susceptible (cld) strain. The values less than 1.0 indicate negatively correlated resistance (i.e. susceptible strain is more resistant).
[c]A23187: Ca^{2+} inophore. EDTA: ethylenediamine tetraacetic acid.
[d]LD_{50} value significantly different from that of cld strain at >95% confidence limit.

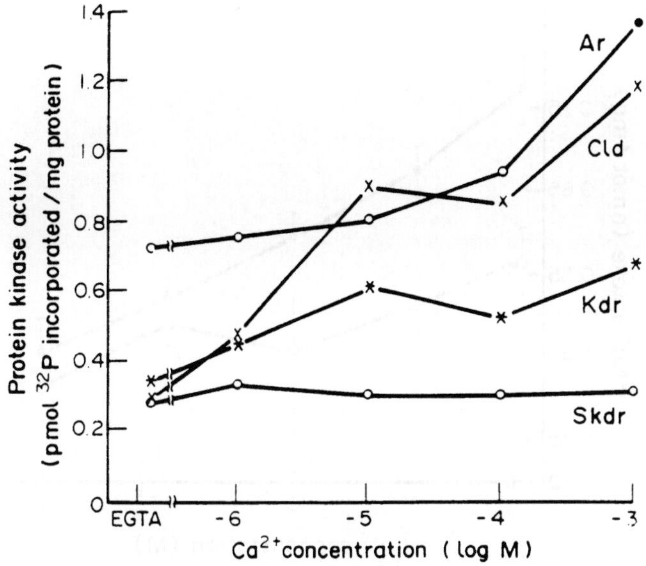

Fig. 5. Calcium sensitivities of synaptosomal Na^+-Ca^{2+} protein kinase from 4 strains of houseflies. Ar and Cld are DDT-susceptible strains. Kdr and Skdr are a moderately resistant and highly DDT-resistant strains, respectively. Note that the differences in phosphorylation levels between the tests at zero Ca^{2+} (i.e. EGTA) and those at high Ca^{2+} (e.g. at 10^{-3}M) constitute the Ca^{2+}-dependent protein kinase activity (Rashatwar et al., 1987, Copyright Pergamon Journals Ltd).

To ascertain the involvement of Ca^{2+} sensitivity changes in the resistant neural components, $^{22}Na^+$ uptake was studied under varying concentrations of Ca^{2+} in the medium. $^{22}Na^+$ uptake is significantly stimulated at low external Ca^{2+} concentrations (Fig. 6) in agreement with the results of previou study using German cockroaches. Such a response to Ca^{2+} was minimal in the synaptosomal preparation from resistant houseflies.

DISCUSSION

In using resistance as a tool for mode of action studies, one must be careful in selecting the right type of resistance, making sure it has very specific characteristics. First, the resistant population should preferably have a single gene mutation (R gene) which specifically confers resistance; second, the R gene is preferably the one conferring "target insensitivity"; and third, the site of alteration is identical or very close to the actual site of the vital insecticidal interaction. The last point is very important, since the primary site of action is not necessarily identical to the site of modification by R genes. For instance, the mechanism of resistance could be reducing penetration of the insecticide into the target tissue. In

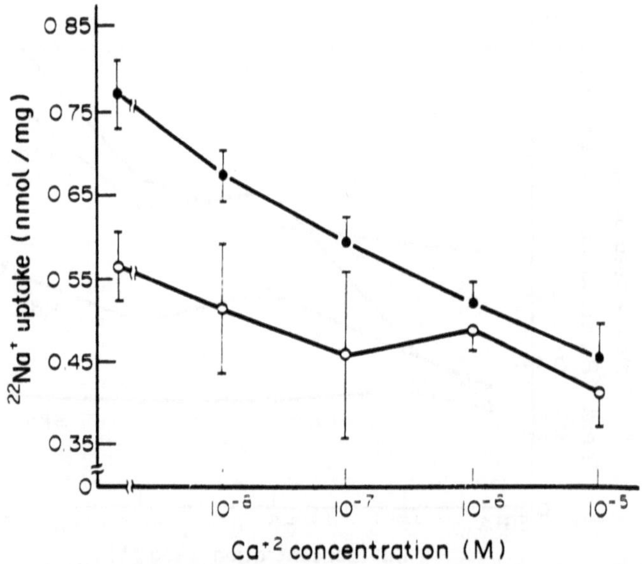

Fig. 6. Calcium sensitivities of $^{22}Na^+$ uptake processes in the synaptosomal preparation from the susceptible Cld (filled circles) and the resistant Skdr (open circles) strain of houseflies. Vertical lines indicate standard deviations (Rashatwar et al., 1987 Pergamon Journals Ltd).

the case of the cyclodiene resistance, the mechanism of resistance happens to involve the actual target site for cyclodienes and gamma-BHC (i.e., the picrotoxinin receptor). This fortunate coincidence has helped us in studying the importance of the roles this receptor plays in the process of cyclodiene poisoning of animals. The same logic may not be applied, however, in the case of the kdr type resistance phenomenon at this stage. The primary site of action of DDT, pyrethrins and pyrethroids is the sodium channel of which the biochemical operation processes are largely unknown. The observations made in this work are (a) kdr insects show cross-resistance *in vivo* to both agents affecting the sodium channel and those affecting calcium homeostasis, and (b) *in vitro* several Ca^{2+}-requiring systems from the kdr insects show lower Ca^{2+} sensitivity than those from their susceptible counterpart. The most logical explanation of the event would be that in the kdr insects those Ca^{2+}-modulated systems including the sodium channel are modified in such a way that they are less dependent on Ca^{2+} (i.e. reduced sensitivity toward Ca^{2+}). It has been well established that the sodium channel operation is affected by Ca^{2+} (Frankenhaeuser and Hodgkin, 1957), and that the effect of DDT and pyrethroids on the sodium channel operations are manifested better at lower external Ca^{2+} concentrations (e.g. Matsumura and Narahashi, 1971; Gammon, 1980). Since the sodium channels from the kdr insects are less dependent on Ca^{2+}, a hypocalcemic condition, which is required to fully express the action of these insecticides, is less likely to occur in kdr insects than in their susceptible counterpart. Such a hypothesis is, however, difficult to prove at this stage, since the biochemical mechanisms by which Ca^{2+} affects the sodium channel operations are not yet fully understood. One interesting observation is that apparently several Ca^{2+}-requiring systems from the kdr insects show the same tendency. Meanwhile, many recent studies indicate that most of these protein amino acid sequences are very similar, containing highly evolutionarily conserved regions of homology particularly in the area of calcium and ATP-binding regions (e.g. Narin et al., 1985, Fliegel et al., 1987). While much more data would be needed to put these two sets of evidence together to construct a working hypothesis, the current research data provide a sufficient base to explore the above feasibility in the future.

ACKNOWLEDGEMENTS

Supported by the California Agricultural Experiment Station, University of California, Davis and research grant ES01963 by National Institute of Environmental Health Sciences, Research Triangle Park, North Carolina.

REFERENCES

Brown, A. W. A., 1958, Insecticide Resistance in Arthropods, World Health Organization. Monograph Ser. 38: Geneva.

Frankenhaeuser, B. and Hodgkin, A. L., 1957, The action of calcium on the electrical properties of squid axons. J. Physiol. 137:218.

Gammon, D. W., 1980, Pyrethroid resistance in a strain of Spodoptera littoralis is correlated with decreased sensitivity of the CNS in vitro. Pestic. Biochem. Physiol. 13:53.

Georghiou, G. and Saito, T., eds., 1982, Pest Resistance to Pesticides, pp. 809, Plenum Press, New York.

Ghiasuddin S. M., Kadous, A. A. and Matsumura, F., 1981, Reduced sensitivity of a Ca-ATPase in the DDT-resistant strains of the German cockroach. Comp. Biochem. Physiol. 68C, 15-20.

Ghiasuddin, S. M. and Matsumura, F., 1982, Inhibition of gamma-aminobytyric acid (GABA)-induced chloride uptake by gamma-BHC and heptachlor eposcide. Comp. Biochem. Physiol. 73C:141.

Griepy, J. and Hodges, R. S., 1983, Location of a trifluoperazine binding site on troponin C. Biochemistry. 22:1586.

Matsumura, F., 1971, Studies on the biochemical mechanisms of resistance in strains of the German cockroach. Proc. Second Intern. Congr. Pesticide Chem. 2:95.

Matsumura, F. and Ghiasuddin, S. M., 1983, Evidence for similarities between cyclodiene type insecticides and picrotoxinin in their action mechanisms. J. Environ. Sci. Health. B18:1.

Matsumura, F. and Narahashi, T., 1983, ATPase inhibition and electrophysiological change caused by DDT and related neuroactive agents in lobster nerve. Bioch. Pharmacol. 20:825.

Matsumura, F., Tanaka, K. and Ozoe, Y., 1986, GABA related system as targets for insecticides. In Sites of Action for Neurotoxic Pesticides, R. M. Hollingworth and M. B. Green, eds., American Chemical Soc. ACS Symposium Series 356, Washington D.C., p. 44.

Matsumura, F. and Tanaka, K., 1984, Molecular basis of neuroexcitatory actions of cyclodiene-type insecticides. Cellular and Molecular Neurotoxicology, T. Narahashi, ed., Raven Press, New York, p. 225.

Narahashi, T., 1981, Modulation of nerve membrane sodium channels by chemicals. In J. Physiol. (Paris). 77:1093.

Narin, A. C., Hemmings, H. C., Jr. and Greengard, P., 1985, Protein kinases in the brain. Ann. Rev. Biochem. 54:931.

Olsen, R. W., Ticku, M. K. and Miller, T., 1978, Dihydropicrotoxinin binding to crayfish muscle sites possibly related to gamma-aminobutyric acid receptor-ionophores. Mol. Pharmacol. 14:381.

Poland, A. and Knutson, J., 1982, 2,3,7,8-Tetrachlorodibenzo-p-dioxin and related halogenated aromatic hydrocarbons: Examination of the mechanism of toxicity. Ann. Rev. Pharmacol. Toxicol. 22:516.

Roberts, E. and Hammerschlag, R., 1976, Basic Neurochemistry, G. J. Siegel, R. W. Albers, R. Katzman and B. W. Agranoff, ed., Little, Brown and Co., Boston p. 218.

Scott, J. G. and Matsumura F., 1981, Characteristics of a DDT-induced case of cross-resistance to permethrin in Blattella germanica. Pestic. Biochem. Physiol. 16:31.

Scott, J. G. and Matsumura, F., 1983, Evidence for two types of toxic actions of pyrethroids on susceptible and DDT-resistant German cockroaches. Pestic. Biochem. Physiol. 19:141.

Tanaka, T. and Matsumura, F., 1986, Membrane receptors and enzymes as targets of insecticidal action, J. M. Clark and F. Matsumura, eds., Plenum Press, New York.

Yamasaki, T. and Narahashi, T, 1958, Resistance of house flies to insecticides and the susceptibility of nerve to insecticides. Studies on the mechanism of action of insecticides. XVIII. Botyu-Kagaku (Scientific Insect Control) 23:146.

INDEX